An Introduction to
Ocean Dynamics

An Introduction to
Ocean Dynamics

Author

Navale Pandharinath

M.Sc. (Maths), M.Sc. (Statistics)

Director (Retd),
India Meteorology Department,
Met. Centre Hyderabad.
Guest Faculty UCESS
(M.Sc. – Earth Ocean &
Atmospheric Sciences UOH, Hyderabad)

Contributing Author

Sreeram P. Navale

B.E (ECE)

Software Engineer

BSP BS Publications

An Imprint of **BSP Books Pvt. Ltd.**

4-4-309/316, Giriraj Lane,

Sultan Bazar, Hyderabad - 500 095

An Introduction to Ocean Dynamics
by Navale Pandharinath

Third Reprint 2024

Published by:

BSP BS Publications
An imprint of **BSP Books Pvt. Ltd.**
4-4-309/316, Giriraj Lane, Sultan Bazar,
Hyderabad – 500 095.
Phone: 040 – 23445688
e-mail: info@bspbooks.net
www.bspbooks.net

Printed at
Repro India Limited.

ISBN: 978-93-8759-307-7 (Hardbound)

Dedicated to,

My Parents

Late Smt. Narasubai R. Navale
Late Shri. Rukmaji Rao Narsoji Rao Navale
Resident of Kalyani, Tq. Yellareddy,
Nizamabad Dist. Telangana

and

To all my Teachers
from primary school Kalyani, middle School Yellareddy,
High School Mahabubnagar and City College,
Nizam College and Osmania University Hyderabad

Navale Pandharinath

FOREWORD

The book "An Introduction to Ocean Dynamics" by Prof. N. Pandharinath and Sri Sreeram P. Navale is intended to enlighten especially the students who are interested to study oceanography.

The study of Oceanography is not the same as the study of a basic science like physics or chemistry, because it is an applied science requiring considerable knowledge of almost all basic sciences. Then only one can have a comprehensive understanding of the subject.

Towards that end, the authors have tried deal first with the basic aspects of oceanography in general and then providing the essential mathematical background in subsequent chapters for advanced and detailed study. At the same time, I have to sound a word of caution in that what the authors have tried is not to give all the required information but to provide only a stepping stone for understanding the more detailed and complicated aspects of oceanography especially its relationship with atmospheric sciences.

The advent of the satellite, the methods of abundant data procurement, the advent of high speed computers have revolutionized the study of ocean science.

Several books on oceanography are available now but only a few deal with the inter relationships of oceanography with mathematics and fluid dynamics from a teachers point of view in Indian conditions.

The senior author is a professional meteorologist dealing with forecasting methods for a long time in the India Meteorological Department and also a teacher in the subject on retirement.

I earnestly hope the book serves the needs of students and professionals alike for the study of oceanography.

B. Rami Reddi

Formerly Reader in Physical Oceanography

Cochin University of Science and Technology

Kochi, Kerala State.

Guest Faculty of U.C.E.S.S.

UOH-Hyderabad

PREFACE

Climate change is not restricted to remote or recent past but it is continuing on all time scales that affect human activities. All atmospheric, geological, environmental changes that are observed at present on various scales like thunderstorms, floods, droughts, tropical cyclones, monsoons, heat waves, snowfall, glaciers, landslides, earthquakes, Tsunamies including human and animal health and behaviour all being attributed to global warming, even by the common man at the market place. These changes are also echoed by IPCC. A number of studies on these topics are flooded over the world and become a household talk.

Atmosphere, ocean, land surface and ecosystems are all interlinked and activities of man that contribute to global warming manifesting in climate change. Atmospheric dynamics and ocean dynamics have much in common being related to the classical Hydrodynamics. In the present book an attempt is made to amalgamate the basic geophysical fluid dynamics to stress the role of basic ocean dynamics on climate change of the past and the present.

Oceans occupy about 71% of surface of the earth and about 80% of the atmospheric water vapour enters from the ocean surface. More than 90% of the earth's CO_2 is dissolved in sea waters. The total mass of CO_2 in atmosphere is about 0.023×10^{17} kg, while the mass of CO_2 in ocean water is about 1.4×10^{17} kg which is more than 60 times the CO_2 in the atmosphere. Sea water plays a peculiar role in respect of dissolved natural gases N, O_2, CO_2 & H_2S. These gases are closely related to living matter on the land and in the sea.

Ocean is a central component of the earth's climate system. Heat, water, momentum, greenhouse gases and a number of substances cross the sea surface. Oceans release about 80% of the water vapour into the atmosphere. In this process of water vapour, heat is transported from oceans to the atmosphere. Ocean currents transport heat and salt around the globe. Oceans huge heat capacity buffers atmospheric temperature changes. Weather and climate had been changing from the origin of earth's atmosphere and life on earth. And no doubt it will continue to change in future. There is no accepted general theory of climate that one-one relates to earths system. The earths system includes atmosphere, land, snow/Ice, oceans and biosphere. Climate never fits into rigid demarcations. It changes from one type to another, one generation to the next, from one century to the next and from one ice age to the next.

According to recent UN report, at present mankind so persisting with thoughtless and extravagant consumption of natural resources and damaging the resources in an unprecedented manner. As a consequence about 30% of amphibians, 23% of mammals and about 12% birds under the threat of extinction. Human activities are pumping GHGs into atmosphere and 70% of this accounts to energy sector. IPCC noted that global warming would melt polar ice, Himalayan glaciers retrea, sea level rise, drops in production of wheat, rice, maize and more than one billion people may face fresh water shortage by 2020.

In relation to climate change of recent two and half centuries, there will be no effect or little effect on western boundary currents, little effect on major wind systems like trade winds, zonal winds, no effect, on Hadley, Ferral and polar cells of wind circulation and little or no effect on Gulf-stream, Ocean Gyres.

Regional climate is highly influenced by thermal properties of the earth's surface, particularly neighbouring ocean. Ocean moderates extreme conditions due to its large heat capacity as compared to atmosphere, land-ice and continents. Ocean circulations are driven by atmospheric wind stress and also by transfer of heat and moisture. Wind stress is responsible for transfer of momentum to the ocean. Poleward transfer of heat achieved by general circulation of earth-atmosphere system. Of this 60% contributed by atmospheric circulation and 40% by ocean.

Ocean dynamics is treated as geophysical fluid Dynamics adopting the fundamentals of Hydrodynamics. Ocean water both in time and space always acted by the same forces, gravity (momentum), Coriolis force, pressure, temperature, wind and salinity. Ocean flow is mainly guided by the sun and moon and Techtronic process. Atmospheric influences that feed to the dynamical process in the oceans wind stress and insolation or heat fluxes.

For various reasons ocean and atmospheric sciences are taught at a very few institutions and universities in India, though these subjects have wide application in ocean navigation, aviation. Presently ocean science is used by coastal fisherman, marine scientists, Coastal Security Agencies, Naval scientists and port authorities. These scientists and other stakeholders are mainly attracted by the ocean phenomena of storm surges, tides, tsunamis, strong winds and ocean waves and currents. The routine users & other stakeholders require forecasts of wave heights, wave direction and speed, period of waves and swell waves, SST, sea surface currents, mixed layer depth, astronomical tides, wind speed and direction. As a routine IMD provides some of these elements in coastal bulletins which are used by fisherman, oil and shipping industries, ports & harbours, maritime boards, coastal tourism department and enforcement and Strategic Agencies (like marine police, Navy & Coastal guards) and oil and Natural Gas Corporation.

Keeping these requirements in view the present book is written. It covers general physical properties of oceans like SST, salinity, ocean stratification, density, potential temperature, neutral density layer, internal and external forces fluid mechanics applicable to oceans. Mathematical treatment of continuity equation, momentum equations, conservation of vorticity, dominant forces of ocean dynamics-gravitation force, buoyancy force, friction force & coriolis force have been presented by simple equations. Types of flows – ocean general circulation, wind driven circulation, ocean gyres, boundary currents, Jets, Mesoscale eddies, planetary waves, internal waves, tidal currents presented.

Hydrodynamical equations of motion (Euler's & Lagranges) including viscosity, turbulence, vertical mixing, horizontal mixing, static & dynamic stability and double diffusion discussed.

Effect of wind on upper ocean layer, inertial motion, Ekman layer at the top surface explained. Ekman's assumptions, Ekman transport equations, coastal upwelling, Ekman pumping, Langmuir, circulation discussed briefly.

Ocean Geostrophic balance, Hydrostatic equilibrium, geostrophic equation derived. Concepts of barotropic and baroclinic flow given.

Wind driven ocean circulation, included Sverdrups theory, Stommels theory of western boundary currents and Munks solution have been explained. At the end ocean waves, sea and swell wave characteristics, motions particals in shallow water and in deep waters, standing waves discussed.

Climate & climate change, geological time scale, earth-atmosphere system with corbon fluxes, evolution of atmosphere, life on earth, role of ocean on climate explained with an emphasis, ocean is responsible for development of life on earth and in moderating climate. In the absence of ocean moderation, there could have extreme weather events.

The treatment of subject is based on notes prepared by the senior author from various books, some of which given in references. The authors hope the book will be well received by beginners in ocean dynamics, teachers, students and scientists alike. The senior author after retirement as Director, IMD, worked as ground instructor meteorology in Andhra Pradesh Aviation

Academy, Flytech Aviation Academy Hyderabad and guest faculty at Centre for Earth and Space Sciences, University of Hyderabad.

The authors express their hearty indebted thanks to M/S BSP Books Pvt., Ltd. Nilkhil Shah and Anil Shah for their patronage and keen interest in publishing this book and also a number of books on meteorology. Profuse thanks are due to Prof. Chakravarthi, Prof. A.C. Narayana, Prof. Jayanand CMSD, UOH and to retired Prof. B. Ramireddy, Cochin University for encouragement and guidance in writing this book. At the end the authors express thanks to the team of BSP Books particularly Naresh Davergave, Jakkani Laxminarayana and S. Sandhya in bringing out this book in a short time.

-Authors

CONTENTS

CHAPTER 5: OCEAN MIXED LAYER AND THERMOCLINE

CHAPTER 6: STABILITY

CHAPTER 7: OCEANIC HEAT BUDGET

CHAPTER 8: FRICTION AND TURBULENCE

CHAPTER 9: FLUID STATICS

CHAPTER 1
THE OCEANS

Introduction

Oceans cover about 71% of the earth's surface area, and it is one major part of the earth system. The total mass of the hydrosphere is about 1.37×10^{21} kg. The oceans help in the processes of the atmosphere by the transfer of mass, momentum and energy through its surface. Oceans receive water and dissolved substances from the land. This dissolved substances settle down as sediments, which ultimately become rocks. At a lower level physical oceanography and meteorology are merging. The ocean provides the feedback leading to slow changes in the atmosphere.

In a broad sense, oceans provide us food (fishes etc.), affects weather and help in ocean transport, marine navigation. Ocean beds are sources for oil and gas extraction and are used for boating, fishing, navigation, surfing, swimming, recreation and extracting energy from waves. Because of these activities we are interested in the study of sea waves, currents, winds and temperature. It is a known fact that of natural disasters tropical cyclones (which form over the oceans) cause highest damage. We study the oceans to predict the formation, intensification, movement and ferocity of tropical systems to mitigate its adverse effects. In addition to tropical cyclones, the studies contribute for transport of heat energy from ocean to atmosphere in tropical latitudes to extra-tropical latitudes. Marine voyages also require the help of ocean studies. We shall study briefly the composition of sea water and exchanges that occur across the sea-air interface and motion of the sea in deep and shallow waters and discuss the effects of the wind on ocean currents.

1.1 Composition of Sea Water

From a very very long time men living near the sea obtained common salt for cooking by evaporating/boiling of sea water. It is now known that there are a number of salts dissolved in sea water. The dipolar nature of water molecule is responsible to dissolve salts by breaking and the ions of the salt are completely free from one another. More than 99% of sea salts contain six ions/atoms which are given below. The common salt (NaCl-Sodium chloride) dissolved material comprises about 85%.

Approximate composition percentage of sea water

Salt	Atoms	Ions
Chloride	55	55
Sodium	30	30
Sulphate	4	8
Magnesium	3	4
Calcium	1	1
Potassium	1	1

The above composition is more or less same throughout the global sea water.

Salinity: The number of grams of dissolved material in 1000 gm of sea water is called salinity. The average salinity of sea water is about 35 gm/kg or about 3.5% by weight.

Dissolved earth material enters ocean waters mainly through rivers on land entering into seas in the form of ions. Besides these ions, rivers carry eroded particles of soil and rock which are deposited as sediments on the sea floor. Dissolved salts in sea water differs from the average composition of the earth's crust, this is because certain elements are dissolved more readily than others and some chemicals are removed from ocean waters by living organisms. The shells and skeletons of living organisms in sea contain calcium (Ca) and silicate (SiO_2) and sea plants and animals too remove some elements from the sea/ocean water.

Sea plants, animals consume very little soluble sodium chloride (NaCl) hence they accumulate in sea water at a faster rate than other ions. Sea waters dissolve most types of ions except calcium. Calcium carbonate ($CaCO_3$) is deposited near sea shores, which form limestone rocks.

1.2 The Exchange of Earth Materials between Sea and Atmosphere

Earth's matter exists in three states viz solid, liquid and gas. The structure of solids is crystalline or amorphous. Liquid state matter exists as molecules or group of molecules. The gaseous state matter exists in molecular form but they are widely separated. The three states of matter generally changes one form to another with temperature (or heat).

As regards to exchange of matter between the oceans/seas and the atmosphere, some salts move from sea to the atmosphere as breaking waves toss the water droplets into the air, which generally evaporate leaving tiny salt crystals into the air. These crystals act as or become condensation nuclei. Sea plants release oxygen near surface, some escape into the air and the remaining goes to the depth of sea by currents. Water is exchanged between the sea and air. More than 80% of water vapour in the atmosphere is pumped from the oceans by way of evaporation. This occurs when the sea is warmer than air. The salinity of sea surface water is affected by the loss or gain of water which takes place in evaporation and precipitation processes. Thus the salinity of sea surface is greater in the vicinity of sub-tropical high pressure belts in both hemispheres.

There is no sharp boundary between the hydrosphere (sea) and the atmosphere. The waters of hydrosphere contains solid materials and gases in solution while the atmosphere contains solid and liquid particles. Salts from sea enters the atmosphere by breaking of sea waves. The evaporated tiny particles of salt act as condensation nuclei to form clouds. Oxygen and carbon dioxide are exchanged across air-sea interface. On land animals inhale oxygen and release (exhale) CO_2 from their bodies. Fish and other marine animals use O_2 that is dissolved in sea water which is replaced by O_2 (oxygen) of the atmosphere.

Plants absorb CO_2 and release O_2. Sea plants release O_2 near sea surface, some of which escape to the atmosphere. The remainder is carried to the depths of the sea by ocean currents. Water is exchanged between sea and the atmosphere as water vapour. About 80% of the water vapour in the atmosphere enters from the ocean by evaporation. This occurs readily when sea is warmer than the air. This water vapour returns to the sea when condensed in the atmosphere (in the form of precipitation) flows through river and falls into the sea. Some water is retained on land. The salinity of sea surface water is affected by evaporation and precipitation processes. By evaporation of water salinity increases, while with precipitation salinity decreases. Because of this salinity of ocean surface water is greater in sub-tropical anticyclone high pressure belts (where evaporation is more than precipitation water content).

Ocean processes are non-linear and turbulent. Ocean (like atmosphere) is a stratified fluid on the rotating earth. The air and water (both fluids) have many similarities in their fluid dynamics but there are some important differences. Water is practically incompressible. Atmospheric moisture plays very important role in water (in terms of latent heat). In case of ocean thermodynamics there is no counter part. All oceans are bounded by countries (laterally). The ocean circulation is forced in a different way as compared to atmosphere. Atmospheric motion is transparent to incoming solar radiation (i.e., insolation) and heated from below (i.e., at surface of the earth). Ocean exchange heat and moisture with atmosphere at the ocean upper surface. Convection in the ocean is by buoyancy loss from above. Wind stress over the surface drives ocean circulation, particularly upper one kilometer of depth. The wind driven and buoyancy driven circulations are inter wind. Ocean circulations affect climate and paleoclimate. As noted earlier about 71% of the earth's surface is occupied by oceans, with an average depth of about 3.7 km. Ocean basins are very complex, bottom topography notched (jagged) much more than land surface. Abyssal ocean currents are comparatively weak and temperature changes are very little. Because of this submarine ocean relief erosion is very very slow as compared to the mountains on land.

The ocean volume is about 3.2×10^{17} m^3, mass is 1.3×10^{21} kg and has huge (enormous) heat capacity, which is 1000 times the heat capacity of the atmosphere. Because of this (reason) it plays an important role in climate. The important features of ocean are given below.

Ocean surface area: 3.61×10^{14} m^2

Mean ocean depth: 3.7 km

Ocean Volume: 3.2×10^{17} m^3

Mean density of ocean water: 1.035×10^3 kg/m^3

Mass of the ocean: 1.3×10^{21} kg

The following table gives the albedo of different surfaces.

Surface	Albedo (%)
Ocean	2-10
Frost	6-18
Cities	14-18
Grass	7-25
Soil	10-20
Grass land	16-20
Desert land (sand)	35-45
Ice	20-70
Cloud (thin, thick status)	30, 60, 70
Snow (old)	40-60
Snow fresh	75-95

1.3 The Cryosphere

About 2% of the water on the surface of the earth is frogen and this is known as Cryosphere . In Greak Kryos - meaning "frost" or "cold". The Cryosphere includes: Ice-sheets, sea-ice, snow, glaciers and frozen ground (permafrost). Most of the ice is contained in the ice sheets over the land masses of Antarctica (89%) and Greenland (8%). These ice sheets store about 80% of the fresh water on the earth.

The Antarctica ice sheet average depth is about 2 km while the Greenland ice sheet is about 1.5 km thick. Climate is affected by the surface area covered by ice (not by the amount of ice). The albedo of ice varies 40-95%, about 70% in the mean, which reflects the incident radiation on it. The perennial (year-round) ice cover 11% of the land area and 7% of the ocean area.

Two forms of ice observed in Antarctica ocean.

(i) Sea ice, which is formed by freezing of sea water and

(ii) Ice bergs, which are broken off pieces of glaciers. Sea ice is important because it regulates the exchange heat, moisture and salinity in polar oceans and insulates relatively warm ocean water from the cold polar atmosphere.

1.3.1 Sea Icing

The physical properties of sea ice depends on the salt content. Salt content is a function of the rate of freezing, age, thermal history. The composition of salts in sea ice more or less same as in brine. For practical purposes the chlorinity and salinity of sea ice have the same meaning as for water (although the salts are not uniformly distributed in the ice).

The sea ice of salinity 10 ‰ at 3 °C is a mush (soft pulp) having 200 gm of brine per kilogram. Sea ice contains small bubbles of gas which changes the properties. The gases occur as small bubbles in the ice. The ice which has been frozen rapidly contain large gas quantity and in this case bubbles represent gases originally in solution in the water or in old ice (that has undergone partial thawing and has been refrozen in which case atmospheric air is trapped in the ice)

Pure water at 0 °C has density 0.9998674 kg/m^3 and pure ice at 0 °C has density 0.91676 kg/m^3. The specific heat of ice depends on temperature and changes in narrow limits. Whereas sea ice varies largely and depends on the salt content and temperature. The change in sea ice temperature depends on either melting or freezing and the amount of heat required depends on the salinity of the ice.

The specific heat of pure ice is less than that of pure water. The very high specific heat of ice of high salinity at the initial (near) freezing point is due to the formation of ice from the enclosed brine or its melting. The latent heat of fusion of pure ice at 0 °C and at atmosphere pressure is 79.67 cal/gm.

The vapour pressure of sea ice has not been determined but it could be very near to that of pure ice, which is given below.

Temp °C	0	−10	−20	−30
Vapour pressure h Pa	6.11	2.61	1.04	0.39

The Latent heat of evaporation of pure ice is variable.

1.4 Some Physical Properties of Sea Water

The most abundant element of hydrosphere is oxygen. It comprises 88.9 % oxygen and 11.1 % nitrogen by weight. Waters of oceans, lakes and rivers contain dissolved elements of earth's crust in small amounts. Sea water has about 3.5% dissolved minerals, of which sodium and chlorine ions are the largest, whose combination is found sodium chloride (NaCl, common salt) and hence the salty taste to the sea water.

Liquid water made up of multiple groups of H_2O molecule having one, two or three elementary molecules called monohydrol, dihydrol and trihydrol. These forms depend on temperature, immediate past history of water and other factors. The degree of polymerization decreases with increasing temperature. Nuetral waters have variable amounts of heavy hydrogen (deutrium – isotope of hydrogen) and oxygen. This modifies the density and other properties of water. Fresh water or rain water have lower heavy isotopes as compared to sea water. The important properties of water are given below.

Density of pure water at 4 °C is 0.999×10^3 kg/m^3

While the average density of sea water is 1.035×10^3 kg/m^3

Specific heat (C_ω) 4.18×10^3 J/kg °K

Latent heat of fusion (L_f) 3.33×10^5 J/kg

Viscosity (μ) 10^{-3} kg/m, sec

$$\text{Kinetic viscosity } (\upsilon) = \frac{\text{vis cos ity of water}}{\text{density}} = 10^{-6}\,\text{m}^2/\text{sec}$$

Thermal diffusivity (K) 1.4×10^{-7} m^2/sec

Heat capacity of water is the highest as compared to all solids and liquids except liquid Ammonia. This does not allow extreme range of temperature but allows large quantity of heat transfer water to atmosphere.

Ocean temperature ranges from about –2 °C to + 30 °C, from $3\,‰$ to $37\,‰$, chlorinity $19.00\,‰$ & froms $34.325\,‰$.

Temperature, salinity of deep ocean and bottom ocean water vary 4 °C to – 1 °C, salinity $34.6\,‰$ to $35\,‰$ and high pressure. (‰ Stands for per thousand or per mile).

Sea water diurnal temperature range is less than 2 °C and maintains uniform (water) body temperature.

1.5 Energy Exchange Processes between Sea and Atmosphere Interface

In the process of evaporation of sea water, energy is transferred from the sea to the atmosphere. 80% of the worlds atmospheric water vapour goes from the sea (oceans). Energy that is required to evaporate sea water is derived from the insolation. This energy is transferred from sea to the atmosphere as latent heat of the water vapour. When water vapour condenses it releases latent heat to the atmosphere, which remains as a heat in the atmosphere.

A tropical cyclone resembles a great heat engine that derives its energy mainly from the transfer of sensible and latent heat from sea to air. The main input is water vapour, a form of latent heat.

The oceans cover about 71% of the earth's surface, and so a large part of insolation (incoming solar radiation) in shortwave radiation is absorbed by the oceans. The oceans then radiate back a great portion of terrestrial long wave radiation. This long-wave radiation by the oceans absorbed by the atmospheric greenhouse gases (like CO_2, O_3, water vapour, CH_4 etc). This energy heats the atmosphere.

In equatorial low latitudes more energy (insolation) is absorbed by the earth-atmosphere system than energy radiated back to the space as terrestrial radiation. Thus ocean surface warms the tropics. In contrast to this, at high latitudes where energy is deficit and sea surface temperatures are low in polar regions.

Most of the insolation (short wave radiation) is absorbed in the top few meters of the oceans. A part of this absorbed heat energy transmitted downwards by vertical mixing by the winds and waves. As a result there is a surface layer with a uniform temperature (in the sea). This layer may extend to about two to three hundred meters depth.

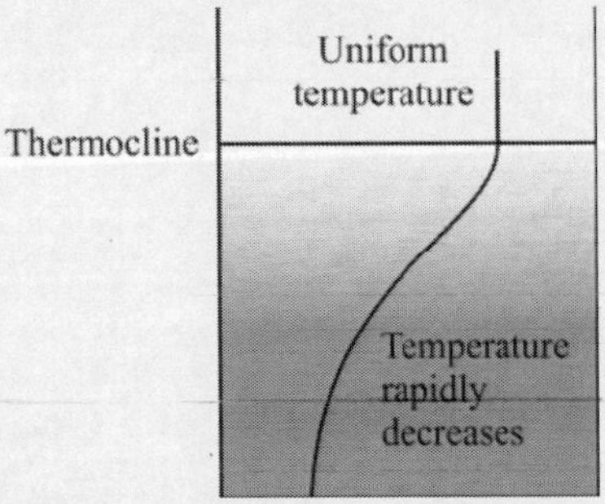

Below this (surface) layer, temperature decreases rapidly for (another) a few hundred meters because warm surface waters do not reach this depth. In boundary (range of depth) where temperature changes rapidly with depth is called the thermocline. Thermocline tends to seal off vertical water movements in many parts of the ocean. This thermocline depth is also a zone of highest density gradient is called pycnocline. Below the thermocline temperature decreases gradually. Even in tropics the temperature of the ocean water at a depth of one kilometer is only a few degrees above freezing point.

The exchanges of matter and energy are related to the events of atmospheric circulation and its waters. Because of this the study of transfer processes that occur across air-sea interface is important, which helps in understanding of weather and its forecasting.

CHAPTER 2
OCEAN CURRENTS

2.1 Introduction

The sun drives the oceanic circulation through the atmospheric circulation (wind), and in turn ocean circulation greately exerts influence on worlds climate by way of winds, temperature, precipitation and humidity (i.e., climatic controls).

Drifts: Movement of ocean water top shallow layer with speed about 3-4 kmph (2-3 mph) is called drift.

Ocean currents: Movement of ocean water with deep effects, with speed exceeding 16 kmph (10 mph) is called oceancurrent.

Ocean water circulation mainly depends on wind-stress on ocean water and different densities within the ocean water itself. (Density depends on temperature and salinity of sea water).

Circulation in both the atmosphere and the oceans is driven by insolation and the earth's rotation (Coriolis force). Radiation of the earth-atmosphere system is positive at low latitudes and negative at higher latitudes. Heat is redistributed from low to higher latitudes through wind system in the atmosphere and ocean current systems. There are two principal components of ocean circulation – wind driven surface ocean currents and density-driven (thermohaline) deep circulation.

Changes in temperature are caused by fluxes of heat across the air-sea boundary. Changes in salinity is brought by the removal or addition of fresh water through evaporation and precipitation. In polar regions freezing and melting of ice lends support to variation of salinity. All those processes are linked to solar radiation directly or indirectly.

2.2 Motion of the Sea and the Effect of Wind on Ocean Currents

Ocean waves are largely produced by the action of wind. The stronger wind and longer it blows it produces larger waves. Even after the wind has stopped, energy transferred to water causes ocean waves continue to travel for hundreds of kilometers. The waters of oceans are always on the move. Mariners have made use of ocean currents in their journey across the sea for many centuries and even now wind systems are the major driving forces of the ocean currents. It may be noted that, even in the absence of wind (speed), temperature differences and the force of gravity sets ocean waters to move.

The energy that drives the atmosphere and ocean is received from the sun's radiation. Solar heating in equatorial region and cooling near the poles create temperature differences coupled with gravity convection develop ocean currents (due to the tendency of cold water to sink and less dense warm water to rise).

The direction of ocean currents are affected by the rotation of earth (i.e., coriolis force = f). The Coriolis force deflects the ocean currents to the right in the northern hemisphere and in the left in the southern hemisphere. The combined effect of winds and gravity, coriolis force produce an inter connected ocean system develop clockwise and anti-clockwise currents, which are the main features of the world ocean currents pattern.

Ocean currents carry heat from equatorial region to polewards. In contrast dense deep currents flowing toward equator transfer cold water from high latitudes to equatorward.

The driving force of ocean currents in equatorial region (east to west) is the trade wind system in each hemisphere. These are north equatorial current and south equatorial currents respectively. Water piles up on the west side of the oceans, sloping upwards towards the west about 1 cm/200 km. The ocean waters flow up this slope as long as the trade winds are blowing.

Within the equatorial trough the winds are generally light or variable. In response to this ocean waters flow in opposite direction from west to east down the sea slope. This flow is called the equatorial counter current. This counter current is seen clearly in eastern parts of Pacific. North equatorial and South equatorial currents flow west wards, which are diverted by land barriers northward along the east coast in the northern hemisphere. The warm Gulf stream in the Atlantic ocean and Kuroshio current in the Pacific ocean are permanent. In the southern hemisphere the south equatorial currents are diverted to southwards along the east coast of the continents. Heat energy is also transported polewards in these ocean currents.

In middle latitudes strong westerlies (called roaring–fifties) force the ocean waters to travel west to east, particularly in southern hemisphere. There are land masses which divert the ocean waters. They drive the currents continuously around the southern hemisphere. This is called West wind Drift. Ocean currents are slow as compared to the wind speed. Narrow currents, like the Gulf stream, flow at about 8 kmph but in mid ocean, the ocean speed is less than 2 kmph.

In some areas of ocean, there is upwelling (upward flow of deep sea water). When winds blow equatorward along the west coasts of continents, the surface sea water is forced to move towards equator. This is coupled with coriolis force deflects the water away from the coast. This results in upwelling of deep cold water to replace surface water which was diverted. The best example is the upwelling off Peru coast, called the Peru current or Humbolt current. When the Peru current moves northward along west coast of south America, and when nears the equator it joins the south equatorial current. The upwelling of sea water, like near Peru, is also observed in Coastal regions of California, Western Australia, Vietnam and South Africa. The upwelling of cold water near northern California produces fog.

2.3 Ocean Currents caused by Density Differences

Density of sea water depends on temperature and salinity. Water becomes denser when cooled or when more salts are dissolved in it. At the air-sea interface transfer of energy or mass takes place which changes the density of sea water. It is hypothised that density differences cause ocean deep water to move in slow currents.

Deep ocean currents can be gauged or assessed indirectly by measuring temperature and salinity of sea water at different depths. Measurements of oxygen content (in sea water) also gives some information because oxygen is transferred to deepest parts of ocean as it is required for the survival of marine life.

In all oceans including in the tropics, cold water lies at the bottom whose temperature would be a few degrees above zero degree Celsius. Very cold (freezing) water forms at surface near polar regions – Antarctica, Greenland, where the water is densest. (i) In Antarctica dense surface water sinks and spreads out at the ocean floor. Freezing (like in evaporation) leaves the salt in the unfrozen water. As ice forms at the surface, the salinity of the remaining water increases. High salinity and very low temperature causes water densest, which sinks the water to ocean floor and moves away north words (from Antartica); (ii) In Arctic region, the cold water is relatively fresh and light (less dense) by river outflows. This water at surface moves south ward and when it meets warmer (less dense) waters of the North Atlantic Drift, it sinks and moves southwards at great depths. Thus in both (cases) hemispheres denser deep water moves towards equatorial region. The following figure shows some important features of deep ocean currents.

A_1, Antartica bottom current moves slowly, crossing the equator. Finally meets North Atlantic deep current A_2, which overrides A_1 on its journey towards Antarctica. The Atlantic intermediate current A_3 which rises near Antarctica. This water returns to the North Atlantic Ocean with the surface currents. A_4 is another Antarctic intermediate current. This moves northwards from Antarctica and crosses the equator flowing at a depth less than 2 km at equator. This current in Atlantic Ocean sends tongues of cool low-saline water under the warm salty surface waters of the Sargasso sea.

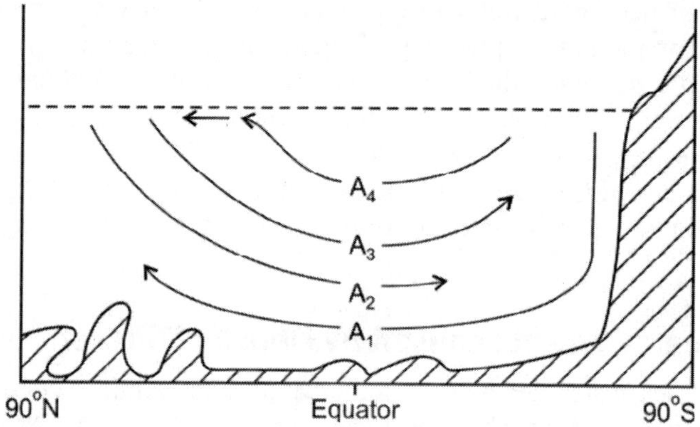

Fig. 2.1 Deep ocean current

Another current exchanges water between the Mediterranean sea and the North Atlantic ocean. Due to the high rate of evaporation Mediterranean water is the saltiest on the earth. Thus dense salty water sinks as it is not made lighter with local precipitation and river flows.

Water exchange takes place through into the Straight of Gibraltar. North Atlantic ocean surface water flows into the Mediterranean sea (about 2×10^9 kg/sec). Under this a heavy salty water flows out via Straight of Gibraltar as a compensating current and spreads out at an intermediate depth into (south-east portion of) the North Atlantic ocean.

The deep waters of other oceans is less known, but they have also layered water masses.

2.4 Ocean Currents

The main ocean currents (shown in Fig. 2.1) are:

(I North Pacific 1. Cool-Oyashivo

2. (Cool) Alaska Currents
3. Warm-California Current
4. Warm North Pacific Current

5. Warn-North Equatorial Pacific Current
6. Warm Equatorial Counter Current

South Pacific

7. Offchina-Kuroshivo Current
8. South Equatorial Pacific Current
9. Cool Peru Current or Humbolt Current

II Indian Ocean

1. North Equatorial Current
2. Equatorial Counter-Current
3. South Equatorial Current
4. Off east Africa-Agulhas Current

III North Atlantic

1. Cool Labrador Current
2. North Atlantic Drift
3. Warm Florida Current
4. Canaries Current (North Africa Coast)
5. Guinea Current (West Africa Coast)

South Atlantic

6. Warm Brazil Current
7. Folkland Cool Current (East South America)
8. Cool Bengula Current (South Africa)

IV Antarctic

1. Antarctic Circumpolar Current Westward drift

2.5 Indian Ocean Currents

In certain polar regions, water is subjected to extreme cooling-sinks and flows equatorward in the thermohaline circulation.

In order to know the net polarward heat transport in the oceans at any latitude, one would require to know the direction and speed of the flow of water and its temperature at all depths.

Relatively a thin layer close to the solid earth has frictional coupling between moving water and the earth and the same holds for air masses. Except these the frictional coupling is very weak. In case of a projectile moving above the earth and thermocline region the frictional coupling is practically zero.

2.6 Some Definitions and Explanation of Terms

1. **Oceanography:** It is the study of ocean in relation to environment. It deals with description quantitatively, which enables to predict future state with some confidence.
2. **Geophysics:** It is the study of physics of the earth.
3. **Physical Oceanograhy:** The study of physical properties and dynamics of the ocean. The main aim is to understand the interaction of the ocean with the atmosphere, water masses formation, ocean currents, the ocean heat budget and coastal dynamics. It is viewed as a sub-descipline of geophysics.
4. **Geophysical Fluid Dynamics (GFD):** It is the study of dynamics of fluid in motion with respect to the scales influenced by the rotation of the earth. In meteorology and oceanography the GFD used to calculate the planatary flow fields.
5. **Hydrography:** The preparation of nautical charts, including charts of ocean depths, currents, internal density field of the ocean and tides.
6. **Earth System Science:** It is the study of the earth as a single system. Which includes many interacting subsystems like the ocean, atmosphere, cryosphere and biosphere and changes in these systems in response to human activity.
7. **Important Physical Properties of Water:**
   - **(i)** Heat Capacity of water is the highest of all solids, liquids except liquid ammonia. This property prevents extreme ranges in temperature. Heat transfer by water movement is very large. Water tends to maintain uniform body temperatures.
   - **(ii)** Latent heat of fusion is highest except Ammonia. Thermostatic effect at freezing point owing to absorption or release of latent heat.
   - **(iii)** **Latent heat and evaporation:** Highest of all substances. Largest latent heat of evaporation is very important in heat and water transfer to atmosphere.
   - **(iv)** **Thermal expansion:** Temperature of maximum density decreases with increasing salinity. For pure water it is at 4 °C. Fresh water and dilute sea water have their maximum density at temperature above the freezing point. This property plays an important role in controlling temperature distribution and vertical circulation in lakes.
   - **(v)** **Surface tension:** Highest in all liquids. This property is important in physiology of the cell. Controls certain surface phenomena and drop formation and behaviour.

(vi) **Dissolving power:** In general water dissolves more substances and in greater quantities than any other liquid. Obvious implications in both physical and biological phenomena.

(vii) **Dielectric constant:** Pure water has highest of all liquids. Of most important in behaviour of inorganic dissolved substances because of resulting high dissociation.

(viii) **Electrolytic dissolution:** Very small. A neutral substance, yet contains both H^+ and OH^- ions.

(ix) **Transparency:** Relatively great. Absorption of radiant energy is large in IR and UV. In visible portion of energy spectrum there is relatively little selective absorption, hence is colourless. Characteristic absorption is important in physical and biological phenomena.

(x) **Conduction of heat:** Highest of all liquids. Although on small scale as in living cells, the molecular processes are far outweighted by eddy conduction.

CHAPTER 3

GEOPHYSICAL PROPERTIES OF OCEANS AND EARTH

3.1 Introduction

The earth is an (oblate spheroid) ellipsoid which rotates about its minor axis. The equatorial bulge of the earth is due to earth's rotation.

The equatorial radius of the earth 6378 km.

The polar radius of the earth 6357 km.

The measurement of the earth is made in units of degrees of latitude and longitudes, nautical miles or meters. Longitude measurements made from Greenwhich meridian and latitude with respect to the earth's great circle.

1° latitude =111 km

1° longitude =111 cos ϕ km

1nautical mile (nm) = length of the arc of 1 minute of a great circle of the earth.

Equator is a great circle midway between the poles from which latitudes are measured.

$$1 \text{ meter} = \frac{P_{axis} \text{ meridian distance between equator to pole}}{10^{7} \text{ (ten million)}}$$

1 nm = 1852 m or 1.852 km

Earth's volume = 1083×10^{12} km^3 or 1.083×10^{21} m^3

Average density of the earth = 5.41 g/m^3

Density of continental crust = 2.7 g/m^3

Density of ocean crust = 3.0 g/m^3

Density of upper mantle = 3.4 g/m^3

Density of outer core =10 g/m^3

Density at the centre of the earth $\simeq$ 13.6 g/m^3

Mass of the earth = 5974 × 10^{18} tonnes $\simeq 6\times10^{24}$ kg

Equatorial circumference of the earth = 40 077 km

Surface area of the earth = 510 100 000 km^2

Earth's dry land area = 149 400 000 km^2 (29.3% of the earth)

Surface area of oceans = 360 700 000 km^2 (70.8% of the earth)

Average oceanic crust = 6 km in thickness

Maximum thickness of the continental crust $\simeq$ 70 km

Age of the earth $\simeq$ 4600 million years or 4.6 × 10^9 years

Volume of the oceans and seas $\simeq$ 1 285 600 000 km^3 or 97.2% of the total worlds water

About 2.15% is frozen in bodies of ice, the remaining lies on or under the land.

0.001% of global water lies as water vapour in atmosphere.

Total Area of oceans $\simeq$ 360 700 000 km^2

Area of Pacific ocean $\simeq$ 179 700 000 km^2 (more than 1.2 times the area of dry earth's land area)

Area of Atlantic ocean $\simeq$ 106 100 000 km^2

Area of Indian ocean $\simeq$ 74 900 000 km^2

Average global ocean depth $\simeq$ 3550 m

The greatest depth of ocean $\simeq$ 11033 m, in the Marianas trench in the Pacific.

The lowest ocean surface temperature = –2 °C in the white sea.

The warmest ocean surface temperature = 35.6 °C in shallow parts of the Persian gulf.

Largest Bay is Hudson Bay, Area $\simeq$ 822 300 km^2

Largest gulf is, gulf of Maxico, Area $\simeq$ 1500 000 km^2

Largest sea area, south china sea, Area $\simeq$ 2 974 600 km^2

Salinity and main constituents of sea water already given.

3.2 Major Oceans

The Pacific, the Atlantic and the Indian ocean. The oceans are interconnected and not separate bodies. Around the north pole a large body of sea water called the Arctic ocean and around the south pole a large body of sea called Antarctic ocean. These two are the polar extensions of the main oceans.

Seas

Parts of oceans are called seas. The major seas of the world are given below.

South china sea, Area $\simeq$ 29 746 00 km^2

Caribbean sea, Area $\simeq$ 2 589 800 km^2

Mediterranean sea, Area $\simeq$ 2 512 150 km^2

Bering sea, Area $\simeq$ 2 273 900 km^2

Sea of Okhotsk, Area $\simeq$ 1 507 300 km^2

The ocean floor: The echo-sounders mapping presents the following features.

The continental shelves: The gently sloping areas around land masses are called continental shelves. The higher parts of continental shelves are islands. The shelves extend outward to a water depth of about 200 m. Geologically shelves are submerged of the coast lines and are the real boundary of the continents. The continental shelves end where the slope (gradient) suddenly changes at the continental slopes. The slopes fall sharply to the Abyss.

The Abyss: It is largely covered by oozes consists of volcanic dust or the remains of marine organisms and large plains and some other interesting features.

Ocean Trenches: The deepest of the abyss are the ocean trenches. The maximum depth of trench is 11033 m in the Pacific ocean, called the Marianas Trench. These trenches are zones of crustal instability being associated with much earthquake activity.

Sea Mounts: Many volcanic sea mounts rise from the abyss and some of them surface as islands (example in Hawaii). The underwater high mountains are the long ocean ridges, extending from about 4000 m deep to about 1000 m with isolated peaks (as in Iceland) emerging above the surface.

Main Ocean Trenches

***In Atlantic Ocean*:** Puerto Rico trench, depth 9212 m, Sandwich island trench, Meteor chasan, depth 8250 m.

In Pacific Ocean

***North Pacific*:** (i) In Japan trench, Rampo chasm, depth 10230 m, (ii) Cyrillian trench,10550 m (in Ilurilp island) (iii) Aleutian trench 7435 m (iv) Marianas trench 11033 m (Ifalik island) (v) Phillippines trench 10497 m (Cape Johnson chasm).

South Pacific: Tonga trench 10888 m (Ata island).

Kermadec trench 10000 m (L'Esperance Rock island).

Atacama trench 8050 m (Clule, Lagarators Promontory).

In Atlantic ocean

Puerto Rico trench (Dominican Republic) Milwankee abyss 8385 m.

Sandwich island trench, Visokoi island, Meteor chasm 8250 m

In Indian ocean: Sound trench, Island of Java 7450 m

TERMINOLOGY OF CRUSTAL DEFORMATION FEATURES

3.3 Sea Floor Features

Ridge: A long and narrow elevation with sides steeper than those of rise.

Rise: A long and broad elevation. The elevation rise is gentle from the ocean bottom.

Isolated rises from the ocean bottom which appear like mountains are called sea mountains. Ridges are curved particularly if parts of them rise above sea level (which are also called arcs). The broad top of a rise is called plateau.

Sill: A submerged elevation dividing or separating two basins is called a sill. The sill depth is greatest depth at which there is free horizontal link or communication between the basins.

Depression: Trough, Trench and Basin are the large scale depressions on the ocean bottom.

Trough: A long, broad depression with gentle sloping sides is called a trough.

Trench: A long and narrow depression with steep sides is called a trench.

(i) ***Basin*:** A large depression, roughly circular or oval shape is called basin. A depression having depth more than 6000 m is called deep depression (3000 fathoms or 5480 m).

(ii) ***Basin shelf*:** The zone extending from the low water line (permanent immersion) to the depth about 120 m, where there is a marked or deep descent toward the great depths.

Continental shelf: The feature bordering the continents is called continental shelf.

Insular shelf: It is used for the feature surrounding island.

Slope: The downward slope (declivity) from the outer edge of the shelf into deeper water. Continental slope and insular slope are used to describe the slopes bordering continents and islands respectively.

Bank, Shoal, Reef: The upper parts of elevations which reveal effects of erosion or depositions (used for Bank, Shoal, Reef).

***Bank*:** Approximately flat-topped elevation over which the depth of water is relatively small but at the same time it is sufficient for surface navigation.

***Shoal*:** A detached elevation with deep depth which is hazardous for surface navigation but that is not composed of rocks or corals.

***Reef*:** An elongated rock or coral elevation which is hazardous to surface navigation.

***Canyon*:** The steep-walled fissures that penetrate the slope and cut across the shelf are called canyon and valley, also called gully, gorge, mock-valley.

3.3.1 Basin

A basin is a depression that is filled with sea water and separated by land or sub-marine barriers from the open ocean (with which horizontal link or communication is restricted to less depth to great depths. in the basin.)

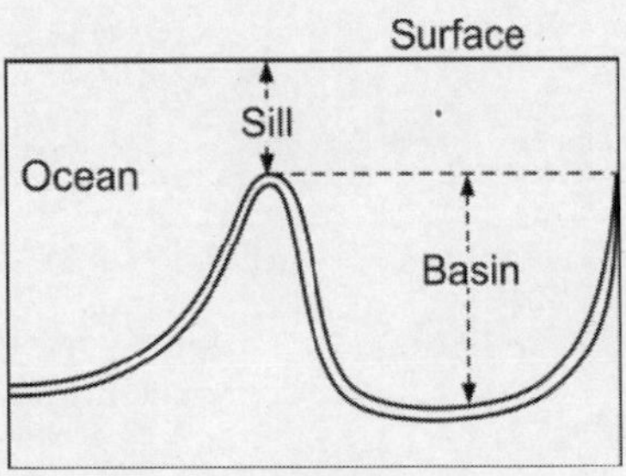

Fig. 3.1

The maximum depth of an entrance from the ocean to a basin is called threshold depth or the sill depth of that entrance.

The water in the basin is restricted horizontal communication with adjacent sea at all levels above that of the lowest sill depth, but below the sill depth renewal of water in the basin can take place by vertical motion only. Below the sill depth water is nearly uniform (homogeneous) and has the character as the water at sill depth.

The Beach: The zone extending from the upper and landward limit of effective wave action to low tide level is called beach.

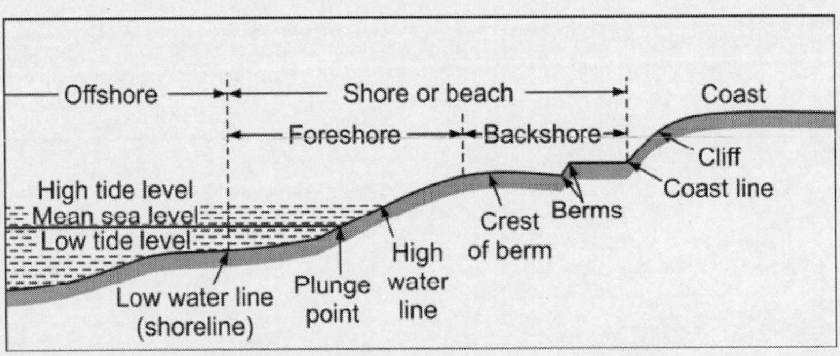

Fig. 3.2 Beach profile (various parts)

The upper part of the beach is covered only during periods of high waves (when storm coincides with spring tides). The various parts of the beach are shown in the figure.

Berms are small impermanent terraces formed by deposition during clam weather and by erosion during storms.

Plunge point is the variable zone where the waves break.

CHAPTER 4

DISTRIBUTION OF TEMPERATURE SALINITY AND DENSITY

Composition of Sea Water

With reference to standard concentration Cl = 19.00‰ (gm/kg), the composition of sea water is given below.

Salt	‰ (gm/kg)	% (percentage)
Chloride	18.98	55.04
Sodium	10.56	30.61
Sulphate	2.65	7.68
Magnesium	1.27	3.69
Calcuim	0.40	1.16
Potassium	0.38	1.10
Bicarbonate	0.14	0.41
Others	0.11	0.30
Overall salinity	34.48	

Water mass: Generally a water mass is defined by Temperature –salinity curve (T–S curve).

Salinity (s): Salinity is defined as the number of grams of dissolved material in one kg of sea water. The average salinity of sea water is about 35 gm/kg or about 3.5% by weight.

4.1 Distribution of Sea Surface Temperature (SST) and Salinity (s)

Sea water density (ρ) is a function of temperature (T), salinity (s) and pressure (p) i.e., $\rho = \rho\ (T, S, P)$. Over the world warmest water is near the equator and the coldest water is near the poles. Sea surface temperature (SST) deviates in general along the zonal (current) is small. While in the meridional direction is considerable. The anomalies of SST (deviations from long term averages) are small, (less than 1.5 °C), except in the equatorial Pacific where the deviations reach up to 3 °C. In tropic the annual range of SST is less than 2 °C while in polar latitudes is higher. Ocean temperature ranges from about – 2 °C to + 30 °C, salinity ranges 3‰ to 37‰ . While chlorinity is about 19‰ . Diurnal temperature range of SST is less than 2 °C and maintains uniform body temperature. The mean temperature of oceans water is 3.5 °C. 50% of the ocean water temperature ranges $1.3\ °C < t < 3.8\ °C$. The average salinity is about 34.7 and ranges $34.6 < s < 34.8$. In general density of sea water is independent of pressure, that is $\rho = \rho\ (s, t)$. Density anomaly $\sigma = \rho - \rho_r$ where

ρ= actual density and ρ_r= reference density = 1000 kg/m^3.

The cooler water equatorward of lat 40° tend to be on the eastern side of the basin while north of lat 40° cooler water tends to be on the western side of the basin. Salty water is denser than fresh water. ρ_{warm} (density of warm water) $< \rho_{cold}$ (density of cold water).

Fresh water has maximum density at 4 °C (assuming salinity S = 0). Fresh water colder than 4 °C is less denser than the fresh water, because of this ice forms on top of fresh water lakes. Cooling at the surface in severe winter forms ice rather than dense water.

At the sea surface $\sigma \simeq 26$ kg/m^3 and this monotonically changes with temperature to bottom.

The Thermal expansion coefficient α_T is given by the relation, $\alpha_T = \dfrac{1}{\rho_r}\dfrac{\partial \rho}{\partial T}$,

where s & p are kept constant

$$\alpha_T = 1 \times 10^{-4}/°C$$

α_T is larger at higher temperatures.

4.2 Dependence of Density, Salinity (β_s)

Dependence of density, salinity (β_s) is defined as $\beta_s = \dfrac{1}{\rho_r}\dfrac{\partial \rho}{\partial S}$, where T, P are kept constant. β_s Varies very little $\beta_s \simeq 7.6 \times 10^{-4}$ psu

psu = practical salinity unit.

The approximate dependence of σ on T, S, P is given as

$$\sigma = \sigma_o + \rho_r \; [-\alpha_T(T - T_o) + \beta_s(S - S_o)]$$

where σ_o (T_o, S_o) is density anomaly at T_o, S_o

$$\alpha_T = \alpha_T\,(p).$$

4.3 Structure of Sea Surface Temperature (SST) Salinity (S) and Temperature

(a) **SST (Sea Surface Temperature):** SSTs are warmest in the tropics (up to 30 °C) and coldest (0 °C) in polar latitudes. Warmest water temperature (29 °C) zone moves north & south with the sun. There are large zonal (east-west) variations in tropics, with relatively cool temperatures in the east and warmest temperatures in the west of the international date line. The later region is the place of greatest atmospheric convection. When SST exceeds 27 °C the location becomes a breeding ground for tropical cyclones (Typhoon/ Hurricanes) in Pacific. In middle latitudes the largest east-west SST difference related to or juxtaposed of cold boundary currents flowing toward the equator and warm boundary currents flowing toward the poles.

(b) **Salinity (SSS Sea Surface Salinity):** In sub-tropics, evaporation from the sea surface is large (vigorus) and exceeds precipitation [E_{VP} > PPtn]. Near the sea surface salinity increases as salts are not evaporated.

In high latitudes and near the equator, precipitations exceed evaporation and sea surface waters here are relatively fresh. High salinity (exceeding 38 psu) found is Mediterranean and Persian gulf. Low salinity (less than 3 psu) found near melting ice edges and at river outflows.

4.4 Equation of State of Water

The equation of state is an equation which gives the relation between density (ρ), temperature (t), salinity (s) and pressure (p) i.e., $\rho = \rho\ (t, s, p)$

Temperature changes by heating, cooling and diffusion.

Eq. (4.1): Temperature change + temperature advection/convection = Heating/ cooling term + diffusion

$$\frac{dT}{dt} = \frac{\partial T}{\partial t} + u\frac{\partial T}{\partial x} + v\frac{\partial T}{\partial y} + \omega\frac{\partial T}{\partial z}$$

where t = time, T = temperature

or $$\frac{dT}{dt}=\frac{Q_H}{\rho_{C_p}}+\frac{\partial}{\partial x}\left(K_H\frac{\partial T}{\partial x}\right)+\frac{\partial}{\partial y}\left(K_H\frac{\partial T}{\partial y}\right)+\frac{\partial}{\partial z}\left(K_v\frac{\partial T}{\partial z}\right)$$

Salinity changes by addition or removal of fresh water (which changes the dilution of salts).

Eq. (4.2): Salinity change + salinity advection/convection = evaporation/precipitation/run-off/brine rejection + diffusion.

$$\frac{dS}{dt}=\frac{\partial S}{\partial t}+u\frac{\partial S}{\partial x}+v\frac{\partial S}{\partial y}+\omega\frac{\partial S}{\partial z}$$

$$=Q_s+\frac{\partial}{\partial x}\left(K_H\frac{\partial S}{\partial x}\right)+\frac{\partial}{\partial y}\left(K_H\frac{\partial S}{\partial y}\right)+\frac{\partial}{\partial z}\left(K_v\frac{\partial S}{\partial z}\right)$$

Density $\rho=\rho(S, T, P)$ is a function of Salinity (S), Temperature (T) & Pressure (P).

Equation of state is dependence of density on salinity, temperature and pressure.

Eq. (4.3): Density change + density advection/convection = density sources + diffusion

$$\frac{d\rho}{dt}=\frac{\partial\rho}{\partial t}+u\frac{\partial\rho}{\partial x}+v\frac{\partial\rho}{\partial y}+\omega\frac{\partial\rho}{\partial z}$$

$$\frac{d\rho}{dt}=\frac{\partial\rho}{\partial S}\frac{dS}{dt}+\frac{\partial\rho}{\partial T}\frac{dT}{dt}+\frac{\partial\rho}{\partial p}\frac{dp}{dt}$$

Note: $$u\frac{\partial\rho}{\partial x}=\frac{\partial x}{\partial t}\frac{\partial\rho}{\partial x}=\frac{\partial\rho}{\partial t}\ \text{etc}$$

where (in all above equations):

Q_H = heat source $\begin{pmatrix}+\text{ for heating}\\ -\text{ for cooling}\end{pmatrix}$ applied only near the sea surface.

C_P = Specific heat of sea water (at constant pressure)

Q_s = Salinity source $\begin{pmatrix}+\text{ve for evaporation, brine rejection}\\ -\text{ ve for precipitation \& run-off}\end{pmatrix}$

applied near sea surface.

K_H = horizontal eddy diffusion

K_v = vertical diffusion

Eq. (4.4): $\rho = \rho$ (S, T, P) is full equation of state, from which evolution of density in terms of T, S change computed in eq. (4.3).

The coefficient for three terms in Eq (4.4) are the haline concentration coefficient, the thermal expansion = coefficient and adiabatic compressibility.

Note: In oceanography the temperature measured in degrees celsius with accuracy ± 0.02 °C, Salinity grams per kilogram of sea water (i.e., in parts per thousand or per mile) denoted as ‰ with accuracy $\pm 0.02\%$, Pressure is defined as the Pressure exerted on one square meter.

1 Atmosphere = 760 mm of Hg (Mercury) column at Temperature of 0 °C, g (acceleration of gravity) 980.665 cm/sec^2

$$\text{Density} = \frac{\text{Mass}}{\text{Volume}}$$

Density of sea water depends on three variables

Temperature, salinity and Pressure

$$\rho = \rho(S, \theta, p)$$

where

S = salinity

θ = Potential temperature

p = Pressure

Density anomaly σ is defined as

$$\sigma = \rho - \rho_r$$

where

ρ = *in situ* density or actual density

ρ_r = reference density 1000 kg/m^3

Density of sea water generally not measured but calculated from the measurements of temperature, salinity, conductivity and pressure using equation of state of sea water. The equation is derived from laboratory observations of T, S, P chlorinity or conductivity. The International Equation of State published: The equation has accuracy of 10% or 0.01 units of $\sigma(\theta)$.

The study of zonal averages shows a warm salty light lens of saline fluid in the sub-tropics (in both hemispheres) shoaling on the equatorial and polar flanks. The lens is floating on a cold, comparatively fresh abyss exists in each ocean basin with local characteristics and variations. In zonal average data plot the lens lies in the middle of the thermocline. The lens is deeper in the Pacific as compared to Atlantic.

The thermocline is clearly seen deep in the subtropics (± 30 degrees) and shallow in equatorial regions. The mean temperature of the ocean's water is T = 3.5 °C and the mean salinity s = 34.7. About 50% of the water is in the range of temperature 1.3 °C to 3.8 °C and salinity 34.6‰ to 34.8‰.

4.5 Water Masses of the Oceans

Water bodies are identified by the combination of physical and chemical properties. These are called water masses. These properties are notably used to identify water temperature and salinity. Temperature and salinity of water bodies are conserved. Temperature and salinity are changed by the processes taking place at the boundaries of the ocean. Within the ocean the changes takes place due to mixing of water masses of different characteristics and biological process like in air mass fronts homogeneity in meteorology. For example deep water leaving the Mediterranian through Gibralter strait has high salinity because of intense cooling in water and more evaporation. The homogeneous water mass has salinity 38.4 ‰ and temperature of about 12.8 °C. In equatorial region, the salinity is low because of high precipitation and high temperature (which causes surface water density low).

According to Defant (1961-2005) we have following graphs of mixing of two water masses that produce a line on T-S plot.

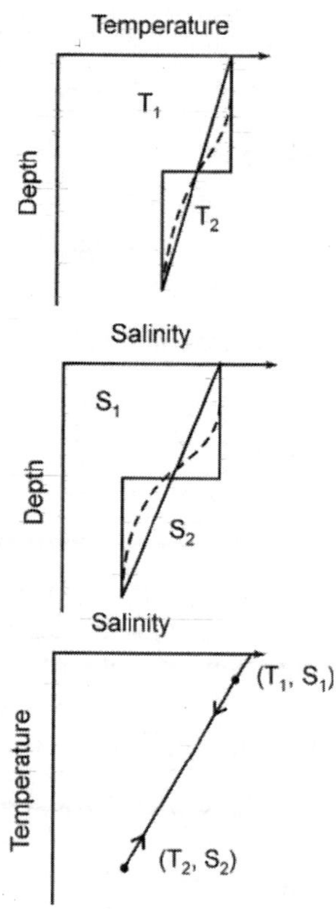

Broadly, a water mass is defined by T-S (Temperature-Salinity) curve.

Ocean water masses generally defined on the basis of their temperature-salinity and not density because two water masses of different temperatures and salinities may have the same density.

Helland-Hansen (1916) introduced T-S diagram to study water masses. At a given place the temperatures and corresponding salinities of the sub-surface water are plotted against each other.

In general the plot shows a well defined curve and gives T-S relationship of the sub-surface water of that place (location/region). This T-S diagram (plot) became the most important tool in physical oceanography and conventionally used to detect the anomalies in the distribution.

Based on Helland-Hansen's suggestion a water mass is defined by a T-S curve. In exceptional cases, a water mass may be defined by a single point in a T-S diagram, that is by means of a single temperature and single salinity value. Such exceptional cases are found in basins where homogenous water is present over a wide range of depth.

A water type is defined by means of a single temperature and salinity values.

A given water type is present along a surface in the area and has no thickness.

4.6 The Water Masses of the Indian Ocean

The Indian Ocean is closed toward the north, and all water masses in the upper layers are characteristic of the middle latitudes and equatorial regions. No Subpolar water mass enters the Indian Ocean. Due to lack of data the types of water in the Indian Ocean is of Preliminary character. In surface water the temperature increases rapidly from sub-tropic convergence to the north and in the equatorial region it is uniformly high during most part of the year (25 to 29 °C).

During August lower temperatures found (down to 22 °C) along southeast coast of Arabia and the east coast of Africa up to the equator due to upwelling under the influence of prevailing southwest Monsoon.

In February lower temperatures are found around the Galf of Oman and the Bay of Bengal.

The salinity of surface water in equatorial region, northern region and west of Australia has subtropical maximum, however it shows variation during monsoon period (because of rain and river water flows).

Temperature and salinity mostly lies between T = 8 °C, S = 34.6‰ to

t = 15 °C and S = 35.5‰.

4.6.1 Distribution of Oxygen in Indian Ocean

Because of paucity of data distribution of oxygen in the Indian Ocean is not clear. Along African coast oxygen content decreases southward to the equator. In Red sea water oxygen content increases in the direction Red Sea water spreads. The Antarctic Intermediate water has comparatively high oxygen ranging 4.4 to 3.6 ml/L.

4.7 Distribution of Temperature and Salinity in Bay of Bengal

In Bay of Bengal the depth of sea increases away from the land boundary and also from north to south. At its centre along 90 °E it has deeper depth exceeding 4000 m. A large number of rivers (Perennial) like Brahmaputra, Ganga Hoogly, Mahanadi, Godavari and Krishna flow into the Bay of Bengal with fresh water during Southwest and Northeast Monsoon period. Canyons exist off the mouth of the most rivers with the deepest one at the Head Bay of Bengal. Along the Southwestern part the Palk Straits between India and Srilanka.

The temperature of the water ranges 24 °C to 28 °C, characteristic of tropical waters. During winter the temperature difference of about 4 °C (south to north), which increases toward the equator. During summer the lowest temperature lies in the central Bay and temperature increase towards the coasts.

The density of water changes with Northeast or Southwest monsoon, seasons with the lightest water at at the surface along the coast and at Head Bay of Bengal tighter water spreads from north to south towards equator. Density changes are largely due to addition of fresh water (rain, river discharge). Thus water mass characteristics mainly depend on salinity and function of river discharge and monsoon activity. As a result water masses classification is based on mostly on density instead of T-S relation.

Average sea surface temperature and salinity in the Indian Ocean between parallels on lat given in the table below.

	Temperature °C	Salinity ‰
40 °N		
30 °N		
20 °N	26.1	35.05
10 °N	27.2	34.92
0 equator	27.90	35.14
10 °S	27.4	34.57
20 °S	25.9	35.15
30 °S	22.5	35.89
40 °S	17.0	35.1
50 °S	8.7	33.87
60 °S	1.6	
70 °S	−1.5	

CHAPTER 5
OCEAN MIXED LAYER AND THERMOCLINE

5.1 Oceanic Mixed Layer and Thermocline

The top thin layer of the ocean is stirred by prevailing winds and convention is called mixed layer. The layer has constant temperature and salinity from surface to a depth about 10-200 m. At the ocean surface there is a mixed layer which is in direct contact with the overlying atmosphere. This layer has relatively uniform properties of temperature & salinity. The next layer is thermocline. The temperature difference in the mixed layer, top and bottom is 0.02 to 0.1 °C (colder than at surface). In tropics and middle latitudes the mixed layer thickness is 10-200 m.

The depth and temperature of the mixed layer changes day to day and from season to season for two reasons. The heat fluxes at surface heat and cool the surface water. The turbulence in the mixed layer mixes heat downward. Turbulence depends on the wind speed and on the intensity of breaking waves. This turbulence also mixes water within the layer and with water in the Thermocline.

Below the mixed layer, water temperature falls rapidly except in high latitudes (polar regions). The range of depth where temperature gradient (slope) is large is called Thermocline. Thermocline is also the pycnocline, a layer where density gradient is greatest.

The slope of the thermocline changes slightly with seasons, called seasonal thermocline. The permanent themocline extend below the seasonal thermocline to depth 1500-2000 m.

Mixing of momentum is frictional process while mixing of properties is diffusion process. Mixing process is weak but it maintains stratification.

The figure 5.1 schematically shows the processes involved in the top/upper mixed layer.

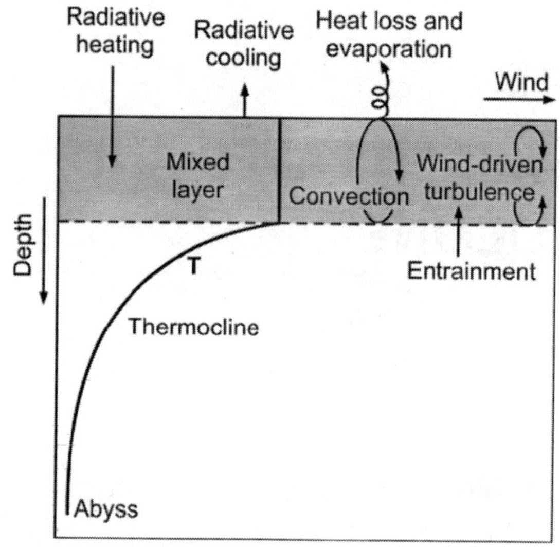

Fig. 5.1

Horizontal eddy viscosity >> vertical eddy viscosity. Isopycnals (along adiabatic surfaces), Diapycnal mixing is across the adiabatic surfaces.

Isopycnal: Horizontal mixing of sea water along the same density surface is isopycnal.

Diapycnal: Vertical mixing of sea water across (or Perpendicular) to two or more isopycnals (different density surfaces) is called diapycnal.

A_H is horizontal eddy viscosity is along the isopycnal and A_V, Vertical eddy viscosity is normal to A_H is along diapycnal.

Note: If density of ocean water changes with depth (vertically) but does not change horizontally then constant pressure surfaces are parallel to the ocean surface and also parallel to the levels of constant density or isopycnal surfaces.

5.2 Mixed Layers

(a) Surface mixed layer: It is the upper boundary layer of the ocean, which is directly forced by atmosphere: (i) Through surface stress by wind and (ii) buoyancy (heat Q, fresh water) exchange through the surface. Wind stress generates strongest motion at the surface, which decreases with depth (vertical shear in the velocity).

The above two motions include waves that add turbulent energy (increasing mixing). Wind driven Langnuir circulations (LC) can promote mixing.

(b) **Bottom mixed layer:** Close to the ocean bottom, turbulence mixing generated by currents or current shear (with interaction of bottom).

At longer scales of time, on the shelf, a bottom ekman layer can develop (where frictional and coriolis forces balance). Eddy viscosity varies both in time and space, which effects the Ekman layer structure. Bottom currents due to density differences also cause mixing. The dense water flows down the continental slope as a plume and mixes vigorously with lighter surrounding water. This turbulent mixing process is called entrainment.

(c) **Internal mixing:** In the interior of the ocean (away from boundaries/Ekman friction) shows vertical profiling of water properties of temperature, salinity and hence density- generally not smooth but they are stepped. The vertical scale of the steps changes (varies) decimeters to many meters. Turbulence and/or double diffusion mix the water column internally.

(d) **Turbulent mixing:** Richardson number (R_i) is a non-dimensional quantity defined as $R_i = \dfrac{N^2}{\left(\dfrac{\partial u}{\partial z}\right)^2}$

where $N^2 = -\dfrac{g}{\rho}\left(\dfrac{\partial \rho}{\partial z}\right)$

N is the Brunt Vaisala frequency

$\dfrac{\partial u}{\partial z}$ = vertical shear of horizontal speed

ρ = density of water.

If R_i is small, stratification is weak but if shear is large, this implies active/vigorous mixing.

If $R_i < \dfrac{1}{4}$, vigorous mixing begins. Vertical stratification stabilizes the mixing Solar Radiation entering the ocean surface is largely absorbed in the top few meters depending on wavelength. IR is absorbed within a few millimerers (mm), blue/green light may penetrate to about 100 m (particularly in clear water). IR radiation heat loss to the atmosphere, latent heat loss through evaporation takes place within top few mm of the surface. With cooling and salinization of surface water, density of water increases. For onset of convection (in incompressible fluid) $\dfrac{\partial \rho}{\partial z} > 0$

The buoyancy frequency of the Thermocline layer is given by

$$\frac{g\,(\rho_1-\rho_2)}{\rho_1} \simeq \frac{-g}{\rho_2}\frac{d\rho_2}{dz}\delta = N^2\delta \qquad \text{.....(5.1)}$$

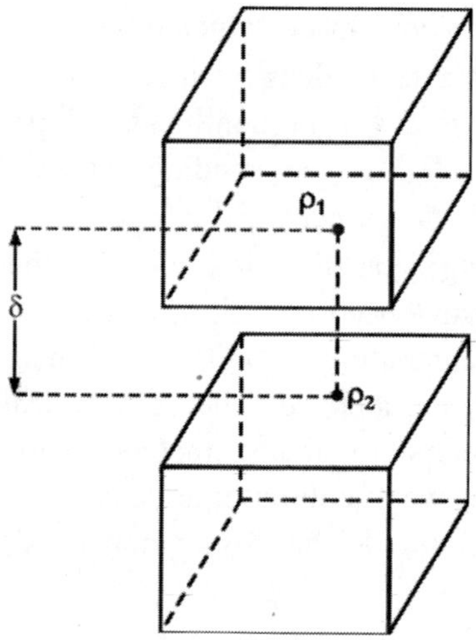

Fig. 5.2

where ρ_1 = density of the parcel

ρ_2 = density of surrounding (environment) water

δ = distance or depth of movement = $\rho_1\rho_2$

This implies $\dfrac{d^2\delta}{dt^2}$ = acceleration = $-g\left(\dfrac{\rho_1-\rho_2}{\rho_1}\right)$

Horizontal currents carry properties 'to' and 'from' the mixed layer in a process called subduction. It may be noted that it is difficult to measure salinity and hence conductivity is used instead of salinity. Density is calculated using temperature, conductivity and pressure.

The uniformly mixed layer depth with constant S & T is generally found to be in the top 1-100 m of the ocean. This depth depends on the wind speed and flux of heat through the sea surface.

For comparing the temperature & density of water masses at different depths in the ocean, potential temperature and Potential density is used (which removes the influence of pressure on density).

As said earlier, water parcels below the mixed layer move along neutral surface.

Solar radiation light is promptly absorbed in the ocean. Generally, about 95% of the solar radiation is absorbed in the top 100 m of the clear sea water and rarely penetrated below a few meters in turbid coastal waters.

Presence of phytoplankton change the colour of sea water, and the change in colour can be measured by satellite based radiometers. Thus water colour is used to measure the phytoplankton concentration from the satellite radiometers.

Reynolds Number-Turbulence

Reynolds number (R_e) is defined as the ratio of non-linear terms to the viscous terms

$$R_e = \frac{\text{Non}-\text{linear terms}}{\text{Viscous terms}} = \frac{\left(u\frac{\partial u}{\partial x}\right)}{\left(\nu\frac{\partial^2 u}{\partial x^2}\right)} = \frac{U\frac{U}{L}}{\nu\frac{U}{L^2}} = \frac{UL}{\nu}$$

where U = velocity,

L = length describing the flow

ν = kinetic molecular viscosity [$\nu = 0.01\ \text{cm}^2/\text{sec}$ for water]

Example 5.1: If $U = 0.1$ m/s, L = mega meter $R_e = 10^{11}$

Note: Molecular viscosity is significant only over very short distances, a few millimeters.

Turbulence plays role from non-linear terms in the momentum equation viz $u\frac{\partial u}{\partial x}$ etc.

Friction is important only over very short distances, a few millimeters in ocean. For most of ocean flows friction is ignored.

5.3 Density, Potential Temperature and Neutral Density

$$\text{Density } (\rho) = \frac{\text{Mass (M)}}{\text{Volume (V)}}$$

Potential temperature: The temperature of a parcel of water that is attained when it is raised adiabatically (*in-situ*) from a depth to the sea surface is called Potential temperature of water and is generally denoted by θ.

Potential density (ρ_θ): It is the density of a parcel of sea water that would attain, if the water parcel raised adiabatically to the sea surface without changing salinity.

σ = density anomaly $= \rho - \rho_r$

where ρ = density of sea water

ρ_r = constant reference density

$$\frac{\sigma}{\rho_r} << 1$$

(i) $\sigma_\theta = \sigma\,(S,\theta,0)$, σ_θ is a function of salinity, potential temperature and pressure (p) = 0

(ii) $\sigma_4 = \sigma\,(s,\theta,4000)$,

Pressure p = 4000 decibars at depth of 4 km

(iii) $\sigma_T = \sigma\,(S,\theta,P,P_r)$

where p = pressure

P_r = pressure at some reference

P_r = 0 at reference

5.4 Neutral Surface and Density

A parcel of water moves locally along a path of constant density. Hence less density water moves above and more density water moves below clearly water moves along a path of constant potential density σ_T referenced to the local depth 'r'. Such a path is called a neutral path. "A neutral surface element is the surface tangent to the neutral paths through a point in the water". No work is required to move a parcel on a neutral surface, because there is no buoyancy force acting on the parcel as it moves (provided friction is ignored).

γ = neutral density is a function of salinity *insitu* temperature (t), Pressure (p), longitude, Latitude

$$\gamma = \gamma\,(s,\ t,\ p\ \text{long/lat})$$

Neutral surface defined above differs slightly from an ideal neutral surface. If a parcel moves around a gyre on a neutral surface and returns to its starting location, its depth at the end will differ by about 10 meters from the depth at the start. If potential density surfaces are used, the difference can be hundreds of meters, a far larger error.

Entrainment: The dense water flows down the continental slope as a plume, mixing vigorously with the lighter ambient water (water around). This turbulent process is called 'entrainment'.

5.5 Observed Mean Ocean Circulation

Ocean currents carry heat from equatorial regions towards poles, while currents flowing towards equator transfer cold water from high latitudes.

The general ocean flow is dominated by closed circulation patterns called "Gyres". Gyres are pronounced in the northern hemisphere (NH) where zonal flow is blocked by coasts. In subtropics of the NH there are anticyclonic gyres, called subtropical gyres, which flow eastward in middle latitudes (like the North Pacific and North Atlantic currents) and westward flow in the tropics (the North equatorial currents in the Pacific and Atlantics).

The driving force for the equatorial currents is the trade wind system (in both hemisphers). The speed of tropical currents in the interior of the gyres are less than 10 cm/sec. At the western edge of these subtropical gyres, there are strong poleward (northward in NH and southward in SH) currents, speed reaching or exceeding 100 cm/sec. The Kuroshivo current in the North Pacific and the Gulf Stream in the Atlantic ocean. These strong currents flow northward along the east coats from the tropics, which turn into interior at about 40° latitude, and move (spread) eastward across the ocean basin. It may be noted that water piles up on the Westside of the oceans (that is eastside of continental coasts) sloping upward towards the west about 1cm in 200 km. The sloping continues as long as the winds are blowing (the ocean waters flow up this slope). Within the Equatorial Trough the winds are light and variable. Because of this, the water then flows in opposite direction from West to East down the slope. This is called the Equatorial counter current.

This counter current is well (strongly) developed in the eastern part of the Pacific Ocean, but also occurs in other oceans. The east to west flow is a striking aspect of the observed current system. All intense boundary currents are also observed in thermal structure at the ocean surface, with strong temperature gradients near the western boundaries of the middle-latitude oceans and in the Antarctic Circum Polar current of the SH (which is called West Wind Drift).

The ocean and atmospheric systems in low latitudes tend to move north and south with migration of equatorial trough position during the year. On the average the meteorlogical equator or thermal equator (or heat equator) tends to be slightly north of the geographical equator with the ocean and atmospheric systems centering around it.

In the polar regions of the North Pacific and North Atlantic basins there are cyclonic gyres called sub-polar gyres with southward flowing western boundary currents (like the Ovashivo current in Pacific and the Labradar currents in the Atlantic). It is noticed that there are no intense eastern boundary currents.

As compared to the wind speed, the ocean currents are slow. Narrow currents like the Gulf Stream sometimes flow at about 8 kmph but in general in mid- ocean the ocean speed is less than 2 kmph.

Upwellings: In some ocean areas there is an upward flow of deep water, called upwelling. Winds that blow equator wards along the west coasts of continents, surface sea water is forced towards low latitudes, which are deflected by the cariolis force away from the coast. This results in cold deeper water rises and replaces surface water which has been diverted. This phenomena is prominent along Peru coast of South America during December, called Peru current or Humbolt current. However, every few years, the Peru current slackens and upwelling disappears along the Peru shore which will be covered by warm waters. This phenomena is called ELNino. ELNino is disasterous for fish population and other marine life. Upwelling of water also occurs in the coastal regions of california, Western Australia, Vietnam and southwest Africa. The upwelling of cold water produces cool waters and fogs off the coast of northern California.

CHAPTER 6

STABILITY

There are several ways of developing instability in the ocean. The following are the important.

(i) Static stability, which is associated with change of density with depth,

(ii) Dynamic stability, which is associated with shear (of flow speed), and

(iii) Double diffusion, which is related with salinity and temperature gradients in the ocean. Molecular diffusion of heat is 100 times faster than molecular diffusion of salt.

6.1 Static Stability

Let $\rho_1 < \rho_2$ (ρ density), implies $\rho_2 > \rho_1$

Consider the two cases as shown in Fig. 6.1 and Fig. 6.2 in which density layers are shown. In these two cases we derive the conclusion of stability in the interface of the levels.

In Fig. 6.1 setting the interface is unstable.

In Fig. 6.2 setting the interface is stable.

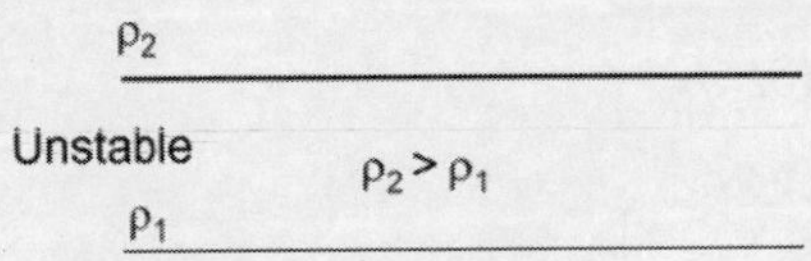

Fig. 6.1

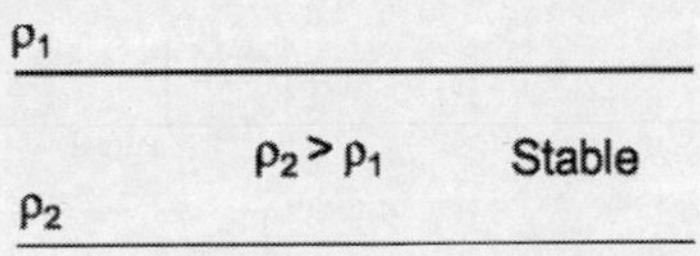

Fig. 6.2

Let V be the volume of water parcel (a cubide shown in Fig. 6.3) displaced vertically. Let ρ_1 be the density of water in the parcel and ρ_2 be the density of surrounding water. Let g be the gravity and let δz (= p_1 p_2) be the vertical displacement. Mass of the parcel = V g ρ_1

Displaced surrounding mass by the parcel = Vg ρ_2

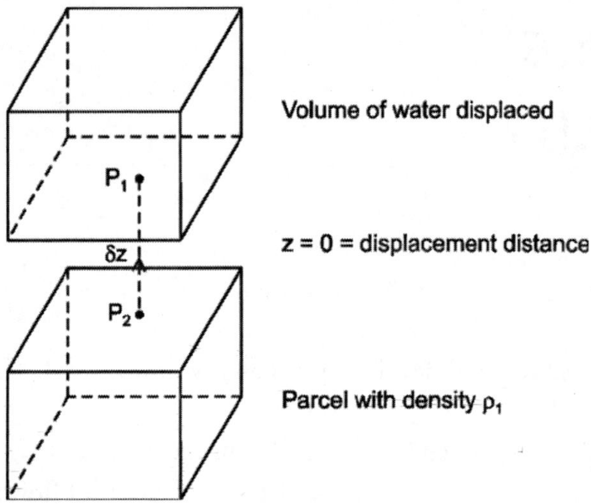

Fig. 6.3

∴ The force of buoyancy F = Vg ($\rho_2 - \rho_1$)

Acceleration of displaced parcel (a) is

$$a = \frac{F}{m} = \frac{Vg(\rho_2 - \rho_1)}{V\rho_1} \qquad \text{.....(6.1)}$$

But (by Taylor's expansion)

$$\left.\begin{aligned} \rho_2 &= \rho + \left(\frac{d\rho}{dz}\right)_{water} \delta z + 0\left(\delta.z^2\right) \\ &\text{higher order terms } [(\delta z)^2, (\delta z)^3 ....] \end{aligned}\right\} \qquad \text{..... (6.2)}$$

$$\rho_1 = \rho + \left(\frac{d\rho}{dz}\right)_{Parcel} \delta z + .0(\delta z^2)$$

$$\therefore \rho_2 - \rho_1 = \left[\left(\frac{d\rho}{dz}\right)_{water} - \left(\frac{d\rho}{dz}\right)_{parcel}\right] \delta z \qquad \text{..... (6.3)}$$

Let E be the stability of water column,

using eq. (6.1),

$$E = -\frac{a}{g\delta z} = -\frac{g(\rho_2 - \rho_1)}{\rho_1} \times \frac{1}{g\delta z}$$

$$= -\frac{\left(\frac{d\rho}{dz}\right)_{water} - \left(\frac{d\rho}{dz}\right)_{parcel}}{\rho + \left(\frac{d\rho}{dz}\right)_{parcel}}$$

$$E \simeq -\frac{1}{\rho}\left(\left(\frac{d\rho}{dz}\right)_{water} - \left(\frac{d\rho}{dz}\right)_{parcel}\right)$$

$$E \simeq -\frac{1}{\rho}\left(\frac{d\rho}{dz}\right) \qquad \text{.....(6.4)}$$

$$\text{since}\left(\frac{d\rho}{dz}\right) >>> \left(\frac{d\rho}{dz}\right)_{parcel}$$

Eq. (6.4) is the stability equation.

Stability is defined

$$\left.\begin{array}{ll} \text{if } E > 0 & \text{stable} \\ E = 0 & \text{neutral stability} \\ E < 0 & \text{unstable} \end{array}\right\} \qquad \text{.....(6.5)}$$

In the above notation,

$\left(\frac{d\rho}{dz}\right)_{water}$ is proportional to the rate of change of density of water column

$\left(\frac{d\rho}{dz}\right)_{Parcel}$ is proportional to the compressibility of sea water, which is generally negligible.

N (= stability frequency), is given by the relation

$$N^2 \equiv -g\,E \qquad \text{.....(6.6)}$$

N is also called buoyancy frequency or the Brunt Vaisala frequency.

The frequency (N) quantifies, the importance of stability. Consequently 'N' is a fundamental variable in dynamics of stratified flow, denotes the vertical frequency of internal waves in the ocean.

6.2 Dynamic Stability and Richardson's Number (R_i)

Let velocity vary with depth in a stable, stratified flow. The flow may become unstable if the change in velocity becomes large with depth, that is current shear develops. For example wind blowing over the ocean surface which creates waves. If the wind is strong the surface becomes unstable and waves break. Here stable fluid has become unstable by velocity shear.

Similarly, if the density difference in a sheared flow is much less than at the sea surface (as in the thermocline or at the top of a stable atmosphere boundary layer) Kelvin Helmholtz instability results. Richardson Number (R_i) is given as

$$R_i \equiv \frac{g\,E}{\left(\frac{\partial u}{\partial z}\right)^2}$$ where g E (numerator) is the strength of static stability,

while the denominator $\left(\frac{\partial u}{\partial z}\right)^2$ is the strength of velocity shear

If $R_i > 0.25$ the flow is stable, and

$R_i < 0.25$ the flow is unstable, velocity shear increases turbulence

Large Reynolds number (R_e) indicates turbulent flow and small Reynolds number indicates smooth or laminar flow.

In case of Richardson number (R_i)

for turbulence $R_i < 0.25$

and stability $R_i > 0.25$.

Large R_e and small R_i criteria is satisfied in some oceanic flows. Under these conditions turbulence mixes the fluid vertically and causes vertical eddy viscosity and eddy diffusivity. Since the ocean tends to be strongly stratified and ocean currents tend to be weak, turbulent mixing intermittent and rare.

6.3 Double Diffusion and Salt Fingers

It is found in some seas (regions) less dense water (ρ_1) overlies more dense water (ρ_2). In such water, even in the absence of currents water column is found to be unstable. The reason of this instability is the molecular diffusion of heat, which is 100 times faster than the molecular diffusion of salt.

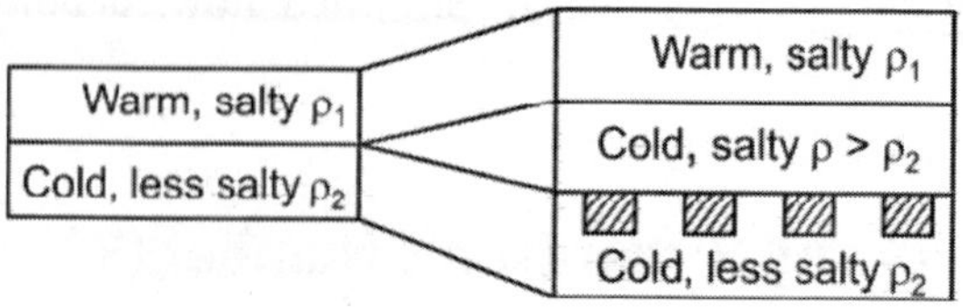

where $\rho > \rho_2 > \rho_1$

Fig. 6.4

Suppose two thin layers (a few meters thick) separated by a sharp interface as shown in the Fig. 6.4. Suppose the upper layer is warm and salty and the lower layer is cold and less salty as compared to the upper layer. Then the interface becomes unstable even if the upper layer is less dense than the lower.

Reason: Heat diffuses across the interface faster than salt. This creates a thin cold salty layer between the two initial layers. The cold salty layer is more dense than the cold less-salty below and the water in this layer sinks. Since the layer is thin, the fluid sinks in fingers one to five centimeter in diameter and about 10 cm long like our fingers (in shape and size). This is salt fingering. Because two constituents diffuse across the interface, the process is termed double diffusion.

There are four variations on this subject. By combination of two variables taken two at a time will have four possibilities which are given below.

1. Warm salty over cold less salty $\left(\frac{\text{WS}}{\text{CLS}}\right)$
2. Cold less salty over warm salty $\left(\frac{\text{CLS}}{\text{WS}}\right)$
3. Warm less salty over cold salty $\left(\frac{\text{WLS}}{\text{CS}}\right)$
4. Cold salty over warm less salty $\left(\frac{\text{CS}}{\text{WLS}}\right)$

The above four combinations result in double diffusion.

Double diffusion changes the regional (local) distributions of temperature and salinity but on large-scale circulation of ocean it will have no effect.

CHAPTER 7
OCEANIC HEAT BUDGET

7.1 Introduction

The main heat source for the ocean and Atmosphere is solar radiation. About 50% of the insolation reaching the earth is absorbed by the ocean and land, which is temporarily stored near the surface. About 20% of the insolation is directly absorbed by the atmosphere. Most of the heat stored by the oceans is released to the atmosphere locally by evaporation and IR radiation and the remaining heat is transported by ocean currents from low latitudes to middle latitudes. Heat lost by the tropical oceans drives the atmospheric circulation. The oceanic heat budget and heat transports are essential in understanding the earth's climate and its variations. Atmospheric convection is triggared by surface warming. Vertical mass transport is confined to few locations of strong updraughts. The ocean is forced from above by air-sea fluxes. Ocean convection is most prevalent in higher (coldest regions) latitudes.

Surface water density increases by two methods.

1. Direct cooling, which decreases temperature and increases density
2. Brine rejection in ice formation increases salinity and hence increases density of water just below the ice.

7.2 Buoyancy (b)

A parcel of water sinks depending on its buoyancy anomaly. Using the oceanographic notation

$$b = \frac{-g(\sigma - \sigma_o)}{\rho} \qquad \text{.....(7.1)}$$

where g = acceleration due to gravity

σ = density anomaly = $\rho - \rho_r$

where ρ = density of parcel

ρ_r = reference density = 1000 kg/m^3

$\sigma - \sigma_o$ = difference between density of the parcel and its surrounding.

Density is a function of temperature (T), salinity (S) and pressure (P) and is denoted as $\rho = \rho$ (T, S, P)

The buoyancy of sea water at the surface depends on distribution of temperature (T) and salinity (S).

The flux of heat and fresh water at the surface induce temperature and salinity which is responsible for buoyancy changes and also ambient pre-existing stratification of the water column.

The equations governing the evolution of T & S are given below.

$$\frac{dT}{dt} = -\frac{1}{\rho_r C_w} \frac{\partial Q}{\partial z} \quad .....(7.2)$$

$$\frac{dS}{dt} = S \frac{\partial \varepsilon}{\partial z} \quad .....(7.3)$$

where

C_w is heat capacity of water

Q turbulent vertical flux of heat

ε fresh water heat flux driven by air-sea exchange convection, ice formation and vertical mixing

At sea surface:

$Q = Q_{net}$ = the net heat flux across the sea surface

$\varepsilon = \varepsilon_{surface}$ = Evaporation – Precipitation = E – P

Q and ε decay with over the mixed layer from their surface values.

'–'sign in eq. (7.2) indicates heat loss from ocean ($Q_{net} > 0$); its temperature decreases.

Note: The dependence of density on salinity is defined as $\beta_s = \frac{1}{\rho_r} \frac{\partial \rho}{\partial s}$ (where T&P are kept constant)

β_s varies very little and it has value about 7.6×10^{-4} PSU^{-1}

$$\sigma = \sigma_\circ + \rho_r \; [-\alpha_T \; (T - T_\circ) + \beta_s \; |S - S_\circ|] \quad .....(7.4)$$

$\sigma_\circ$ $(T_\circ, S_\circ)$ is density anomaly at Point T_o, S_o

$$\sigma = \rho - \rho_r, \; \alpha_T = -\frac{1}{\rho_r} \frac{\partial \rho}{\partial T} \quad .....(7.5)$$

α_T = coefficient of thermal expansion of water

$$\frac{db}{dt} = \frac{-g}{\rho_r}\left(\frac{\alpha_T}{C_W}\frac{\partial Q}{\partial z} + \rho_r\beta_s\ S\frac{\partial \varepsilon}{\partial z}\right) \quad \text{.....(7.6)}$$

$$= -\frac{\partial B}{\partial z}\text{; where B = vertical buoyancy flux}$$

$$B_{surface} = \frac{g}{\rho_r}\left[\frac{\alpha_T}{C_W}\ Q_{net} + \rho_r\beta_s S(E - P)\right] \quad \text{.....(7.7)}$$

The first term in the brackets of RHS of eq (7.7) represents thermal flux and second term haline flux components of buoyancy.

$$\alpha_T = -\frac{1}{\rho_r}\frac{\partial \rho}{\partial T}$$

$$\beta_s = \frac{1}{\rho_r}\frac{\partial \rho}{\partial S}$$

The units of buoyancy flux = $m^2\ s^{-3}$ = velocity × acceleration

Buoyancy flux comprises both thermal and haline components

7.3 Ocean Heat Budget Equation

$$Q_{net} = Q_{SW} + Q_{LW} + Q_S + Q_L + Q_{AV} \text{ units W/m}^2 \quad \text{.....(7.8)}$$

where

Q = Total heat gain or loss = Q_{net}

Q_{SW} = shortwave flux (insolation into the sea)

Q_{LW} = longwave flux (Net IR radiation from the sea)

Q_S = sensible heat flux (out of the sea by condensation)

Q_L = lalent heat flux by evaporation of water

Q_{AV} = heat advection (carried away by currents)

Albedo of different surfaces (%)

Surface	Albedo (%)	Surface	Albedo (%)
Ocean	2-10	Desert (land)	35- 45
Forest	6-18	Ice	20-70
Cities	14-18	Cloud	30, 60-70
Grass	7-25	Snow old	40-60
Soil	10-20	Snow fresh	75-95
Grass land	16-20		

$$\Delta E = C_p m \ \Delta t$$

where

ΔE = Energy change due to change in temperature Δt

C_p = specific heat of water at constant pressure

$= 4.0 \times 10^3$ J/kg °C

m = mass of water

7.4 Heat Flux

The transfer of heat across or through surface is called heat flux.

Heat flux and water flux changes the density of surface water and also buoyancy. The sum of the heat and water fluxes is also called the buoyancy fluxes.

The flux of heat at the surface is quick, while at the deeper layers is slow and smaller. Since the temperature of the ocean and land over a long period is observed to be steady and hence the total heat fluxes into and out of the ocean must be zero (Budget). Equation (7.8) represents the heat budget (with major components) of ocean.

The shortwave flux (Q_{sw}) in the insolation that reaches the sea surface and penetrates the ocean, warms it down to a depth of 100-200 m depending on the transparency of water.

The longwave flux (Q_{LW}) is the next flux of longwave (IR) radiation at the sea surface due to the radiation emitted out by the ocean, according to black body law

Emitted radiation per unit area = σT_e^4 .....(7.9)

where $\sigma = 5.67 \times 10^{-8}$ W/m²/°K⁴

7.5 Sensible Heat Flux (Q_s), Heat through Sea Surface

$$Q_s = \rho_{air} C_p C_s U_{10} [SST - T_{air}] \quad .....(7.10)$$

where

ρ_{air} = density of air at surface

C_s = stability dependent bulk transfer coefficient for heat

C_P = specific heat of air at constant pressure

T_{air} = temperature of air

U_{10} = wind speed at height of 10 m above surface

SST = sea surface temperature

If SST > T_{air}, $Q_s > 0$ and the sensible heat flux is out of the ocean, which cools the sea.

The global average temperature of the surface ocean is one or two degrees warmer than the atmosphere, the sensible heat is transferred from ocean to the atmosphere.

Latent heat flux (Q_L) is the flux of heat carried by evaporated water.

$$Q_L = \rho_{air} L_e C_L U_{10} [q_* (SST) - q_{air}] \quad .....(7.11)$$

where $\quad Q_s = \rho_{air} C_p C_s U_{10} [SST - T_{air}]$

C_L= stability dependent bulk transfer coefficient forwater vapour ($\simeq 10^{-3}$)
L_e = latent heat of evaporation
q_{air} = specific humidity (in kg vapour/kg air)
q_* = specific heat at saturation, which depends on SST

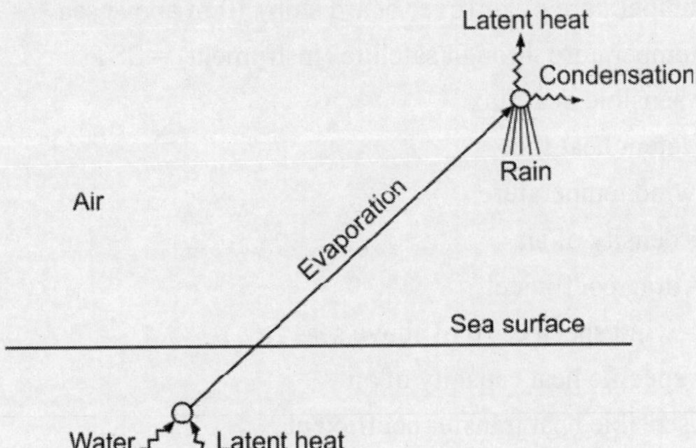

Fig. 7.1

The net outward transfer of water in evaporation, (which is) = Precipitation, that is upward flux of energy in latent form is

$$Q_L \sim L_e \frac{dm}{dt}$$

where

L_e = latent heat of evaporation

m = mass of water falling /m^2

The rate of precipitation $P = \frac{1}{\rho_r}\frac{dm}{dt}$

$$\Delta T = \frac{Q_{net}\ \Delta t}{h\ \rho_r\ C_w}$$

where

ΔT = warming of seawater (by an amount)

Q_{net} = air-sea flux magnitude to a depth 'h' over a time period Δt

7.6 Indirect Calculation of Fluxes – Bulk Formulas

Variables that can be measured globally by use of observed correlations between fluxes and variables. For fluxes of sensible and latent heat and momentum are termed bulk formulas, which are given below.

$$T = \rho_{air} C_D U_{10}^2$$

$$Q_s = \rho_{air} C_p C_s U_{10} (t_c - t_a)$$

$$Q_L = \rho_{air} L_e C_L U_{10} (q_s - q_a)$$

Where

t_a= temperature of air (over board ship) 10 m above sea

t_c = temperature through satellite (instrument) = SST

Q_s = sensible heat flux

Q_L= latent heat flux

T = wind temperature

ρ_{air} = density of air

C_D = drag coefficient

U_{10} = wind speed at 10 m above sea

C_P = specific heat capacity of air

C_s = sensible heat transfer coefficient

L_e = latent heat of evaporation

C_L = latent heat transfer coefficient

q_s = specific humidity of air at the sea surface

q_a = specific humidity of air at 10 m above the sea

Latent heat flux (Q_L) is influenced by wind speed and relative humidity. Strong winds and dry air evaporate more water from ocean surface than light winds with relative humidity 100%. In polar regions evaporation from ice covered ocean is much less than from open water. This implies the open water (percent area) plays an important role in Arctic heat budget.

The average annual latent flux is

$$-130 \text{ W/m}^2 < Q_L < -10 \text{ W/m}^2$$

The sensible heat flux (Q_s) is dependent on wind speed and air-sea temperature difference. Strong winds with large temperature difference cause high flux and wind chill factor. The annual value of sensible heat flux is

$$-42\ \text{W/m}^2 < Q_s < -2\ \text{W/m}^2$$

7.6.1 Meridional Ocean Heat Transport ($H^{-\lambda}$)

$H^{-\lambda}_{\text{ocean}}$ represents the flux across a vertical plane extending from the bottom of the ocean to the top (surface) and from western coast of an ocean basin to the eastern coast of the basin.

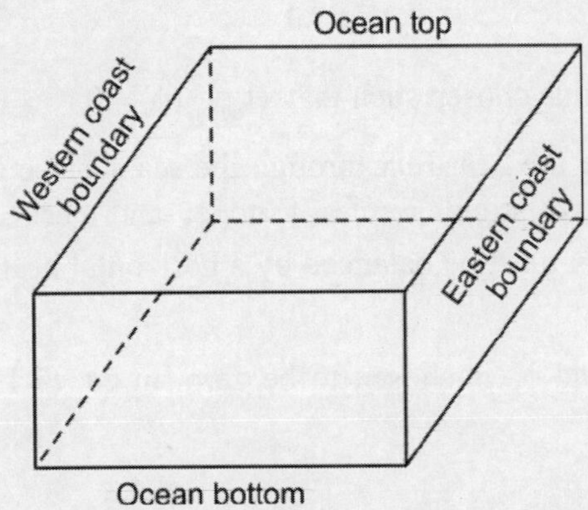

Fig. 7.2

Mathematically

$$H^{-\lambda}_{\text{ocean}} = \rho_r\ C_w\ a\cos\phi \int_{\lambda_{\text{west coast}}}^{\lambda_{\text{east coast}}} \int_{\text{ocean bottom}}^{\text{ocean top}} vTdzd\lambda \qquad \text{.....(7.12)}$$

where

a = radius of the earth

ϕ = latitude

λ = longitude

$a\cos\phi\, d\lambda$ = distance along a latitude circle over an arc d_λ

T = Potential temperature

Meridional ocean heat transport at latitude ϕ is given by

$$H^{-\lambda}_{\text{ocean}}(\phi) = -a^2\cos\phi \int_{\phi_1}^{\phi_2} \int_{\lambda_{\text{west coast}}}^{\lambda_{\text{east coast}}} Q_{\text{net}}\, d\lambda\, d\phi \qquad \text{.....(7.13)}$$

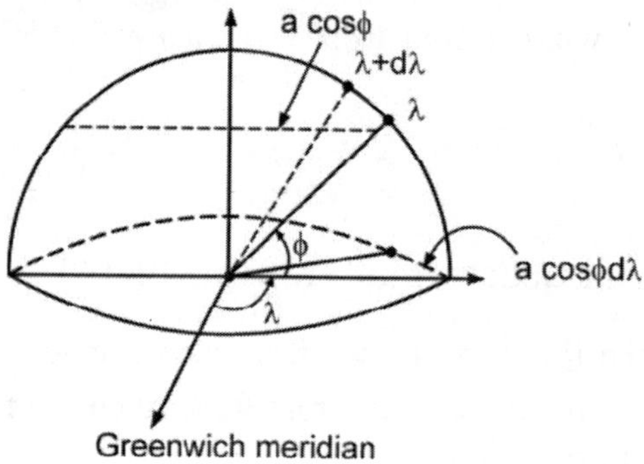

Fig. 7.3

where ϕ_1 = latitude chosen such that $H_{ocean}^{-\lambda}(\phi) = 0$

For steady state the heat flux through the sea surface integrated over the area bounded by two latitude circles ϕ_1 to ϕ_2 and meridional coast λ_{west} to λ_{east}. This heat flux must be balanced by a horizontal heat flux into or out of the basin region.

For $H_{ocean}^{-\lambda}(\phi_1)$ at ϕ_1 is chosen to be zero, in eq. (7.13) if $Q_{net} < 0$, then $H^{-\lambda} > 0$.

7.7 The Ocean Heat Budget

Ocean circulation plays a role in meridional heat transport. Over a long period, a steady climate is observed. This means the ocean and atmosphere transport excess heat received in equatorial region (tropics) to the polar regions. The atmosphere transports the larger part of the poleward heat transport and the remainder part transported by the ocean currents.

Consider the heat budget for a column of ocean, which is obtained by integrating the equation

$$\frac{dT}{dt} = -\frac{1}{\rho_r C_w}\frac{\partial Q}{\partial Z} \quad .....(7.14)$$

Vertically over the depth of ocean.

$$\frac{\partial}{\partial t}(Q) = -Q_{net} - \nabla_h.\vec{H}_{ocean} \quad .....(7.15)$$

where heat content $(Q) = \rho_r C_w \int_{bottom}^{top} T dz$

and $\vec{H}_{ocean}$ is by ocean current.

The heat stored in the column Q_{net} is given by the equation

$$Q_{net} = Q_{SW} + Q_{LW} + Q_S + Q_L \qquad .....(7.16)$$

and the $\vec{H}_{ocean} = \rho_r C_w \int_{bottom}^{top} \vec{U}\, T\, dz$ (is a vector)

Horizontal heat flux by ocean currents integrated over the vertical column, $\nabla_h = \left(i\frac{\partial}{\partial x} + j\frac{\partial}{\partial y} \right)$, horizontal divergent operator.

Eq. (7.15) above gives the changes in heat stored in a column of the ocean are induced by fluxes of heat through the sea surface and the horizontal divergence of heat carried by ocean currents.

In order to maintain steady state, the global integral of the air-sea flux must be zero (since ocean currents carry heat from one place to another and redistribute it around the globe).

1. The magnitude of $H_{ocean}^{-\lambda} \simeq \frac{1}{4}$ to $\frac{1}{3}$ of pole-equator heat transport

2. $H_{ocean}^{-\lambda}$ contributions from the Atlantic, the Pacific & the Indian oceans are significantly different.

7.7.1 Meridional Heat Transport - Note

At the top of the tropical atmosphere, earth gains heat while at the top of the polar atmosphere earth loses heat. The atmosphere and ocean circulation together transport heat from Low latitudes to high latitudes to balance the gain and loss because over a long period neither the low latitudinal regions warming nor the high latitudinal regions cooling but remaining at the same temperatures. The north-south transport of heat is called meridional heat transport.

7.8 Adiabatic Process

If the volume and temperature of a gas (or fluid) changes without the addition or subtraction of heat (energy) from the gas (or fluid) it is called an adiabatic process.

In an adiabatic compression the temperature of the gas (or fluid) will rise, while in an adiabatic expansion the gas (or fluid) will cool.

CHAPTER 8

FRICTION AND TURBULENCE

In fluids mainly two types of friction plays part,

(i) Shearing stresses: This plays part when fluid layers are slipping relative to each other.

(ii) Normal stresses: Shearing stresses acting on a unit area are proportional to the rate of shear normal to the surface on which stress is exerted.

$$\tau = \mu \frac{d\nu}{dt} \qquad .....(8.1)$$

where μ = dynamic viscosity, a proportional factor

$$\left.\begin{array}{l}\text{Frictional forces}\\ \text{Per unit volume}\end{array}\right\} = \begin{array}{c}\text{Difference between shearing stresses exerted on}\\ \text{oppsite sides of a cube of unit dimension}\end{array}$$

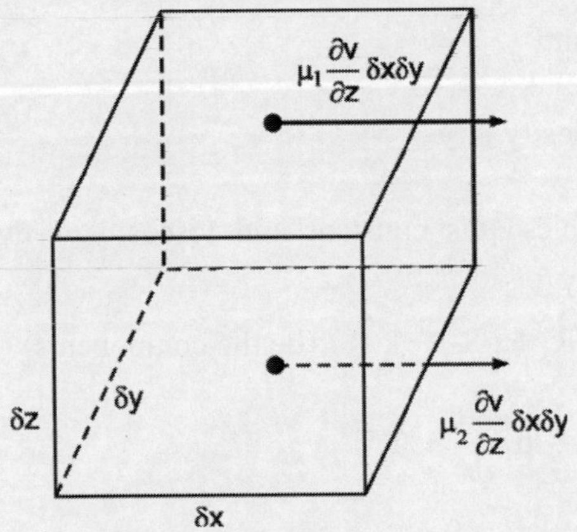

Fig. 8.1

Let the rate of shear at the top of the cube $\left(\frac{dv}{dz}\right)_1$ and at the bottom of the cube be $\left(\frac{dv}{dz}\right)_2$.

Shearing stress per unit area at the top surface $\tau_1 = \left(\mu\frac{dv}{dz}\right)_1$

∴ Total stress acting on the upper surface is

$$\tau_1\ \delta x\ \delta y = \left(\mu\frac{dv}{dz}\right)_1 \delta x\ \delta y$$

Similarly the shearing stress exerted on the bottom surface is

$$\tau_2\ \delta x\ \delta y = \left(\mu\frac{dv}{dz}\right)_2 \delta x\ \delta y$$

∴ The force acting on the cube $(\tau_1 - \tau_2)\ \delta x\ \delta y$ and the force per unit volume is $\frac{(\tau_1 - \tau_2)\ \delta x\ \delta y}{\delta x\ \delta y\ \delta z}$

Volume V= $\delta x\ \delta y\ \delta z$

The above expression can be put in terms of differential as

$$R = \frac{d\tau}{dz} = \mu\frac{d^2v}{dz^2} \qquad \text{using (8.1)}$$

where μ is constant

Kinematic viscosity is $\nu = \frac{\mu}{\rho}$

In hydrodynamics μ is constant and viscosity = dynamic viscosity + kinematic viscosity

Let $R = iR_x + j\ R_y + k\ R_z$ (be the components)

Then
$$R_x = \frac{1}{3}\ \mu\ \frac{\partial\theta}{\partial x} + \mu\ \nabla^2\ v_x$$

$$R_y = \frac{1}{3}\ \mu\ \frac{\partial\theta}{\partial y} + \mu\ \nabla^2\ v_y$$

$$R_z = \frac{1}{3}\ \mu\ \frac{\partial\theta}{\partial z} + \mu\ \nabla^2\ v_z$$

where $\theta = \text{div}\ \nu = \dfrac{\partial \nu_x}{\partial x} + \dfrac{\partial \nu_y}{\partial y} + \dfrac{\partial \nu_z}{\partial z}$

and $\nabla^2 = \dfrac{\partial^2}{\partial x^2} + \dfrac{\partial^2}{\partial y^2} + \dfrac{\partial^2}{\partial z^2}$ $\left(\nabla = i\dfrac{\partial}{\partial x} + j\dfrac{\partial}{\partial y} + k\dfrac{\partial}{\partial z}\right)$

When the fluid moves in layers or laminae, then the motion is called laminar flow. This theory and neglecting divergence of velocity is used in Navier-Strokes equation. Navier-Strokes equations are not applicable in oceanographic problems but used in fluid mechanics. Laminar flow is not applicable in local fluctuations of velocity.

Turbulence: When fluid moves in sheets in an orderly manner it is called Laminar flow. However ocean currents have various eddies of different dimensions. As a result small fluid masses have different velocities in regions. This irregular motion is called turbulent flow and the process is called turbulence. This is similar to the entrainment of mass in convective clouds.

Let $V = = \overline{V} + V'$

where

V = Actual velocity

$\overline{V}$ = Average velocity

V' = Turbulent velocity

$$\left.\begin{aligned} \text{If } V &= iu + jv + kw \\ \overline{V} &= i\,\overline{u} + j\,\overline{v} + k\,\overline{w} \\ V' &= i\,u' + j\,v' + k\,w' \end{aligned}\right\} \quad \begin{aligned} \text{implies } u &= \overline{u} + u' \\ v &= \overline{v} + v' \\ w &= \overline{w} + w' \end{aligned}$$

That is, velocity at any point (given) in space is assumed to be the vector sum of two different velocities $\overline{V}$ and V'.

The whole flow pattern = Average flow (which may steady or accelerated) + turbulent flow.

In ocean currents this type of flow is very common. This superimposed flow has importance because the turbulent flow modifies the apparent viscosity.

8.1 Frictional Stresses due to Turbulent Motion

The whole flow is composed of (1) average flow and + (2) Irregular turbulent flow. In turbulent motion, small fluid masses (parcels) are carried back and forth across any given surface. Because of different fluid masses a velocity gradient develops. As a result turbulent motion causes transport of momentum across the surfaces normal to the gradient and this stress is equal to the rate of momentum transport across the surface. A frictional force per unit volume exists; if the transport momentum into a given volume changes from the transport out of the volume.

The ocean has stable stratification. Vertical displacement works against buoyancy force. Turbulence in the ocean causes mixing. Vertical mixing requires more energy than horizontal mixing. Vertical mixing across surfaces of constant density (isopycnal) is called diapycnal mixing. Diapycnal changes the vertical structure of the ocean and controls the rate at which deep waters ultimately reaches in mid and low latitudes.

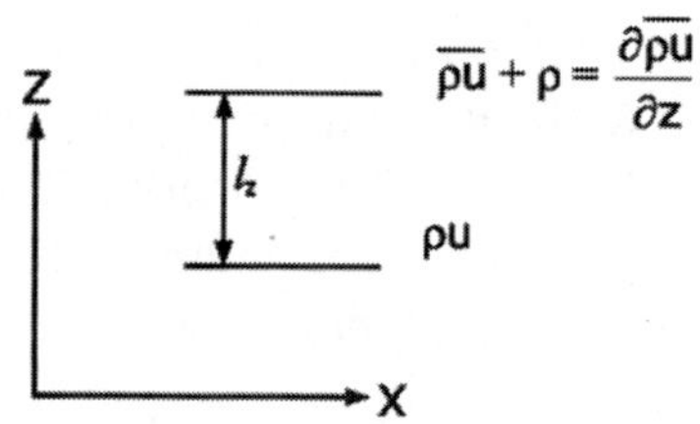

Fig. 8.2

Suppose the average flow in x-direction $\bar{u} > 0$ and that $\bar{v}, \bar{w}$ (in y, z, directions) is zero. Then $u = \bar{u} + u'$, $v = v'$, $w = w'$

Let the average velocity in z-direction changes and u = constant (in x - direction)

Let the water moves in z - direction a distance 'l_z'

The momentum of mean motion is ρ

∴ The rate of momentum (u,v) transport M will be

$$M = -k\,|\overline{w}'|\,l_z\,\frac{d(\rho\bar{u})}{dz}$$

where k is a factor of proportionality. Neglecting changes in density, k = 1, we can write $A_z = \rho|\overline{w}'|l_z$ .....(8.2)

Then the rate of momentum transport (M)

$$M = -A_z\frac{d\bar{u}}{dz} \qquad .....(8.3)$$

Here '–' sign, because the transport takes place towards regions of lower velocity.

According to Prandtl's theory

$$A_z = \rho\,L_z^2\left|\frac{d\bar{u}}{dz}\right| \qquad .....(8.4)$$

Then

$$\tau_{xz} = A_z\,\frac{d\bar{u}}{dz} \qquad .....(8.5)$$

where τ_{xz} = The stress in the direction of x-axis along a surface that is normal to the z-axis.

Eq. (8.5) is similar to $\tau = \mu \frac{dv}{d\eta}$ (definition of viscosity)

where μ = coefficient of dynamic viscosity

The eddy viscosity $\tau = A \frac{d\bar{u}}{dz}$

where $\bar{u}$ = mean motion, A depends on turbulence.

Let the stress in y-direction, a surface perpendicular z axis be τ_{yz}. (See Fig. 8.3).

$$\tau_{yz} = A_Z \frac{d\bar{v}}{dz} \qquad \text{.....(8.6)}$$

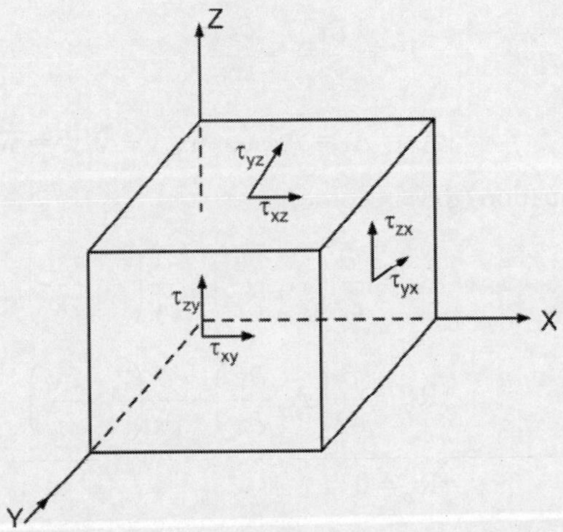

Fig. 8.3

If $\bar{u} \neq 0$, then we have four shearing stresses

$$\left.\begin{aligned} \tau_{xy} &= A_y \frac{\partial \bar{u}}{\partial y} \\ \tau_{zy} &= A_y \frac{\partial \bar{w}}{\partial y} \\ \tau_{zx} &= A_x \frac{\partial \bar{w}}{\partial x} \\ \tau_{yx} &= A_x \frac{\partial \bar{v}}{\partial x} \end{aligned}\right\} \qquad \text{.....(8.7)}$$

u = where flow (speed) in x direction, ν in y-direction & w in z-direction

Here the coefficients of eddy viscosity is defined as

$$A_x = \rho|\bar{u}|l_z$$

$$A_y = \rho|\bar{v}|l_z$$

8.2 Frictional Forces due to Turbulent Motion

As shown in fig 8.3, the frictional force acting in the direction of x-axis is derived from the stresses τ_{xz}, τ_{xy}

$$\therefore \quad R_x = \frac{\partial \tau_{xz}}{\partial z} + \frac{\partial \tau_{xy}}{\partial y} \quad \text{.....(8.8a)}$$

Similary

$$R_y = \frac{\partial \tau_{yz}}{\partial z} + \frac{\partial \tau_{yx}}{\partial x} \quad \text{.....(8.8b)}$$

$$R_z = \frac{\partial \tau_{zx}}{\partial x} + \frac{\partial \tau_{zy}}{\partial y} \quad \text{.....(8.8c)}$$

If $A_x = A_y = A_h$, $u = \bar{u}$, $v = \bar{v}, w = \bar{w}$

The above equation (8.8) become

$$R_x = \frac{\partial}{\partial z}\left(A_z \frac{\partial u}{\partial z}\right) + \frac{\partial}{\partial y}\left(A_h \frac{\partial u}{\partial y}\right)$$

$$R_y = \frac{\partial}{\partial z}\left(A_z \frac{\partial v}{\partial z}\right) + \frac{\partial}{\partial x}\left(A_h \frac{\partial v}{\partial x}\right)$$

$$R_z = 0$$

If horizontal turbulence is neglected the transport of momentum leads to the frictional terms

or

$$\left.\begin{aligned} R_x &= \frac{d}{dz}\left(A\frac{du}{dz}\right), \; R_y = \frac{d}{dz}\left(A\frac{dv}{dz}\right), \\ R_x &= A\frac{d^2u}{dz^2} \text{ and } R_x = A\frac{d^2u}{dz^2} \end{aligned}\right\} \quad \text{.....(8.9)}$$

where A may be variable

8.3 Influence of Stability (E) on Turbulence

According to Taylor, (1) turbulence expressed by eddy viscosity, decreasing with increasing stability.

Rate of turbulent energy per unit volume

$$= A_v\left(\frac{d\nu}{dz}\right),$$

where A_v = Eddy viscosity

The rate at which potential energy increases = gEA_s

where $$E = \text{stability} = \lim_{\delta z \to 0} \frac{1}{\rho}\frac{d\rho}{dz},$$

A_s = eddy diffusivity

This implies

$$A_v\left(\frac{dv}{dz}\right)^2 > gEA_s$$

or $$\frac{\left(\frac{dv}{dz}\right)^2}{gE} > \frac{A_s}{A_v}$$

1. Eddy viscosity 100 to 200 times greater than ordinary viscosity of seawater.
2. According to Taylors theory, turbulence is always present irrespective of stability. But the condition, the rate of Reynolds stresses energy must be greater than the rate at which the potential energy of that region increases. Ocean observations show that $A_s < A_v$. It is theorized that:
   (i) Turbulence is reduced to smaller values of the eddy viscosity and
   (ii) The type of the turbulence alters depending on accompanying eddy diffusivity is smaller than eddy viscosity.

8.4 Horizontal Turbulence

Suppose the field of pressure is balanced by the Coriolis force associated with the mean motion, i.e., pressure gradient = Coriolis force or deflecting force.

$$\left.\begin{aligned} \alpha\frac{\partial p}{\partial x} &= fv \\ \alpha\frac{\partial p}{\partial y} &= -fu \end{aligned}\right\} \text{from equations of motion}$$

If $u = \bar{u} + u', \quad v = \bar{v} + v'$

The turbulent horizontal velocities are given by

$$\frac{du'}{dt} = fv' \qquad \text{.....(8.10)}$$
$$\frac{dv'}{dt} = -fu'$$

where friction is neglected

Eq. (8.10) represents the equations of motion in a circle of inertia.

$\therefore$ The horizontal turbulent motion = motion in a circle of inertia.

However the above conclusion is not observed in practical aspect. Therefore the turbulent motion must reflect in the distribution of pressure.

Writing $p = \bar{p} + p'$, the above equation take the form

$$\left.\begin{aligned} f\bar{v} = \alpha\frac{\partial \bar{p}}{\partial x} \quad &\text{and } fv' = \alpha\frac{\partial p'}{\partial x} \\ -f\,\bar{u} = \alpha\frac{\partial \bar{p}}{\partial y} \quad &\text{and } -fu' = \alpha\frac{\partial p'}{\partial y} \end{aligned}\right\} \qquad \text{.....(8.11)}$$

In eq. (8.11) $-\frac{\partial \bar{p}}{\partial x}$ & $\frac{-\partial \bar{p}}{\partial y}$ represent mean pressure gradient and $\frac{-\partial p'}{\partial x}$ & $\frac{-\partial p'}{\partial y}$ represent local pressure gradient associated with turbulent motion.

8.5 Friction

Friction is of two types, external and internal. External friction is due to the interaction between the bodies at their surfaces of contact and obstructs their relative displacement. Internal friction or viscosity, which develops as tangential forces that obstruct the displacement of portions of fluids, liquids or gases with respect to each other. Friction between moving surfaces is called Kinetic friction, while between relatively static (motion less bodies) is called static friction. The coefficient of static friction in general is greater than coefficient of kinetic friction.

8.5.1 Boundary Friction

According to Prandtl, turbulent motion extends to the surface, that is mixing length has a definite value at the surface.

Flow will be in general, laminar within thin layer near the surface (the laminar boundary layer).

Let the mixing length be $l = k_0(z + z_0)$ .....(8.12)

where z_0 = roughness length, related to the average height z of the roughness elements. Prandtl definition of A_z is

$$A_z = \rho l_z^2 \left|\frac{d\bar{u}}{dz}\right|$$

$$= \rho k_0^2 (z + z_0)^2 \left|\frac{d\bar{u}}{dz}\right|, \quad \text{[using eq. (8.12)]} \quad .....(8.13)$$

and $$\tau = A\frac{d\nu}{dz}$$

$$= \rho k_0^2 (z + z_0)^2 \left(\frac{d\nu}{dz}\right)^2 \quad \text{[using eq. (8.13)]} \quad .....(8.14)$$

From eq (8.14)

$$\therefore \quad \frac{\tau}{\rho} = k_0^2 (z + z_0)^2 \left(\frac{d\nu}{dz}\right)^2$$

or $$\frac{d\nu}{dz} = \sqrt{\frac{\tau}{\rho}}\ \frac{1}{k_0(z + z_0)} \quad .....(8.15)$$

$$A = \rho k_0^2 (z + z_0)^2 \frac{d\nu}{dz} \quad \text{[def eq. (8.12)]}$$

$$= \rho k_0^2 (z + z_0)^2 \times \sqrt{\frac{\tau}{\rho}}\frac{1}{k_0(z + z_0)} \quad \text{[using (8.15)]}$$

$$= \rho k_0 (z + z_0)\sqrt{\frac{\tau}{\rho}} \quad ..... (8.16)$$

Integrating eq. (8.15) w.r.t z with boundary conditions $v = 0$ at $z = 0$ and v at z

$$\int_0^z \frac{dv}{dz}\, dz = \int_0^z \sqrt{\frac{\tau}{\rho}}\frac{1}{k_0(z + z_0)}\, dz$$

$$v = \frac{1}{k_0}\sqrt{\frac{\tau}{\rho}}\left\{\ln\frac{z + z_0}{z_0}\right\}$$

or $$v^2 = \frac{1}{k_0{}^2}\frac{\tau}{\rho}\left[\ln\left(\frac{z + z_0}{z_0}\right)\right]^2$$

or $$\tau = \frac{\rho k_0^2 v^2}{\left[\ln\left(\dfrac{z + z_0}{z_0}\right)\right]^2} \quad .....(8.17)$$

At the sea surface the boundary conditions differ. At the sea surface the wind stress τ_a (that exerts on water) must balance the stress that the water exerts on the air.

$$\tau_a + \left(A\frac{dv}{dz}\right)_0 = 0$$

or $$\tau_a = -\left(A\frac{dv}{dz}\right)_0 \qquad \text{.....(8.18)}$$

Eq. (8.18) implies A_0 and $\left(\frac{dv}{dz}\right)_0$ must be different from zero.

i.e., $$A_0 \neq 0 \ \& \left(\frac{dv}{dz}\right)_0 \neq 0$$

8.6 The Energy Equation when Friction is not Zero

We have proved $R_x = \frac{d}{dz}\left(A\frac{du}{dz}\right)$ & $R_y = \frac{d}{dz}\left(A\frac{dv}{dz}\right)$

Neglecting horizontal turbulence effect, the power of the frictional force per unit volume is

$$R_x u + R_y v = u\frac{d}{dz}\left(A\frac{du}{dz}\right) + v\frac{d}{dz}\left(A\frac{dv}{dz}\right)$$

where $A \simeq A_z$

Consider differentiation of product

$$\frac{d}{dz}\left(uA\frac{du}{dz}\right) = uA\frac{d^2u}{dz^2} + \frac{d(uA)}{dz}\frac{du}{dz}$$

or $$u\frac{d}{dz}\left(A\frac{du}{dz}\right) = \frac{d}{dz}\left(uA\frac{du}{dz}\right) - \frac{d(uA)}{dz}\frac{du}{dz}$$

Integrating w.r.t z between 0 to d

$$\int_0^d u\frac{d}{dz}\left(A\frac{du}{dz}\right)dz = \int_0^d \frac{d}{dz}\left(uA\frac{du}{dz}\right)dz - \int_0^d \frac{d(uA)}{dz}\frac{du}{dz}dz$$

$$= uA\frac{du}{dz} - \left(uA\frac{du}{dz}\right)_0 - \int_0^d \frac{d(uA)}{dz}\frac{du}{dz}dz$$

$$= v_{ox}\tau_{ax} - \int_0^d \frac{d(uA)}{dz}\frac{du}{dz}dz \qquad \text{(using definitions)} \qquad \text{.....(8.19)}$$

At the bottom velocity is zero, where surface stress is equal to the stress exerted by wind. The integral on the RHS represents the loss of KE per unit time and unit surface area, due to vertical turbulence (the dissipation).

If we take A = constant, the integral (RHS of 8.19) would reduce to $A\int_0^d \left(\frac{du}{dz}\right)^2 dz$, which is the form of general dissipation function.

The first term on the RHS of eq. (8.19), $\nu_o \tau_a$ represents the power per unit area which is exerted by the stress of the wind.

The energy equation can be written as

$$\frac{d}{dt}\int_0^d \frac{1}{2}\rho v^2 dz = \nu_{ox}\tau_{ax} + \nu_{oy}\tau_{ay}$$

$$-\int_0^d 10\left(u\frac{\partial p}{\partial x} + v\frac{\partial p}{\partial y}\right)dz - \int_0^d \left(\frac{d(uA)}{dz}\frac{du}{dz} + \frac{d(vA)}{dz}\frac{dv}{dz}\right)dz \quad ..... (8.20)$$

The above equation means KE of a water column per unit area cross section (surface to bottom) changes with time = The power of the stress of wind on the surface – the total work per unit time due to motion across the isobars –The dissipation of energy.

8.7 Viscous Force (Dissipation)

Fluids have viscous molecular process, which smoothout velocity variations and also slowdown the flow. The molecular processes are very weak, hence fluids are treated mostly as 'inviscid' instead of viscous. The effect of turbulent mixing/stirring of the fluid is eddy viscosity. Eddy viscosity empirically determined from observations or indirectly from models.

For ocean circulation, the turbulent motions are meso-scale eddies [fine vertical structure]

Eddy viscosity (proportional) $\propto$ turbulent speed $\times$ path length

Horizontal eddy viscosity (A_H) is >> vertical (A_v) eddy viscosity

The direction of A_H is along isopycnals (adiabatic surfaces) and A_v is across (perpendicular) to isopycnals (called diapycnal mixing). Isopycnals are natural coordinates for quasi-lateral motion and diapycnals for quasi-vertical motion.

In case of molecular viscous stress, the x-momentum dissipation is $= \nu\left(\frac{\partial^2 u}{\partial x^2} + \frac{\partial^2 v}{\partial y^2} + \frac{\partial^2 w}{\partial z^2}\right)$

where ν = coefficient of molecular viscosity.

The x-momentum of eddy viscosity (dissipation)

$$=A_H\left(\frac{\partial^2 u}{\partial x^2}+\frac{\partial^2 u}{\partial y^2}\right)+A_v\left(\frac{\partial^2 u}{\partial z^2}\right)$$

swhere A_v is perpendicular to A_H

The eddy viscosity coefficients are generally used as below.

$$\text{x- momentum dissipation}=\frac{\partial}{\partial x}\left(A_H\frac{\partial u}{\partial x}\right)+\frac{\partial}{\partial y}\left(A_H\frac{\partial u}{\partial y}\right)+\frac{\partial}{\partial z}\left(A_v\frac{\partial u}{\partial z}\right)$$

The momentum balance with eddy viscosity and rotation (coriolis force) is

(i) $$\frac{du}{dt}-fv\equiv\frac{\partial u}{\partial t}+u\frac{\partial u}{\partial x}+v\frac{\partial u}{\partial y}+w\frac{\partial u}{\partial z}-fv$$

$$=-\frac{1}{\rho}\frac{\partial p}{\partial x}+\frac{\partial}{\partial x}\left(A_H\frac{\partial u}{\partial x}\right)+\frac{\partial}{\partial y}\left(A_H\frac{\partial u}{\partial y}\right)+\frac{\partial}{\partial z}\left(A_v\frac{\partial u}{\partial z}\right)$$

(ii) $$\frac{dv}{dt}+fu\equiv\frac{\partial v}{\partial t}+u\frac{\partial v}{\partial x}+v\frac{\partial v}{\partial y}+w\frac{\partial v}{\partial z}+fu$$

$$=-\frac{1}{\rho}\frac{\partial p}{\partial y}+\frac{\partial}{\partial x}\left(A_H\frac{\partial v}{\partial x}\right)+\frac{\partial}{\partial y}\left(A_H\frac{\partial v}{\partial y}\right)+\frac{\partial}{\partial z}\left(A_v\frac{\partial v}{\partial z}\right)$$

(iii) $$\frac{dw}{dt}\equiv\frac{\partial w}{\partial t}+u\frac{\partial w}{\partial x}+v\frac{\partial w}{\partial y}+w\frac{\partial w}{\partial z}$$

$$=\frac{-1}{\rho}\frac{\partial p}{\partial z}-g+\frac{\partial}{\partial x}\left(A_H\frac{\partial w}{\partial x}\right)+\frac{\partial}{\partial y}\left(A_H\frac{\partial w}{\partial y}\right)+\frac{\partial}{\partial z}A_v\frac{\partial w}{\partial z}$$

Table 8.1 (at salinity 35) Molecular, Eddy horizontal (along isopycnal), Eddy vertical (along diapycnal)

	At 0 °C	At 20 °C	
Viscosity	1.83 × 10^{-6} m^2/sec	1.05 × 10^{-6} m^2/sec	100-10000 m^2/sec/1cm^2/sec
Thermal diffusivity	1.37 × 10^{-7} m^2/sec	1.46 × 10^{-7} m^2/sec	100-10000 m^2/sec/$10^{-1}$$cm^2$/sec
Haline diffusivity	1.37 × $10^{-9}$$m^2$/sec	-	100-10000 m^2/sec/10^{-1} cm^2/sec^2

CHAPTER 9

FLUID STATICS

A fluid is a substance that can flow. Fluids include both liquids and gases.

Fluid mechanics: Fluid mechanics deals with the laws of equilibrium and motion of fluids (both liquids & gases). It also deals with the interaction of moving liquids and gases with the solid bodies that come across the flow. Thus it deals with fluid statics and fluid dynamics.

Fluid statics: The study of fluid at rest is called fluid statics. It deals with the laws of equilibrium of liquids and gases under the action of applied forces.

Fluid dynamics: The study of fluids in motion is called fluid dynamics. It deals with the laws of motion of liquids and gases and of other interaction with solids.

The main difference of liquids and gases is that liquids are practically incompressible while gases are highly compressible under the action of pressure.

Incompressible fluid: A fluid, that is liquid or gas, in which its density is not a function of pressure. This amounts that the dependence of fluid density on pressure can be ignored or neglected.

Compressible fluid: The fluid is a gas in which density is a function of pressure – denoted as $\rho = \rho(p)$.

i.e ., the dependence of density on pressure cannot be ignored or neglected.

An Ideal fluid: An ideal fluid is one in which there is no internal friction (viscosity) and treated to be incompressible.

A Viscous fluid: A fluid in which there exists internal friction is called viscous fluid.

A Barotropic fluid: A fluid in which its density depends on pressure. $\rho = \rho(p)$.

In fluid statics we have two types of external forces acting on an element of fluid volume (dv) of the fluid. They are called mass force and surface force.

9.1 Mass Force or Body Force

It is the force acting on the element (of mass or volume) under consideration is independent of other portion of the fluid.

Mass force is proportional to the mass of the fluid element. Mass force is the force of gravity.

If $\vec{F}$ = mass force per unit mass of fluid

ρ = density of fluid

dv = element of fluid volume

Then, Mass force = $\vec{F}\,\rho dv$.

The vector $\vec{F}$ is called the intensity of the mass force field ($\vec{F} = \vec{g}$ in case of gravity, it is the free fall acceleration)

If $\vec{F} = -\text{grad}\,\phi = -\nabla\phi$, then we say, mass force $\vec{F}$ is conserved,

where $\phi = \phi$ (x, y, z, t) is called mass force potential.

$$\nabla\phi = i\frac{\partial\phi}{\partial x} = j\frac{\partial\phi}{\partial y} + k\frac{\partial\phi}{\partial z} = \text{gradient of scalar function } \phi.$$

9.2 Surface Force

Surface forces are those which are applied to the fluid element (dv) by the adjoining particles of the remaining fluid (V). These forces act on the surface (ds) of the elements (dv) under consideration.

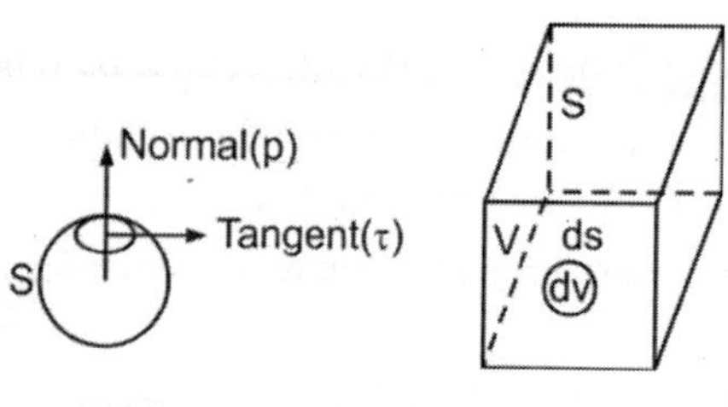

Fig. 9.1

The surface force per unit of surface on which it acts is called average pressure intensity. Any surface force can be resolved into two components.

One is along the tangent (τ) and another is perpendicular to the surface (p).

τ is called tangential pressure intensity

p is called pressure intensity or simply pressure.

p is constant does not depend on the orientation of the surface (ds).

9.3 Equilibrium Equations of Fluid

Let F be the resultant intensity of mass force field acting on the fluid element 'dv'.

The equation under equilibrium is given by

$$\vec{F} = \frac{1}{\rho}\text{grad } p = \frac{1}{\rho}\nabla p \qquad .....(9.1)$$

where ∇p = Pressure gradient

$\vec{F}$ = Mass force field

ρ = density of the fluid

or in component form of eq (9.1)

$$iF_x + jF_y + kF_z = \frac{1}{\rho}\left(i\frac{\partial p}{\partial x} + j\frac{\partial p}{\partial y} + k\frac{\partial p}{\partial z}\right)$$

$$\Rightarrow F_x = \frac{1}{\rho}\frac{\partial p}{\partial x}, \quad F_y = \frac{1}{\rho}\frac{\partial p}{\partial y}, \quad F_z = \frac{1}{\rho}\frac{\partial p}{\partial z}$$

where F_x, F_y, F_z are components of $\vec{F}$ are the projections of vector $\vec{F}$ along x, y, z axis respectively.

These equations are valid only when the velocity of the fluid flow is zero (i.e., stationary or static fluid).

9.4 Equilibrium Condition under the Action of Mass Forces

If a fluid is subjected to the action of mass forces it will be in equilibrium when the field intensity $\vec{F}$ of these forces satisfy.

The scalar products of $\vec{F}$ and the rotation of $\vec{F}$ is zero

i.e., $\vec{F} \cdot \text{Curl } \vec{F} = 0$

or $\vec{F}.(\nabla \times \vec{F}) = 0$

i.e., $\vec{F}.\left[\left(i\frac{\partial}{\partial x} + j\frac{\partial}{\partial y} + k\frac{\partial}{\partial z}\right) x \left(iF_x + jF_y + kF_z\right)\right] = 0$

or $(iF_x + j F_y + kF_z).\left[i\left(\frac{\partial F_z}{\partial y} - \frac{\partial F_y}{\partial z}\right) + j\left(\frac{\partial F_x}{\partial z} - \frac{\partial F_z}{\partial x}\right) + k\left(\frac{\partial F_y}{\partial x} \frac{-\partial F_x}{\partial y}\right)\right] = 0$

$$F_x\left(\frac{\partial F_z}{\partial y} - \frac{\partial F_y}{\partial z}\right) + F_y\left(\frac{\partial F_x}{\partial z} - \frac{\partial F_z}{\partial x}\right) + F_2\left(\frac{\partial F_y}{\partial x} - \frac{\partial F_x}{\partial y}\right) = 0$$

If ρ = constant, then $F = \frac{1}{\rho}$ grad p

$$= \text{grad}\left(\frac{p}{\rho}\right) = \nabla\left(\frac{p}{\rho}\right)$$

This is possible only in a potential force field, with a potential of $\phi = -\frac{p}{\rho} +$ constant.

The equipressure surfaces (or isobaric surfaces) coincide with the equipotential surfaces.

9.5 Hydrostatic Equation

Consider the equilibrium of a fluid in uniform gravitational field.

$$\vec{F} = -\vec{g} = \text{constant} \qquad \text{.....(9.2)}$$

i.e., $(i F_x + j F_y + k F_z) = -kg$

$\therefore$ $F_x = F_y = 0$ and $F_z = -g$

This implies $F_z = \frac{1}{\rho}\frac{dp}{dz} = -g$ [(since $F_x = F_y = 0$)]

or $$\frac{1}{\rho}dp = -gdz \qquad \text{.....(9.3)}$$

integrating (9.3) under boundary conditions

$p = p_0$ at $z = 0$ and $p = p$ at $z = z$

$$\int_{p_0}^{p} \frac{dp}{\rho} = -\int_{0}^{z} gdz$$

$$\left.\begin{aligned} \frac{1}{\rho}(p-p_0) &= -gz \\ p &= p_0 - \rho g z \end{aligned}\right\}$$

Or .....(9.4)

eq. (9.3) & eq. (9.4) are called the basic hydrostatic equation in incompressible fluid.

Note: The quantity $p - p_0$ does not depend on p_0. This implies the pressure exerted on the fluid by external forces is transmitted equally in all direction, which is the famous Law of Pascal.

Other Equilibrium Conditions

1. **Chemical equilibrium:** In an hetrogeneous system, the equilibrium is defined the chemical potential of any component should be same in all phases containing the component.
2. **Thermal equilibrium:** The temperature of the system should be same in all portions in the system.
3. **Mechanical equilibrium:** The pressure should be same in all position of a system under equilibrium, in which no other forces act except an uniform external pressure.
4. **The thermodynamic identity:** In a cycle, change in entropy is equal to zero and the algebraic sum of reduced amounts of heat delivered to the system equal to zero in a reversible process.

$$\oint_{\text{reversible process}} \frac{\delta Q}{T} = 0 \quad \text{(in a reversible process.)}$$

and Less than zero in an irreversible process

$$\oint_{\text{irreversible process}} \frac{\delta Q}{T} < 0$$

9.5.1 Gibbs Function

Let G = Gibbs function of unit mass of the fluid,

and gz = the potential energy of unit mass in gravitational field.

In a compressible fluid, which is in thermal and mechanical equilibrium, then G + gz should be same throughout the whole volume.

9.5.2 Mechanical Equilibrium

Mechanical equilibrium of a fluid in a gravitational field in the absence of thermal equilibrium (that is when the temperature of the fluid varies along the vertical axis 0z).

This equilibrium is stable if (under no convection condition) $\left(\frac{\partial v}{\partial T}\right)_p \frac{ds}{dx} > 0$

where v = specific volume

s = entropy of unit mass of fluid

For most fluids $\left(\frac{\partial v}{\partial T}\right)_p > 0$ and no convection condition takes the form

$$\frac{dT}{dz} > -\frac{gT}{C_p v}\left(\frac{\partial v}{\partial T}\right)_p$$

where C_p= specific heat of fluid at constant pressure.

For ideal gas $\frac{dT}{dz} > \frac{-g}{C_p}$

9.5.3 Archimede's Principle

It states that, if a body is wholly or partly immersed in a fluid, it is buoyed up with force which is equal to the weight of the volume of the fluid displaced by the body, and it acts vertically upwards through the CG (centre of gravity) of the immersed part of the body.

Note: A fluid is a substance that has both definite mass and volume but has no definite shape. A fluid cannot sustain a shear stress under the equilibrium condition.

9.5.4 Ideal and Real Fluids

The compressibility of most of the liquids is very very small and that its influence on flow is negligible. A fluid that is both incompressible and in viscid (non-viscous) is called perfect or ideal fluid.

Real fluids are those in which tangential or shear forces always come into play whenever motion takes place and this gives rise to fluid friction. Such fluids are also compressible.

9.5.5 Inviscid Fluid

An inviscid fluid is one in which there is no friction, that is the fluid viscosity is zero. Even during motion, the internal forces at any internal section are always normal to the section and are purely pressure forces. Such fluid does not exist in reality.

9.6 Relative Density or Specific Gravity of a Material

It is the ratio of density of a material to that of water.

$$\begin{matrix}\text{Relative density}\\ \text{[or]}\\ \text{Specific gravity}\end{matrix} = \frac{\text{Density of the material}}{\text{Density of water}}$$

pressure in water (ocean/sea/lake) increases with depth below the surface.

$$\text{Pressure at a point (p)} = \frac{\text{Normal force (dF) exerted}}{\text{Area(dA)}}$$

In the limit,

$$\text{Pressure (p)} = \frac{F}{A},$$

where F = force

P = pressure

A = Area

If P_1, P_2 are pressures in a fluid at elevations h_1, h_2 (respectively) above some reference level, then

$$P_2 - P_1 = -\rho g (h_2 - h_1)$$

where ρ = density of the fluid

g = gravitational force.

9.7 Fluid Kinematics

It deals with the geometry of fluid motion in terms of a fluid particle qualitatively expressed in terms of velocity vector.

CHAPTER 10

FLUID DYNAMICS

Fluid dynamics is the study of fluids in motion. A fluid particle or element is a minute volume of fluid whose linear dimensions are negligible and it may be assumed as a geometrical point for the mathematical discussion and investigations (its velocity and acceleration).

10.1 Velocity of a Fluid Particle

Let P be the position of a particle at any time 't', where $\vec{OP} = \vec{r}$ and Q be the position at time $(t + \delta t)$ where $= \vec{OQ} = \vec{r} + \delta\vec{r}$

The velocity of the particle at P is defined by the vector

$$\vec{V} = \lim_{\delta t \to 0} \frac{(\vec{r} + \delta\vec{r}) - \vec{r}}{(t + \delta t) - (t)}$$

$$= \lim_{\delta t \to 0} \frac{\delta\vec{r}}{\delta t} = \frac{d\vec{r}}{dt}$$

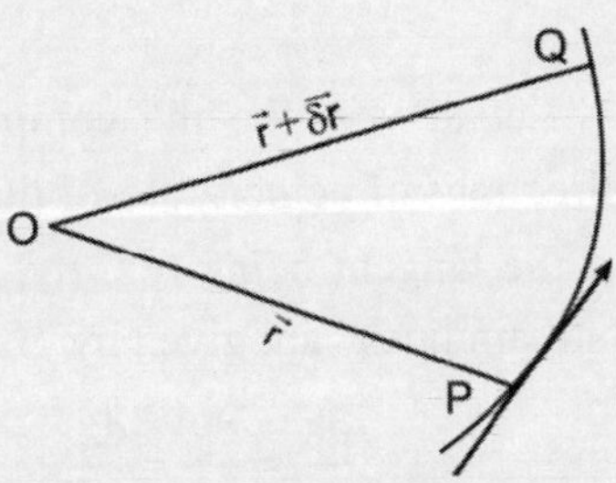

The quantity $\vec{V}$ is a function of $\vec{r}$ and t (also denoted as $\vec{V} = \vec{V}(\vec{r}, t)$)

$$\vec{V} = f(\vec{r}, t)$$

The form of the function f gives the motion of the fluid. The velocity of the particle (or element) varies both in magnitude and direction along its line of flow.

If the motion of individual particle form the parts of the regular system of curves, the motion is called streamline motion, and if the tracks are widely irregular, the motion is turbulent.

Streamlines: A streamline is a curve drawn in the fluid such that, at any time, the direction of the tangent to the curve (at any point) coincides with the direction of the velocity of the fluid particle at that point.

The Fig. 10.1 shows the streamline flow around obstacles of various shapes.

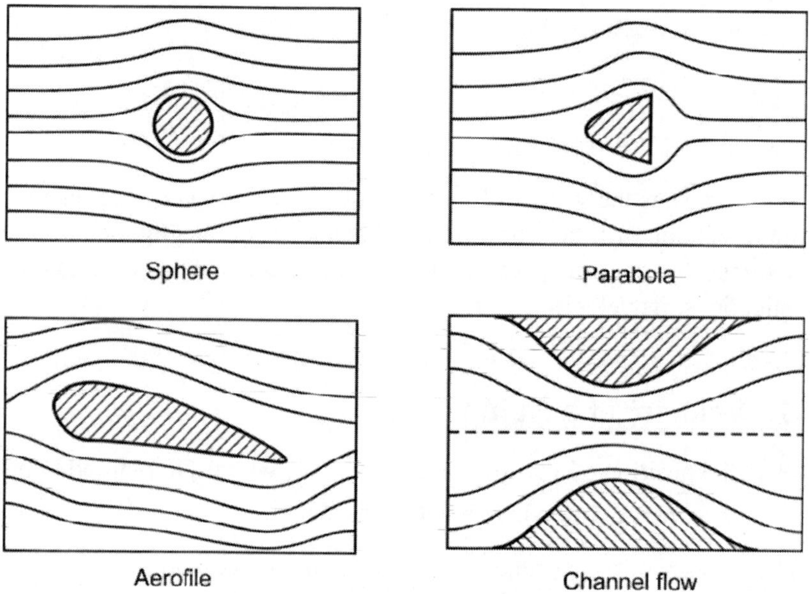

Fig. 10.1

Let u, v, w be the components of velocity (or projections in x, y, z directions of velocity) $\vec{V}$ of the fluid particle at P(x, y, z), the direction ratios of the tangent being (dx, dy, dz) at that point, the differential equations of "stream lines" are given by

$$\frac{dx}{u} = \frac{dy}{v} = \frac{dz}{w}$$

where $\vec{V} = iu + jv + kw$

and $u = \frac{dx}{dt}, v = \frac{dy}{dt}, w = \frac{dz}{dt}$

Acceleration components in x, y, z directions are $\frac{\partial^2 x}{\partial t^2}, \frac{\partial^2 y}{\partial t^2}, \frac{\partial^2 z}{\partial t^2}$

10.2 Path Lines or Trajectories

Path line is a curve which a particular fluid particle described (traced) during its motion, that is the path followed by an element (or parcel of water) of a moving fluid during its motion.

The differential equations of the path lines are given by $\frac{dx}{dt} = u, \ \frac{dy}{dt} = v, \ \frac{dz}{dt} = w$

(where the differentiation is with respect to time 't')

10.3 Difference between the Streamlines and Path Lines

Streamlines show how each fluid particle is moving at a given instant (time), while the path lines show how a given particle is moving at each instant of time.

In a steady flow, the streamlines coincide with the path lines.

Stream tube

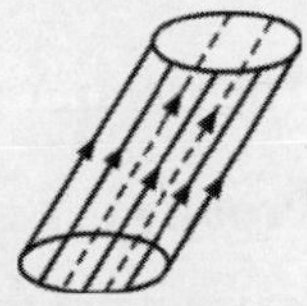

The streamlines drawn through each point of a closed curve enclose a tubular surface in the fluid. This tubular surface is called a stream tube or simply tube flow.

A stream tube of infinitesimal cross-section is called a stream filament.

10.4 Velocity, Acceleration and Speed

Let P ($\vec{r}$), $P_1(\vec{r}_1)$ be the positions of a particle at time 't' and $t_1 = t + \Delta t$, which a particle moved in a small time interval Δt, as shown in Fig. 10.2.

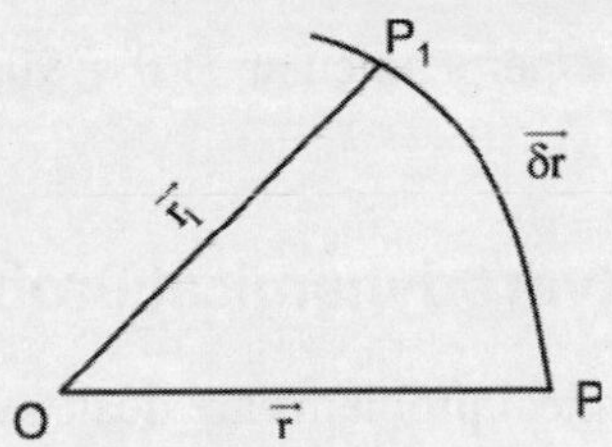

Fig. 10.2

then, velocity $\vec{V} = \frac{\Delta \vec{r}}{\Delta t}$

or $\vec{V} = \frac{d\vec{r}}{dt}$ (in the limit as $\Delta t \to 0$)

Acceleration $\vec{a} = \frac{d\vec{V}}{dt} = \frac{d^2\vec{r}}{dt^2}$

If $\quad \vec{r} = ix + jy + kz$, then

$$\vec{V} = i\frac{dx}{dt} + j\frac{dy}{dt} + k\frac{dz}{dt} \text{ or } iu + jv + kw$$

Speed $\quad V = |\vec{V}| = \sqrt{\left(\frac{dx}{dt}\right)^2 + \left(\frac{dy}{dt}\right)^2 + \left(\frac{dz}{dt}\right)^2}$

Note $\quad V = |\vec{V}| = \sqrt{\vec{V}.\vec{V}},\ \dot{\vec{r}} = \frac{d\vec{r}}{dt},\ \ddot{\vec{r}} = \frac{d^2\vec{r}}{dt^2}$

If 's' is the path length of the particle traced in time period t_0 to t_1, then

$$s = \int_{t_0}^{t_1} |\vec{V}|\, dt = \int_{t_0}^{t} \sqrt{\left(\frac{dx}{dt}\right)^2 + \left(\frac{dy}{dt}\right)^2 + \left(\frac{dz}{dt}\right)^2}\, dt$$

Theorem: A vector of constant length is normal to its time derivative.

Proof: Let $\vec{b} = \vec{b}(t)$ is a vector function of time 't'. Then, $\sqrt{\vec{b}\vec{b}} =$ constant (= length of the vector)

i.e., $\quad \vec{b}^2 = \vec{b}.\vec{b} = \text{const} \qquad .....(10.1)$

Differentiating eq. (10.1) w. r. t to 't' (time), we have $2\vec{b}\cdot\frac{d\vec{b}}{dt} = 0$

i.e., $\vec{b}\cdot\frac{d\vec{b}}{dt} = 0$ implies $\vec{b}$ and $\frac{d\vec{b}}{dt}$ are at right angles.

[$\because$ $\vec{A}\cdot\vec{B} = ab\cos\theta$ $\vec{A}$ is perpendicular $\vec{B}, \theta = 90$, $\vec{A}\cdot\vec{B} = 0$]

10.5 Methods of Hydrodynamical Problems

In general the hydrodynamical problems are dealt by two methods.

1. The Eulerian Flux or Local time–rate method.
2. Lagrangian method or individual time – rate of change.

1. **Euler's Method:** In this method some point is selected in space occupied by fluid, and observe the changes that occur in velocity, density and pressure as the fluid passes through this point. As the

point is fixed x, y, z, t are independent variables. Consequently the derivatives $\frac{dx}{dt}$, $\frac{d^2x}{dt^2}$ etc., have no significance.

Let $f(x,y,z,t) = f(\vec{r},t)$ be any scalar point function related with a fluid motion.

Keeping the point 0(x,y,z) as fixed, the change is $f(x,y,z,\ t+\delta t) - f(x,y,z,t)$

Or $\quad f(\vec{r},t+\delta t) - f(\vec{r},t)$

The local time-rate of change is $\frac{\partial f}{\partial t}$ is given by

$$\frac{\partial f}{\partial t} = \underset{\delta t \to 0}{\text{lmt}} \frac{f(\vec{r},t+\delta t) - f(\vec{r},t)}{\delta t}$$

A similar expression can be determined for a vector point function.

2. **Lagrangian Method:** In this method, a particle (or an infinitesimal volume parcel) of fluid selected and it is pursued on its onward journey, making observations of changes in velocity, density and pressure at each instant and at each point.

In this method the expression $\frac{dx}{dt}$, $\frac{d^2x}{dt^2}$ etc., have definite significance. In this method, to specify a particular fluid – particle, we require its initial position coordinates, say (x_0, y_0, z_0) or $(\vec{r}_0)$ so that there are altogether four independent variables (x_0, y_0, z_0, t) in Cartesian system and $(\vec{r}_0, t)$ in vector treatment.

Here $x = f_1(x_0, y_0, z_0, t)$

$y = f_2(x_0, y_0, z_0, t)$

$z = f_3(x_0, y_0, z_0, t)$

where x_0, y_0, z_0 are called lagrangian variables

Note: Sometimes (a, b, c,) are used instead of (x_0, y_0, z_0). Consider any scalar point function f (x, y, z, t) or $f(\vec{r},t)$ associated with a fluid in motion.

Keeping particle fixed, the change is

$$f(\vec{r}+\delta\vec{r},t+\delta t) - f(\vec{r},t).$$

The change $\delta\vec{r}$ in the position of the particle during the time 'δt' depends upon $\vec{V}$, the velocity of the particle at time 't'.

Thus $\delta\vec{r} = \vec{V}.\delta t$

$$\therefore \quad \frac{df}{dt} = \underset{\delta t \to 0}{\text{lmt}} \frac{f(\vec{r}+\delta\vec{r}, t+\delta t) - f(\vec{r}, t)}{\delta t}$$

$\frac{df}{dt}$ is the individual time–rate of change.

10.6 Relation between the Local and Individual Time-rate

Let u, v, w, be the components (or projections) of velocity $\vec{V}$ along the coordinate axes.

$\vec{V} = iu + jv + kw$, where i, j, k are unit vectors in x, y, z directions respectively.

And $\frac{dx}{dt} = u, \ \frac{dy}{dt} = v, \ \frac{dz}{dt} = w$

Now $f = f(x, y, z, t)$

$$\frac{df}{dt} = \frac{\partial f}{\partial t} + \frac{\partial f}{\partial x}\frac{dx}{dt} + \frac{\partial f}{\partial y}\frac{dy}{dt} + \frac{\partial f}{\partial z}\frac{dz}{dt}$$

or
$$\frac{df}{dt} = \frac{\partial f}{\partial t} + u\frac{\partial f}{\partial x} + v\frac{\partial f}{\partial y} + w\frac{\partial f}{\partial z}$$

$$= \frac{\partial f}{\partial t} + (iu + jv + kw).\left(i\frac{\partial f}{\partial x} + j\frac{\partial f}{\partial y} + k\frac{\partial f}{\partial z}\right)$$

$$= \frac{\partial f}{\partial t} + \vec{V}.\nabla f, \qquad \text{where } \nabla = i\frac{\partial}{\partial x} + j\frac{\partial}{\partial y} + k\frac{\partial}{\partial z}$$

i.e.,
$$\frac{df}{dt} = \frac{\partial f}{\partial t} + \vec{V}(\nabla f) \qquad \text{....(10.2)}$$

this is the required relation.

A similar expression for a vector point function $\vec{U}$ can be derived in the form

$$\frac{d\vec{U}}{dt} = \frac{\partial \vec{U}}{\partial t} + (\vec{V}.\nabla)\vec{U} \qquad \text{....(10.3)}$$

Note: Measurement of Velocity at a point by Flux Method

Hold a small plane surface at the point perpendicular to the direction of flow. Then the velocity at the point is measured by the time rate of flow of volume of fluid per unit area across the surface.

10.7 Acceleration and Advection

Eulerian Method: It consists in determining the motion or flow of field by specifying velocity field of the fluid in space at each instant of time.

That is $\vec{V} = f(\vec{r}, t)$

or

Projections (components) on the rectangular Cartesian axes,

$$u = f_1(x, y, z, t)$$

$$v = f_2(x, y, z, t)$$

$$w = f_3(x, y, z, t)$$

where $\vec{V} = iu + jv + kw$

$\vec{V}$ is the velocity of the fluid at any instant of time 't' at the point $P = (\vec{r} = ix + jy + kw)$ in space. The quantities (x, y, z) are known as the Eulerian variables. Instead of (x, y, z), cylindrical, spherical, natural or other kinds of coordinates can be used.

The components (or projections) of acceleration $\vec{a} = ia_x + ja_y + ka_z$ of the fluid in Cartesian coordinates are given by:

$$a_x = \frac{du}{dt} = \frac{\partial u}{\partial t} + u\frac{\partial u}{\partial x} + v\frac{\partial u}{\partial y} + w\frac{\partial u}{\partial z}$$

$$a_y = \frac{dv}{dt} = \frac{\partial v}{\partial t} + u\frac{\partial v}{\partial x} + v\frac{\partial v}{\partial y} + w\frac{\partial v}{\partial z}$$

$$a_z = \frac{dw}{dt} = \frac{\partial w}{\partial t} + u\frac{\partial w}{\partial x} + v\frac{\partial w}{\partial y} + w\frac{\partial w}{\partial z}$$

$$\vec{a} = \vec{a}_{local} + \vec{a}_{\substack{convection\\ \text{or advection}}}$$

where $\vec{a}_{local} = i\frac{\partial u}{\partial t} + j\frac{\partial v}{\partial t} + k\frac{\partial w}{\partial t}$

and

$$\vec{a}_{\substack{convection\\ \text{or advection}}} = i\left(u\frac{\partial u}{\partial x} + v\frac{\partial u}{\partial y} + w\frac{\partial u}{\partial z}\right)$$

$$+ j\left(u\frac{\partial v}{\partial x} + v\frac{\partial v}{\partial y} + w\frac{\partial v}{\partial z}\right)$$

$$+k\left(u\frac{\partial w}{\partial x}+v\frac{\partial w}{\partial y}+w\frac{\partial w}{\partial z}\right)$$

The components of $\vec{a}_{\substack{\text{convection}\\\text{or advection}}}$ are due to the non-conformity of the velocity field.

Note: $\frac{dU}{dt}=\frac{\partial u}{\partial t}+u\frac{\partial u}{\partial x}+v\frac{\partial u}{\partial y}+w\frac{\partial u}{\partial z}$, this indicates change in U at a point (Eulerian frame work).

In a lagrangian frame work, following the particle of water (fluid), only the time derivative appears, the advection terms do not appear because they are involved in the movement of the particle.

10.8 Lagrangian Method

The components (or projections) of velocity vector

$$\vec{V}=iu+jv+kw \text{ is given by}$$

$$u=\frac{\partial x}{\partial t},\ v=\frac{\partial y}{\partial t},\ w=\frac{\partial z}{\partial t}$$

and acceleration components in x, y, z directions of $\vec{a}=ia_x+ja_y+za_z$ is given by

$$a_x=\frac{\partial^2 x}{\partial t^2},\ a_y=\frac{\partial^2 y}{\partial t^2},\ a_z=\frac{\partial^2 z}{\partial t^2}$$

10.9 Some Definitions of Hydrodynamics

Vorticity: If $\vec{V}$ be the velocity vector of a fluid particle, then the vorticity vector $\vec{W}$ is defined as

$$\vec{W}=\frac{1}{2}\nabla\times\vec{V}=\frac{1}{2}\text{curl }\vec{V}\text{ or }\frac{1}{2}\text{ rot }\vec{V}$$

Vorticity is angular velocity (the rotational velocity) of an infinitesimal element (of volume).

Vortex lines: A vortex line is a curve drawn in the fluid such that the tangent at any point of the curve is in the direction of the velocity vector $\vec{W}$ at that point.

The differential equations of the system of vertex lines are given by

$$\frac{dx}{\xi}=\frac{dy}{\eta}=\frac{dz}{\varsigma}$$

where $\vec{W}=i\xi+j\eta+k\varsigma$

Vortex tube: The vortex lines drawn through each point of closed curve enclose a tubular space in the fluid called a vortex tube.

A vortex tube of infinitesimal cross section is called vortex filament or simply a vortex.

Irrotational motion: The motion of a fluid is called irrotational if the vorticity $\vec{W}$ of every fluid particle (or infinitesimal volume of fluid) is zero. i.e., $\vec{W}=0$.

If $\vec{W}\neq 0$, the motion is called rotational.

10.10 Velocity Potential

Let $\phi=\phi\,(x, y, z, t)$ be a scalar function.

If $\nabla\phi=-\vec{V}$ or grad $\phi=-\vec{V}$, then ϕ is called velocity potential.

In other words, if the expression

$$u\,dx+vdy+wdz=-\,d\phi\text{, an exact differential}$$

$$\text{viz}\quad -d\phi=-\frac{\partial\phi}{\partial x}dx-\frac{\partial\phi}{\partial y}dy-\frac{\partial\phi}{\partial z}dz$$

So that $u=-\frac{\partial\phi}{\partial x}, v=-\frac{\partial\phi}{\partial y}, w=-\frac{\partial\phi}{\partial z}$, then ϕ is called velocity potential or velocity function. In vectors, this is expressed as

$$\vec{v}=-\text{grad}\,\phi=-\nabla\phi$$

or

$$iu+jv+kw=-\left(i\frac{\partial}{\partial x}+j\frac{\partial}{\partial y}+k\frac{\partial}{\partial z}\right)\phi$$

this gives

$$u=-\frac{\partial\phi}{\partial x}, v=-\frac{\partial\phi}{\partial y}, w=-\frac{\partial\phi}{\partial z} \qquad \text{.....(10.4)}$$

From the theory of differential equations, that the system of curves (here stream lines)

$$\frac{dx}{u}=\frac{dy}{v}=\frac{dz}{w} \qquad \text{(A)}$$

meets the surfaces given by the differential equation

$$udx + vdy + w\,dz = 0 \tag{B}$$

then the curves (A) meet orthogonally to (B), (if) provided (B) admits a solution of the form $\phi(x, y, z) = \text{constant}$.

The necessary and sufficient condition being

$$u\left(\frac{\partial v}{\partial z}-\frac{\partial w}{\partial y}\right)+v\left(\frac{\partial w}{\partial x}-\frac{\partial u}{\partial z}\right)+w\left(\frac{\partial u}{\partial y}-\frac{\partial v}{\partial x}\right)=0 \qquad .....(10.5)$$

Using eq (10.4), we have

$$\frac{\partial v}{\partial z}-\frac{\partial w}{\partial y}=\frac{\partial}{\partial z}\left(\frac{-\partial\phi}{\partial y}\right)-\frac{\partial}{\partial y}\left(\frac{-\partial\phi}{\partial z}\right)$$

$$=\left(\frac{-\partial^2\phi}{\partial z\partial y}\right)+\frac{\partial^2\phi}{\partial y\partial z}$$

$$=0$$

that is, $\dfrac{\partial v}{\partial z}=\dfrac{\partial w}{\partial y}$

Similarly we get $\left.\dfrac{\partial w}{\partial x}=\dfrac{\partial u}{\partial z} \text{ and } \dfrac{\partial u}{\partial y}=\dfrac{\partial v}{\partial x}\right\}$ .....(10.6)

Thus, there exists the surfaces which cut the stream lines orthogonally whenever velocity potential exists.

Alternatively, if the expression

$$udx + vdy + wdz = -d\phi\text{, an exact differential,}$$

that is $-d\phi=-\dfrac{\partial\phi}{\partial x}dx-\dfrac{\partial\phi}{\partial y}dy-\dfrac{\partial\phi}{\partial z}dz$

where $u=\dfrac{\partial\phi}{\partial x}, v=\dfrac{\partial\phi}{\partial y}, w\dfrac{\partial\phi}{\partial z},$ .....(10.7)

then ϕ is called velocity potential or simply velocity function.

If $\vec{V} = iu + jv + kw$

$$= i\frac{\partial\phi}{\partial x}+j\frac{\partial\phi}{\partial y}+k\frac{\partial\phi}{\partial z} \qquad \text{[using eq. (10.7)]}$$

then $\vec{V} = -\text{grad }\phi = -\nabla\,\phi$ .....(10.8)

where ϕ is a scalar quantity.

The curves traced out by the moving points of the equation $\frac{dx}{u}=\frac{dy}{v}=\frac{dz}{w}$ are orthogonal to the curves traced by the moving points

u dx + v dy + w dz = 0, if there is an equation which admits a solution of the form $\phi(x, y, z)$ = constant.

The necessary and sufficient condition being

$$u\left(\frac{\partial v}{\partial z}-\frac{\partial w}{\partial y}\right)+v\left(\frac{\partial w}{\partial x}-\frac{\partial u}{\partial z}\right)+w\left(\frac{\partial u}{\partial y}-\frac{\partial v}{\partial x}\right)=0 \qquad .....(10.9)$$

Proof : $\frac{\partial v}{\partial z}-\frac{\partial w}{\partial y}=\frac{\partial}{\partial z}\left(\frac{\partial \phi}{\partial y}\right)-\frac{\partial}{\partial y}\left(\frac{\partial \phi}{\partial z}\right)$ [using eq. (10.7)]

$$=\frac{\partial^2\phi}{\partial z\partial y}-\frac{\partial^2\phi}{\partial y dz}=0$$

implies $\frac{\partial v}{\partial z}=\frac{\partial w}{\partial y}$

Similarly we get $\frac{\partial w}{\partial x}=\frac{\partial u}{\partial z}$ & $\frac{\partial u}{\partial y}=\frac{\partial v}{\partial x}$

∴ The equation (10.9) is satisfied.

Thus the surface exists which cut the streamlines orthogonally whenever velocity potential ϕ exists.

10.11 Equipotential Lines

Let $\phi=\phi(x, y)$ and $u=\frac{\partial\phi}{\partial x}, v=\frac{\partial\phi}{\partial y}$

That is the derivative of ϕ in any direction gives the velocity in that direction. This implies ϕ is a velocity potential function (two dimensional flow)

The lines of constant potential function are called equipotential lines

If $d\phi=\frac{\partial\phi}{\partial x}dx+\frac{\partial\phi}{\partial y}dy$

$$= u\,dx + v\,dy$$

if ϕ = constant, then $d\phi$=0, this implies udx + vdy =0

∴ for an equipotential line udx + vdy = $d\phi$ = 0,

or $\frac{dy}{dx}=-\frac{u}{v}$

Equation of continuity in two dimensional flow

is $$\frac{du}{dx}+\frac{\partial v}{\partial y}=0$$

or $$\frac{\partial}{\partial x}\left(\frac{\partial \phi}{\partial x}\right)+\frac{\partial}{\partial y}\left(\frac{\partial \phi}{\partial y}\right)=0 \qquad \text{using eq. (10.7)}$$

or $$\frac{\partial^2 \phi}{\partial x^2}+\frac{\partial^2 \phi}{\partial y^2}=0=\nabla_H^2 \phi \quad \nabla_H = i\frac{\partial}{\partial x}+j\frac{\partial}{\partial y}$$

This means ϕ satisfies the laplacian equation

$$\nabla^2 \phi = 0.$$

Note:

1. Potential function (ϕ) exists only for irrotational flow.
2. Stream function (ψ) applies to both rotational and non-rotational flow, provided the flow is steady and incompressible.
3. For irrotational flow, both stream function (ψ) and velocity potential function (ϕ) satisfy the laplace equation

   i.e., $\nabla^2 \psi = 0$ and $\nabla^2 \phi = 0$.

10.12 Stream Function (ψ)

Stream function (ψ) is a mathematical postulation such that its derivative with respect to x gives velocity in y- direction with negative sign, and its differentiation w. r. t - y gives the velocity (+) in x-direction.

That is $$\frac{\partial \psi}{\partial x}=-v, \quad \frac{\partial \psi}{\partial y}=u \qquad .....(10.10)$$

$\psi = \psi$ (x, y), its total derivative is

$$d\psi = \frac{\partial \psi}{\partial x}dx + \frac{\partial \psi}{\partial y}dy$$

or $$d\psi = -v\,dx + u\,dy \qquad \text{[using eq. (10.10)]} \qquad .....(10.11)$$

In case of two dimensional incompressible flow, the equations of stream lines are given by

$$\frac{dx}{u}=\frac{dy}{v}$$

or $$vdx - udy = 0 \qquad .....(10.12)$$

From eq. (10.11) & (10.12) we have

$$d\psi = 0, \text{ which implies } \psi = \text{constant}.$$

That is, stream function is constant along a streamline.

∴ Streamlines are the lines of constant stream function.

Thus, stream function (ψ) varies from one streamline to another.

The equation of continuity in two dimensions (x, y) is given by

$$\frac{\partial u}{\partial x} + \frac{\partial v}{\partial y} = 0 \qquad .....(10.13)$$

Substituting the values of u, v from eq (10.10) in eq (10.13), we have

$$\frac{\partial}{\partial x}\left(\frac{\partial \psi}{\partial y}\right) + \frac{\partial}{\partial y}\left(-\frac{\partial \psi}{\partial x}\right)$$

$$= \frac{\partial^2 \psi}{\partial x \partial y} - \left(\frac{\partial^2 \psi}{\partial y \partial x}\right) = 0.$$

i.e., eq (10.13) is satisfied.

This means stream function ψ satisfies the equation of continuity.

10.13 Some Properties of Streamlines

1. Streamlines do not intersect with one another. However they intersect at isolated points of zero velocity or infinite velocity.
2. The fluid mass cannot move across the streamlines or the flow is along or parallel to the streamlines.
3. Streamline spacing varies inversely as the velocity – converging of streamlines in any particular direction.

Note: A path line represents the trace or trajectory of a fluid particle during a period of time. Path lines show the direction of the velocity of the same fluid particle (or parcel of infinite fluid volume) at successive instants of time. A path line can intersect itself at different times.

Streak line: A streak line is the instantaneous picture of the positions of all the fluid particles that have passed through a fixed point or location in the fluid flow. Ex. A line formed by the smoke particles that ejected into the atmosphere from a fixed chimney, or nozzle forms the streak line.

For a steady flow there is no difference (geometrical difference) between streamlines, path lines and streak lines. If they have the same origin, they all coincide.

Types of flows

Steady flow: If the fluid parameters in the fluid flow remains constant, at any point in the fluid field (region), then the flow is termed steady flow.

$$\frac{d\vec{V}}{dt} = 0 \text{ or } \vec{V} = \text{constant}$$

Unsteady flow: If the fluid parameters change in the fluid field with time or the conditions change from time to time, then the flow is termed unsteady flow.

$$\frac{d\vec{V}}{dt} \neq 0 \text{ or } \vec{V} \neq \text{constant.}$$

Steadiness refers to no change in flow parameters (P, v, ρ, η, T) with time, while uniformity refers to no change in space.

Incompressible flow

If the density variation in the flow field is negligible due to changes in pressure and temperature, then the flow is called incompressible. Liquids under moderate pressure changes may be treated as incompressible.

Compressible flow: In a fluid flow if the density changes are applicable the fluid flow is called compressible. The flow of gases is compressible.

Rotational flow: If the fluid particles rotate about their own mass centres while moving along a stream line, then it is called rotational flow.

$$\nabla \times \vec{V} \neq 0.$$

Irrotational flow: If the fluid particles do not rotate about their own mass centres while moving along a streamline, it is called irrotational flow.

$$\vec{W} = \frac{1}{2} \nabla \times \vec{V} = 0$$

Note: In steady fluid flow the pressure field P = P(x, y, z, t) and density field $\rho = \rho(x, y, z, t)$ are independent of time i.e., P (x, y, z, t) = P (x, y, z) and $\rho(x, y, z, t) = \rho(x, y, z)$ or pressure field and density field do not change with time.

10.14 Dimensional Flow Representation

	steady	**unsteady**
one dimensional flow	V = f (x)	V = f (x, t)
two dimensional flow	V = f (x, y)	V = f (x, y, t)
three dimensional flow	V= f (x, y, z),	V = f (x, y, z, t)

Isothermal flow: If the fluid particles (very small parcels) do not experience any change in temperature during its flow (journey) it is called isothermal flow.

Adiabatic flow: If the heat content of the fluid mass (mass of very small volume or parcel) does not change (neither it absorbs heat nor gives out, from one particles to another) during its flow is termed Adiabatic.

Reversible adiabatic flow is called isentropic flow.

Ideal frictionless flow: In this flow shear stress is taken to be negligible between adjacent fluid layers and between fluid layers and the boundary.

In this case $\mu = 0$ (viscosity = 0) or if the velocity gradient normal to the direction of flow is zero.

Real flow: If the frictional resistance to the motion due to viscosity of real fluids is not negligible (that is, it has to be taken into account) the flow is called Real flow.

Laminar flow: If the fluid laminae (thin layers) smoothly slide one over the other, the flow is termed Laminar flow. Laminar flow is also called viscous or streamline flow.

Smooth pipes with low fluid velocities generally assume Laminar flow.

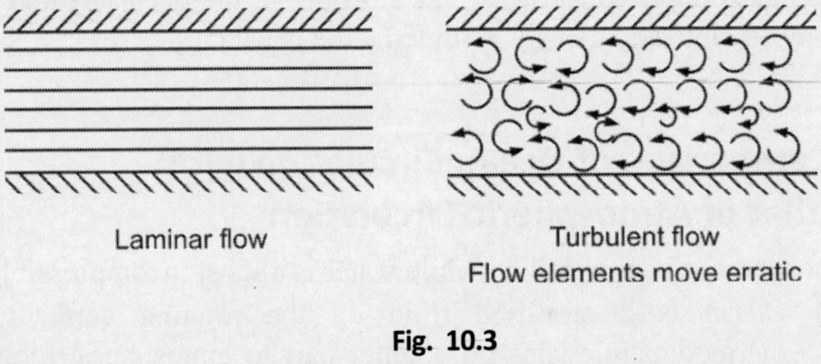

Fig. 10.3

Ideal fluid: An ideal fluid is one in which there is no internal friction (viscosity) and treated to be incompressible.

Viscous fluid: A viscous fluid is one in which there exists internal friction.

Barotropic fluid: A barotropic fluid is one whose density depends on pressure $\rho = \rho(P)$

Definitions

(a) Potential fields

(i) Let $\vec{A}$ be a vector field and ϕ a scalar field.

If $\vec{A} = \nabla\phi = \text{grad}\,\phi$, then $\vec{A}$ is called potential and ϕ is called the potential of the field $\vec{A}$

(ii) If $\vec{A} = \nabla\phi = \text{grad}\,\phi$, then rot $\vec{A}$ (= curl $\vec{A}$) = 0.

ie., every potential field is irrotational.

(b) Solenoidal fields: Let $\vec{A}$ be a vector field and $\vec{\phi}$ is another vector field.

If $\vec{A} = \text{rot}\,\phi$, then $\vec{A}$ is called solenoidal and ϕ is called vector potential of the field $\vec{A}$.

10.15 Principal Forces in Ocean Dynamics

Principal forces acting in the ocean are: Gravity, Buoyancy and Wind.

(i) Earth's rotation develops a psuedo force called coriolis force or drifting force. Oceanic flow depends on conservation of salt, volume and other quantities.

(ii) Equations of motion based on conservation laws applied to oceanic flow. The linear (or first order) differential equations relating to Newton's dynamics of mass acceleration by force changes to non-linear partial differential equations of fluid mechanics.

(iii) Boussinesq approximation is that the flow in the ocean treated to be incompressible and density assumed to be constant.

10.16 Comparison of Ocean Circulation with that of Atmospheric Circulation

1. Atmosphere is compressible, while water is almost incompressible.

   (i) Ocean is a stratified fluid on the rotating earth. Ocean thermodynamics has no counter part to atmospheric moisture, which is a source of latent heat.

   (ii) Atmosphere is not bounded laterally, while all oceans are bounded laterally by continents. Atmosphere is almost transparent to insolation and heated from below by convection, while ocean is heated from top. Ocean exchanges heat and moisture with atmosphere from its top (upper) surface. Ocean convection is driven by buoyancy loss from above.

(iii) Winds blowing over the ocean surface exert a stress on it (ocean surface) which is a principal driver of ocean circulation, particularly in the top (upper) one kilometer depth.

(iv) In ocean dynamics, wind-driven and buoyancy driven ocean circulations are closely interwined.

10.17 Ocean Flows (Currents and Waves)

(i) General circulation: It is a semi permanent feature, based on time averaged circulation.

(ii) Abyssal or deep circulation: It is the circulation of mass in the meridional plane in the deep ocean. This circulation is driven by mixing.

(iii) Wind driven circulation: It is the ocean circulation in top one kilometer of the ocean which is forced by the wind. The circulation may be caused by the local wind forcing or by the winds in other region.

(iv) Gyres: Gyres are the wind-driven cyclonic or anti-cyclonic currents which have dimension of ocean basins.

(v) Jets or squirts: Jets are long narrow currents, with dimensions of a few hundred kilometers, depth of few kilometers. These currents roughly move perpendicular to west coasts.

(vi) Boundary currents: These are the ocean currents flowing parallel to the coasts. Two important boundary currents observed are : Western boundary currents and Eastern boundary currents.

Western boundary currents move on the western edge of the ocean, tend to be fast, narrow jets.

Ex. The Gulf stream, the Kuroshivo current

Eastern boundary currents move on the eastern edge of the ocean, which are weak.

Ex. The California current.

(vii) Meso-scale ocean eddies

Meso-scale ocean eddies are turbulent or spinning flows.

Scale: A few hundred kilometers.

10.18 Oscillatory Flows Generated by Waves

(i) **Planetary waves:** Planetary waves depend on the rotation of the earth for a restoring force. They include Rossby waves, Kelvin waves, Equatorial waves and Yanai waves.

(ii) **Surface waves or gravity waves:** These waves break on the beach. The restoring force is due to the density difference between air and water at the sea surface.

(iii) **Internal waves:** Internal waves are sub-sea waves. These are similar in some respects to surface waves. The restoring force is due to change in density with depth.

(iv) **Tsunamis:** Tsunamis are surface waves with periods near about 15 minutes. These waves are generated or triggered by earthquakes (in ocean).

(v) **Tidal currents:** These are horizontal ocean currents associated with internal waves, and the waves are driven by the tidal potential.

(vi) **Edge waves:** These are surface waves with periods of a few minutes, confined to shallow regions near shore. The amplitude of the waves decay (drop down) exponentially with distance from shore.

CHAPTER 11

EQUATION OF CONTINUITY OR CONSERVATION OF MASS

The equation of continuity expresses the conservation of mass, that is, mass is neither created nor destroyed. It expresses that the rate of change of generation of mass within a given volume is balanced by an equal outflow of mass from the volume, or it means that there are no sources or sinks of mass anywhere in the field.

For simplicity we consider a box of volume element $\delta v = \delta x \delta y \delta z$ in the fluid (as shown in fig.11.1), where $\delta x,\ \delta y,\ \delta z$ are the sides of the box, ρ is the density of the fluid.

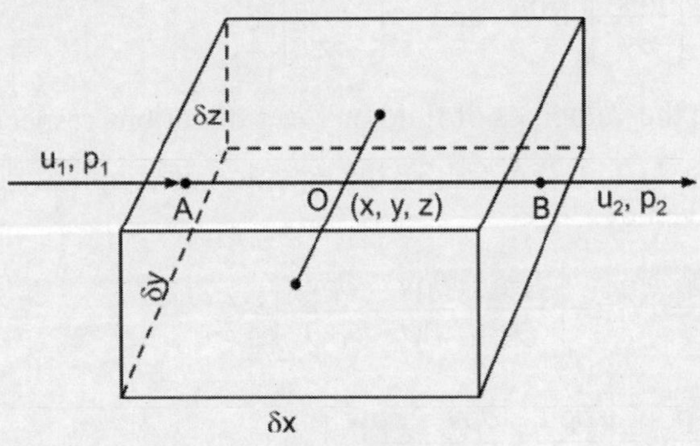

Fig. 11.1

The mass of the fluid in the box = ρ dV (mass = volume × density)

The change of mass per unit time is $\dfrac{\partial \rho}{\partial t}\delta V$ .....(11.1)

where $\dfrac{\partial \rho}{\partial t}$ = the local density change.

1. Let the coordinates of the centre of the box O be (x, y, z). where the velocity of the fluid u and density ρ.
2. Let A, B be the centre points on the left and right faces of the box, where let the velocities of the fluid be u_1, u_2 and density ρ_1 , ρ_2 respectively.
3. The coordinates of the centre of the left side face of the box A is $\left[x-\frac{\delta x}{2}, y, z\right]$ and the centre of right side box B is $\left[x+\frac{\delta x}{2}, y, z\right]$
4. The net flow into the box in x-direction is

$$= \rho_1 u_1 \delta y \delta z - \rho_2 u_2 \delta y \delta z$$

$$= \left[\rho_0 u_0 - \frac{1}{2}\left(\frac{\partial \rho u}{\partial x}\right)_0 \delta x\right]\delta y \delta z - \left[\rho_0 u_0 + \frac{1}{2}\left(\frac{\partial \rho u}{\partial x}\right)_0 \delta x\right]\delta y \delta z$$

$$= -\left[\frac{\partial \rho u}{\partial x}\right]_0 \delta x \delta y \delta z$$

$$= -\left[\frac{\partial \rho u}{\partial x}\right]_0 \delta V \qquad .....(11.2)$$

Similarly the net inflow into the box in y direction and z direction respectively are given by.

$$-\left[\frac{\partial \rho v}{\partial y}\right]_0 \delta V \quad \text{and} \quad -\left[\frac{\partial \rho w}{\partial z}\right]_0 \delta V \qquad .....(11.3)$$

where v, w are the velocities of fluid in y & z directions respectively.

From eq (11.1), (11.2) & (11.3), the equation of continuity is (omiting the subscript 0) given by

$$\frac{\partial \rho}{\partial t}\delta V = -\left[\frac{\partial \rho u}{\partial x} + \frac{\partial \rho v}{\partial y} + \frac{\partial \rho w}{\partial w}\right]\delta V$$

or

$$\frac{\partial \rho}{\partial t} = -\left[\frac{\partial \rho u}{\partial x} + \frac{\partial \rho v}{\partial y} + \frac{\partial \rho w}{\partial z}\right]$$

or

$$\frac{\partial \rho}{\partial t} = -\nabla \cdot \left(\rho \vec{V}\right) \qquad .....(11.4)$$

where

$$\nabla = i\frac{\partial}{\partial x} + j\frac{\partial}{\partial y} + k\frac{\partial}{\partial z}$$

$$\vec{V} = iu + jv + kw$$

$$\rho\vec{V} = i\rho u + j\rho v + k\rho w$$

$$\nabla \cdot \left(\rho\vec{V}\right) = \left(i\frac{\partial}{\partial x} + j\frac{\partial}{\partial y} + k\frac{\partial}{\partial z}\right) \cdot \left(i\rho u + j\rho v + k\rho w\right)$$

$$= \frac{\partial \rho u}{\partial x} + \frac{\partial \rho v}{\partial y} + \frac{\partial \rho w}{\partial z}$$

$\nabla \cdot \left(\rho\vec{V}\right)$ = mass divergence in three dimensional fluid space.

Eq. (11.4) can be written as

$$\frac{\partial \rho}{\partial t} + \nabla \cdot \left(\rho\vec{V}\right) = 0$$

or $$\frac{\partial \rho}{\partial t} + \vec{V} \cdot \nabla\rho + \rho\nabla \cdot \vec{V} = 0$$

or $$\frac{d\rho}{dt} + \rho\nabla \cdot \vec{V} = 0 \text{ or } \frac{1}{\rho}\frac{d\rho}{dt} + \nabla \cdot \vec{V} = 0 \qquad \text{.....(11.5)}$$

since $$\frac{d\rho}{dt} = \frac{\partial \rho}{\partial t} + \vec{V} \cdot \nabla\rho \text{ (implied } \frac{d}{dt} = \frac{\partial}{\partial t} + \vec{V} \cdot \nabla \text{)}$$

eq. (11.5) is the required equation of continuity. The term $\nabla \cdot \vec{V}$ is called dialation or expansion.

$$\frac{d}{dt} = \frac{\partial}{\partial t} + \vec{V} \cdot \nabla$$

denotes that the individual time derivative = local time derivative + the convective acceleration

$\frac{d}{dt}$ = rate of change or individual time derivative

$\frac{\partial}{\partial t}$ = local change or local time derivative

$\vec{V}, \nabla$ = velocity advection or convective acceleration

11.1 Equation of Continuity – Alternate Method

Consider an isolated box of volume δv of sides $\delta x, \delta y, \delta z$ in fluid.

Let δM be the mass of the fluid in the box, which moves with the fluid is constant [see (Fig. 11.2).]

Let ρ be the density of the fluid, then we have

$$\delta V = \delta x \delta y \delta z \quad .....(11.6)$$

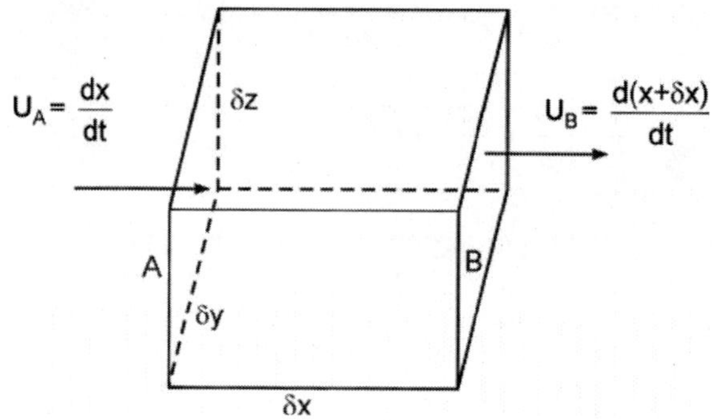

Fig. 11.2

and $$\delta M = \rho \delta v = \rho \delta x \ \delta y \ \delta z \quad .....(11.7)$$

Taking logarithm of (11.7), we have

$$l_n \delta M = l_n \rho + l_n \delta x + l_n \delta y + l_n \delta z$$

Differentiating w. r. t t (time) we have

$$\frac{1}{\delta M}\frac{d\delta M}{dt} = \frac{1}{\rho}\frac{d\rho}{dt} + \frac{1}{\delta x}\frac{d\delta x}{dt} + \frac{1}{\delta y}\frac{d\delta y}{dt} + \frac{1}{\delta z}\frac{d\delta z}{dt} \quad .....(11.8)$$

Let $U_A = \frac{dx}{dt}$,

$U_B = \frac{d(x+\delta x)}{dt}$ be the speed of the fluid to the face A and B respectively.

Then the difference of speed in x – direction in these faces is

$$\delta u = U_B - U_A$$

$$= \frac{d(x+\delta x)}{dt} - \frac{d(x)}{dt}$$

$$\delta u = \frac{d(\delta x)}{dt}$$

Similarly in, y & z directions we have

$$\left.\begin{aligned} \delta v &= \frac{d(\delta y)}{dt}, \\ \delta w &= \frac{d(\delta z)}{dt} \end{aligned}\right\} \quad .....(11.9)$$

Substituting these values eq. (11.9) in eq. (11.8) we have

$$\frac{1}{\delta M}\frac{d(\delta M)}{dt}=\frac{1}{\rho}\frac{d\rho}{dt}+\frac{\delta u}{\delta x}+\frac{\delta v}{\delta y}+\frac{\delta w}{\delta z} \qquad .....(11.10)$$

In the limit as $\delta x, \delta y, \delta z$ tends to zero, and conservation of mass (δM) = constant,

we have $\frac{d\cdot(\delta M)}{dt}=0$

The above equation (11.10) reduces to

$$0=\frac{1}{\rho}\frac{d\rho}{dt}+\frac{\partial u}{\partial x}+\frac{\partial x}{\partial y}+\frac{\partial w}{\partial z}$$

or $$\frac{1}{\rho}\frac{d\rho}{dt}+\nabla\cdot\vec{V}=0 \qquad .....(11.11)$$

where $$\vec{V}=iu+jv+kw$$

$$\nabla=i\frac{\partial}{\partial x}+j\frac{\partial}{\partial y}+k\frac{\partial}{\partial z}$$

Eq. (11.11) is the required equation of continuity.

11.2 Conservation of Mass in Fluid Mechanics

Conservation of mass or equation of continuity expresses that the rate of generation of mass within a given volume (or liquid or fluid parcel) is balanced by an equal net outflow of mass from the volume (parcel).

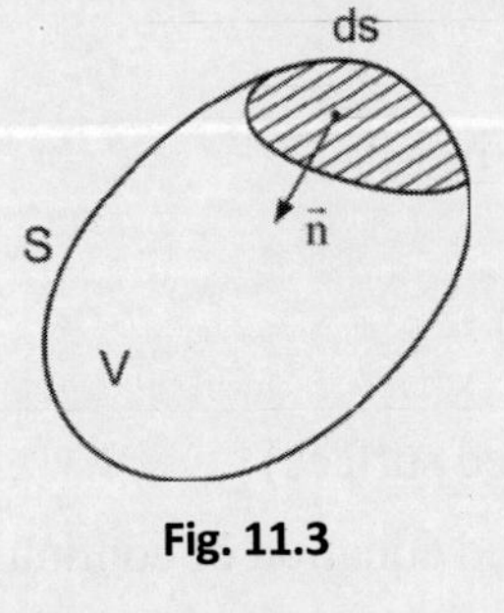

Fig. 11.3

Consider a fixed enclosed parcel of surface area S, enclosing volume V in the region occupied by moving fluid. Let $\vec{n}$ be the unit outward normal drawn to the parcel element ds and $\vec{V}$ be the velocity of the fluid, then the inward normal velocity of the fluid would be $\left(-\vec{V}\vec{n}\right)$.

Volume of fluid entering through the surface ds would be $\left(-\vec{V}\vec{n}\right)$ ds

The mass of fluid entering per unit time through the element ds would be

$$= -\rho\vec{v}\cdot\vec{n}ds$$

∴ the mass of fluid entering in unit time across the whole surface S of the parcel

$$= \int_S (-\vec{V}\cdot\vec{n})ds$$

$$= -\int_V \nabla\cdot(\rho V)dv \quad \text{(by Gauss theorem)} \qquad .....(11.12)$$

[Gauss theorem $\iint_S Fnds = \iiint_V Div\ Fdv$

$$= \iiint_V \nabla\cdot Fdv$$

where V = volume, S = surface area]

The mass of fluid within volume V, enclosed by the surface S is

$$= \iint_V \rho\cdot dv \qquad .....(11.13)$$

If we assume that there are no sinks or sources within the surfaces, then the mass can be increased only through the inflow from the boundary surface S.

Equating the local time – rate of change of eq. (11.13) to eq. (11.12), we have

$$\frac{\partial}{\partial t}\left[\int_V \rho dV\right] = -\int_V \nabla\cdot(\rho\vec{V})dv$$

or
$$\int_V \left[\frac{\partial\rho}{\partial t} + \nabla\cdot\left(\rho\vec{V}\right)\right]dv = 0 \qquad .....(11.14)$$

∴
$$\frac{\partial\rho}{\partial t} + \nabla\cdot\left(\rho\vec{V}\right) = 0 \qquad .....(11.15)$$

(since S is an arbitrary closed surface)

Eq. (11.15) is the required equation of continuity.

Note 1. For incompressible fluid, the equation of continuity is $\nabla\cdot\vec{V} = 0$

or
$$\left(i\frac{\partial}{\partial x} + j\frac{\partial}{\partial y} + k\frac{\partial}{\partial z}\right)\cdot(iu + jv + kw) = 0$$

or
$$\frac{\partial u}{\partial x} + \frac{\partial v}{\partial y} + \frac{\partial w}{\partial z} = 0$$

i.e., the incompressible flow is non-divergent.

2. For compressible flow [equation of continuity]

$$\frac{d\rho}{dt} + \rho\nabla \cdot \vec{V} = 0 \text{ for } \rho \text{changes.}$$

It is convenient to use pressure coordinates (x, y, P), z = (x, y, p)

The elemental volume $\delta V = \delta x \delta y \delta p$ .....(11.16)

$$\delta z = \frac{\partial z}{\partial p}\delta p \qquad .....(11.17)$$

Mass $\delta M = \rho \delta x \ \delta y \ \delta z$

$$= \rho\ \delta x \delta y \frac{\partial z}{\partial p}\delta p \qquad \text{[using eq. (11.17)]}$$

$$\delta M = -\frac{1}{g}\delta x\ \delta y\ \delta p$$

[since $g\ \rho\ \delta z = -\delta p$, mass per unit volume = $\frac{1}{g}$

$$\rho\frac{\partial z}{\partial p} = -\frac{1}{g}]$$

i.e., the mass of elemental fixed volume in pressure coordinates cannot change

The equation of continuity is

$$\nabla_p \cdot \vec{V}_p = \frac{\partial u}{\partial x} + \frac{\partial v}{\partial y} + \frac{\partial w}{\partial p} = 0$$ (is the equation of continuity in pressure coordinates)

Subscript p denotes pressure coordinates. In this equation there is no term representing rate of change of density, as the mass per unit volume is $\frac{1}{g}$ = constant.

11.2.1 An Incompressible Fluid

An ideal fluid is inviscid and non-thermo-conducting, incompressible fluid. Incompressibility means that during fluid flow, the density $\rho = \rho(t,x)$ of an arbitrarily isolated fluid parcel is a function of time t and location X, remains constant, particle (parcel) in turn depend on time (t). The density constancy condition is expressed by the equation

$$\frac{d\rho}{dt} = \frac{\partial \rho}{\partial t} + (\vec{u}\nabla)\rho = 0 \qquad .....(11.18)$$

$$\vec{u} = \vec{u}(t, x) = x'(t)$$

where $\vec{u} = \frac{dx}{dt} \frac{d\rho}{dt} = \frac{\partial \rho}{\partial t} + (\vec{u}\nabla)\rho = 0 \Rightarrow \frac{d}{dt} = \frac{\partial}{\partial t} + (\vec{u}\nabla)$

Direction of flow = derivative of $\vec{u}\nabla$ in the direction of the velocity vector $\vec{u}$.

The quantities of the state of fluid [i.e., its weight and velocity of any point of the space occupied by the fluid is field characterizing scalar field,] satisfying equation (11.18) are called Lagrangian invariants.

Lagrangian invariant is passively transported by fluid motion i.e., its value remains constant for each individual particle, and it is only its location that changes.

11.3 Equation of Continuity for Incompressible Fluid Flow (another method)

Equation of continuity for compressible fluid is given by

$$\frac{d\rho}{dt} + \rho(\nabla \cdot \vec{V}) = 0 \qquad \text{.....(11.19)}$$

According to the Boussinesq approximation, we assume sea water is incompressible.

By definition of coefficient of compressibility (β)

$$\beta = -\frac{1}{V}\frac{\partial V}{\partial p} = -\frac{1}{v}\frac{dV}{dt} \cdot \frac{dt}{dp}$$

$$= -\frac{1}{V}\frac{\frac{dV}{dt}}{\frac{dp}{dt}}$$

where V = volume; p = pressure.

For compressible flow $\beta = 0$, from the above equation we have

$$\frac{1}{V}\frac{\partial V}{\partial t} = 0 \qquad \left(\text{since } \frac{dp}{dt} \neq 0\right)$$

Density is mass per unit volume, and since mass is constant,

$$\frac{1}{V}\frac{dV}{dt} = -V\frac{d}{dt}\left(\frac{1}{V}\right) = -\frac{V}{m}\frac{d}{dt}\left(\frac{m}{V}\right)$$

$$= -\frac{1}{\rho}\frac{d}{dt}(\rho) = 0$$

This implies $\frac{d\rho}{dt} = 0$ .....(11.20)

Using eq. (11.20), in eq. (11.19), we have

$$\rho(\nabla \cdot \vec{V}) = 0$$

or
$$\rho\left[\left(i\frac{\partial}{\partial x} + j\frac{\partial}{\partial y} + k\frac{\partial}{\partial z}\right).(iu + jv + kw)\right] = 0$$

i.e.,
$$\frac{\partial u}{\partial x} + \frac{\partial v}{\partial y} + \frac{\partial w}{\partial z} = 0.$$

This is the required equation of continuity for incompressible fluid flow.

11.4 Equivalent Forms of Equation of Continuity

(i) The equation of continuity is

$$\frac{\partial \rho}{\partial t} + \nabla \cdot (\rho \vec{V}) = 0 \quad .....(11.21)$$

Expanding ∇ operator, we have

$$\frac{\partial \rho}{\partial t} + \rho(\nabla.\vec{V}) + \vec{V} \cdot \nabla\rho = 0 \quad ....(11.22)$$

$$\frac{\partial \rho}{\partial t} + \rho\left[\left(i\frac{\partial}{\partial x} + j\frac{\partial}{\partial y} + k\frac{\partial}{\partial z}\right) \cdot (iu + jv + kw)\right] +$$

$$(iu + jv + kw) \cdot \left(i\frac{\partial}{\partial x} + j\frac{\partial}{\partial y} + k\frac{\partial}{\partial z}\right)\rho = 0$$

$$\frac{\partial \rho}{\partial t} + \rho\left(\frac{\partial u}{\partial x} + \frac{\partial v}{\partial y} + \frac{\partial w}{\partial z}\right) + u\frac{\partial \rho}{\partial x} + v\frac{\partial \rho}{\partial y} + w\frac{\partial \rho}{\partial z} = 0 \quad .....(11.23)$$

Eq. (11.23) represents another form of continuity equation.

(ii) From eq. (11.22) above we have

$$\frac{\partial \rho}{\partial t} + u\frac{\partial \rho}{\partial x} + v\frac{\partial \rho}{\partial y} + w\frac{\partial \rho}{\partial z} + \rho\nabla \cdot \vec{V} = 0$$

Combining the first two terms we have

$$\frac{d\rho}{dt} + \rho\nabla \cdot \vec{V} = 0 \quad .....(11.24)$$

where $\rho = \rho\ (x, y, z, t)$ density of the fluid

$\vec{V} = \vec{V}(x, y, z, t) = iu + jv + kw$ Velocity of the fluid

$$\nabla \cdot \vec{V} = \left(i\frac{\partial}{\partial x} + j\frac{\partial}{\partial y} + k\frac{\partial}{\partial z} \right) \cdot (iu + jv + kw)$$

$$= \frac{\partial u}{\partial x} + \frac{\partial v}{\partial y} + \frac{\partial w}{\partial z} = \text{divergence of vector } \vec{V}$$

$\nabla \cdot \vec{V}$ gives the rate of volume expansion of a fluid element (or parcel). It is also called dialation.

Eq. (11.24) is another form of equation of continuity.

(iii) Continuity equation for incompressible fluid.

In case of incompressible fluids $\frac{d\rho}{dt} = 0, \because \rho = \text{constant}$

From eq. (11.24), we have $\left[\frac{d\rho}{dt} + \rho \nabla \cdot \vec{V} = 0 \right]$

$\nabla \cdot \vec{V} = 0 \quad$ since $\rho \neq 0$

or $\quad$ div $\vec{V} = 0$

$$\left(i\frac{\partial}{\partial x} + j\frac{\partial}{\partial y} + k\frac{\partial}{\partial z} \right) . (iu + jv + kw) = 0$$

$$\frac{\partial u}{\partial x} + \frac{\partial v}{\partial y} + \frac{\partial w}{\partial z} = 0 \qquad \text{.... (11.25)}$$

Eq. (11.25) is equation of continuity for incompressible fluid flow.

In eq. (11.25) velocity vector $\vec{V}$ is solinoidal.

(iv) Equation of continuity for steady motion

Equation of continuity is $\frac{\partial \rho}{\partial t} + \nabla \cdot (\rho \vec{V}) = 0$

For steady motion $\frac{\partial \rho}{\partial t} = 0$

$\therefore$ equation of continuity for steady motion becomes $\nabla \cdot \rho \vec{v} = 0$ or $\text{div}(\rho \vec{v}) = 0$

$$\text{or} \left(i\frac{\partial}{\partial x} + j\frac{\partial}{\partial y} + k\frac{\partial}{\partial z} \right) \cdot (i\rho u + j\rho v + k\rho w) = 0$$

or $\frac{\partial \rho u}{\partial x}+\frac{\partial \rho v}{\partial y}+\frac{\partial \rho w}{\partial z}=0$ .....(11.26)

eq. (11.26) is the equation of continuity for steady motion.

(v) With steady motion $\frac{\partial \rho}{\partial t}=0$, the fluid flow through a cross section of a stream filament does not depend on the location of the cross section.

For two arbitrary cross sections ds_1, ds_2 of an elemental filament, the following condition holds, $\rho_1 v_1 d s_1 = \rho_2 v_2 d s_2$.

11.5 The Equation of Continuity in Lagrangian Method

Consider a fluid particle (element/parcel) of an infinitesimal volume dv and density ρ at time 't',

$\frac{d}{dt}(\rho dv)=0$, since the mass of the fluid particle (ρ dv) is invariant as it move about

$\therefore \quad \rho\, dv = \text{constant} = \rho_o d v_o \text{ (say)}$ .....(11.27)

where $\rho_o d v_o$ = mass of the particle/element/parcel in the initial stage.

In rectangular Cartesian coordinates

$$dv = dx\, dy\, dz\,,\ dv_o = da\, db\, dc$$

and $\quad dx\, dy\, dz = \mathcal{F}\, da\, db\, dc$

$$\mathcal{F} = \frac{\partial(x,y,z)}{\partial(a,b,c)} \quad .....(11.28)$$

$$\rho dv = \rho_o\, d v_o$$

$$\rho dx\, dy\, dz = \rho_o\, da\, db\, dc$$

$$\rho \mathcal{F} = \rho_o \quad .....(11.29)$$

which is the equation of continuity

In summary, equation of continuity expresses that

(i) The local loss of mass $\left(\frac{-\partial \rho}{\partial t}\right)$ = the divergence of the specific momentum

$$-\frac{\partial \rho}{\partial t}=\frac{\partial \rho u}{\partial x}+\frac{\partial \rho v}{\partial y}+\frac{\partial \rho w}{\partial z}$$

(ii) In another form, equation of continuity states that, the rate of expansion of the moving element = the divergence of the velocity

$$-\frac{1}{\rho}\frac{d\rho}{dt} = \frac{\partial u}{\partial x} + \frac{\partial v}{\partial y} + \frac{\partial w}{\partial z}$$

(iii) In case of stationary distribution of mass $\left(\frac{\partial \rho}{\partial t} = 0\right)$

The equation of continuity reduces to

$$\frac{\partial \rho u}{\partial x} + \frac{\partial \rho v}{\partial y} + \frac{\partial \rho w}{\partial z} = 0$$

11.6 Conservation of Mass and Salt in an Ocean Basin

Conservation of mass and salt in an ocean basin provides us with useful information about the flow in the ocean.

Consider an arbitrary ocean basin (as shown in fig.11.4)

P↓ = Precipitation

E ↑= Evaporation

R→ = River flow in

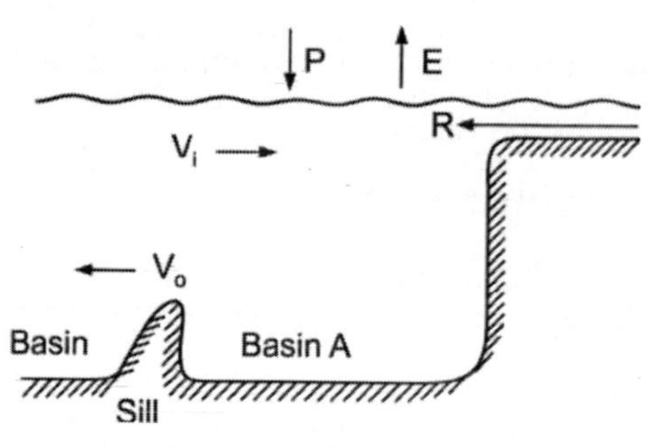

Fig. 11.4

Let S_i = salinity of inflow water into the basin

S_o = salinity of outflow water from the basin

ρ_i = density of inflow water into the basin

ρ_o = density of outflow water from the basin

V_i = volume of inflow water into the basin

V_o = volume of outflow water from the basin

P = precipitation over the basin (in)

E = Evaporation (out) over the basin

R = River inflow into the basin

Mass of inflow water = $\rho_i V_i$ (into the basin)

Mass of outflow water = $\rho_o V_o$ (out from the basin)

In case of no change in volume of basin, the conservation of mass implies

$$\rho_i V_i = \rho_o V_o \qquad .....(11.30)$$

If we assume $\rho_i = \rho_o$, the above equation reduces to

$$V_i = V_o \qquad .....(11.31)$$

Volume of inflow into the basin = $V_i + P + R$

Volume of outflow from the basin = $V_o + E$

From the conservation principle we have

$$V_i + P + R = V_o + E$$

or

$$V_o - V_i = R + P - E \qquad .....(11.32)$$

eq. (11.32) is the required equation of conservation of mass.

If we assume that there is no deposit or removal of salt from the basin, we have

$$\rho_i V_i S_i = \rho_o V_o S_o \qquad .....(11.33)$$

or

$$V_i S_i = V_o S_o \qquad [\text{if we assume } \rho_i = \rho_o]$$

eq. (11.33) is equation for conservation of basin salt.

CHAPTER 12

EQUATIONS OF MOTION

Newton's second law of motion can be expressed in the following different ways.

1. The first derivative of the momentum of a particle ($\overline{P}_i$) is equal to the force $\vec{F}_i$ acting on the particle.

$$\frac{d\overline{P}_i}{dt} = \vec{F}_i \text{ or } \frac{d(m_i \ \vec{V}_i)}{dt} = \vec{F}_i \qquad .....(12.1)$$

where

$\overline{P}_i$ = momentum = $m\vec{V}_i$,

m_i = mass of the particle,

$\vec{V}_i$ = velocity of the particle

2. A small change in momentum of a particle is equal to the impulse of the force acting on it.

Impulse = Force × time

$$J = \vec{F}_i \ t$$

$$d\vec{p}_i = \vec{F}_i \, d\, t \text{ or } d\ (m_i \ \vec{V}_i) = \vec{F}_i dt \qquad .....(12.2)$$

where

m_i = mass

$\vec{F}_i dt$ = impulse of force $\vec{F}_i$

$J = \vec{F}\ (t_2 - t_1)$, [impulse during the period t_1 to t_2.]

3. The acceleration ($\vec{a}_i$) of a particle is directly proportional to the force ($\vec{F}_i$) acting on it and inversely proportional to the mass (m_i) of the particle and coincides with the direction of the force.

$$\vec{a}_i = \frac{d\vec{V}_i}{dt} = \frac{\vec{F}}{m_i}$$

It follows from eq. (12.1),

$$\frac{d(mi\vec{V}_i)}{dt} = \vec{F}i \quad \text{or} \quad m_i \frac{d\vec{v}_i}{dt} = \vec{F}_i$$

$$\vec{a}_i = \frac{d\vec{V}_i}{dt} = \frac{\vec{F}_i}{m_i} \qquad .....(12.3)$$

where $\vec{a}_i = \frac{d\vec{V}_i}{dt}$ = acceleration.

The principal forces acting in ocean dynamics are:

(i) Gravity, (ii) Friction and (iii) Coriolis force.

Force is a vector quantity and hence gravity, friction, coriolis force have both magnitude and direction.

- **Gravity:** The weight of the water in ocean develops pressure.

  Changes in gravity results due to the motion of the sun and moon. This produces tides, tidal currents and tidal mixing of ocean water, which mixes interior of the ocean.

  Buoyancy is the upward or downward force due to gravity on (element) a particle of water.

  Horizontal pressure gradients in the fluid are developed because of varying water weights in different parts (regions) of the ocean.

- **Friction:** When the surface of one body moves (or slides) over another body, each body exerts a frictional force on the other, parallel to the surfaces. The force exerted on each body is opposite to the direction of its motion (relative to the other).

  Wind stress is the friction due to wind blowing across the sea surface. It transfers horizontal momentum to the sea and is responsible for creating currents.

- **Coriolis:** Force is a pseudo force acts due to motion (rotation) of the earth.

  Other forces that play part in ocean dynamics are atmospheric pressure and seismic (due to earthquakes). In ocean dynamics, conservation of mass leads to equation of continuity, conservation of momentum leads to momentum equations, conservation of angular momentum leads to conversation of vorticity and conservation of energy leads to heat energy to heat budget, mechanical energy to wave equation.

12.1 Euler's Equation of Motion for a Perfect Fluid

An ideal or perfect fluid is a continuous fluid which has no shearing stress.

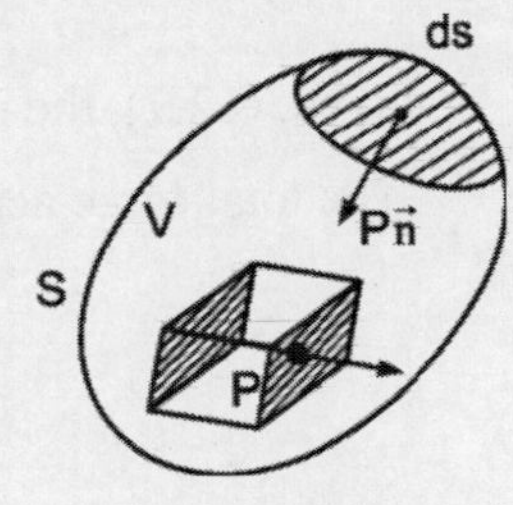

A perfect fluid is frictionless or non-viscous. Consider any arbitrary closed surface 's' drawn in the fluid region occupied by a non-viscous fluid moving with it. It contains the same fluid particles at every time.

Consider a fluid particle P within S, with density ρ and volume δv (surrounding P).

The mass of the element ρ δv is always remains same.

Let $\vec{V}$ be the velocity of P.

The momentum of the volume v is given by

$$M_{\circ} = \int_{v} \rho \vec{V} dv \quad \text{.....(12.4)}$$

The rate of change of momentum is

$$\frac{dM_{\circ}}{dt} = \int_{v} \left(\frac{d\rho\vec{v}}{dt} \right) dv$$

$$= \int_{v} \rho \frac{d\vec{V}}{dt} dv + \int_{v} \vec{V} \frac{d(\rho dv)}{dt}$$

$$\frac{dM_{\circ}}{dt} = \int_{v} \frac{d\vec{V}}{dt} \rho dv \quad \text{.....(12.5)}$$

Since ρ d v = constant and hence its derivative is zero.

If $\vec{F}$ = external force (that is body force, like gravity) per unit mass acting on the fluid. Then the total body force acting on the fluid volume v contained in S at time t

$$= \int_{v} \vec{F} \rho dv \quad \text{.....(12.6)}$$

Let p be the pressure at a point on the surface element ds, having $\vec{n}$ unit outward normal,

Then, the total surface force is

$$= \int_{v} -p\, \vec{n} ds$$

$$= -\int_{v} \nabla p dv \quad \text{.....(12.7)}$$

[by Gauss theorem $\iiint_V \nabla\vec{F}dv = \iint_S Fnds$]

In eq. (12.7), the negative sign for the surface force acts inwards.

The total force acting on v

$$= \int_V \vec{F}\rho dv - \int_V \nabla p dv$$

$$= \int_V (\rho\vec{F} - \nabla P)dv \qquad \text{.....(12.8)}$$

Using Newtons second law of motion, namely, the rate of change of linear momentum = total force acting on the mass of same fluid.

Thus eq. (12.5) = eq. (12.8)

i.e.,

$$\int_V \frac{d\vec{V}}{dt}\rho dv = \int_V (\rho\vec{F} - \nabla p)dv$$

or

$$\int_V \left(\rho\frac{d\vec{V}}{dt} - \rho\vec{F} + \nabla p\right)dv = 0 \qquad \text{.....(12.9)}$$

Since the surface S is arbitrary, from eq. (12.9) it follows

$$\rho\frac{d\vec{v}}{dt} - \rho\vec{F} + \nabla p = 0$$

or

$$\frac{d\vec{V}}{dt} = \vec{F} - \frac{1}{\rho}\nabla p \qquad \text{.....(12.10)}$$

The above eq. (12.10) is the Euler's equation of motion.

We have proved $\frac{df}{dt} = \frac{\partial f}{\partial t} + \vec{V}\cdot\nabla f$

or

$$\frac{d}{dt} = \frac{\partial}{\partial t} + \vec{V}\cdot\nabla$$

The above eq. (12.10) can be written as

$$\frac{\partial\vec{V}}{dt} + \left(\vec{V}\cdot\nabla\right)\vec{V} = \vec{F} - \frac{1}{\rho}\nabla p \qquad \text{.....(12.11a)}$$

In Cartesian coordinates the Euler's equation of motion eq. (12.10) is

$$\frac{d\vec{V}}{dt} = \vec{F} - \frac{1}{\rho}\nabla p \qquad \text{(is given).}$$

$$\frac{d}{dt}(iu + jv + kw) = (iF_1 + jF_2 + kF_3) - \frac{1}{\rho}\left(i\frac{\partial p}{\partial x} + j\frac{\partial p}{\partial y} + k\frac{\partial p}{\partial z}\right)$$

Equating like terms, we have

$$\left.\begin{aligned} \ddot{x} = \frac{d^2x}{dt^2} \text{ or } \quad \frac{du}{dt} &= F_1 - \frac{1}{\rho}\frac{\partial p}{\partial x} \\ \frac{dv}{dt} &= F_2 - \frac{1}{\rho}\frac{\partial p}{\partial y} \\ \frac{dw}{dt} &= F_3 - \frac{1}{\rho}\frac{\partial p}{\partial z} \end{aligned}\right\} \quad \text{.....(12.11b)}$$

If we use the eq. (12.11a) viz

$$\frac{\partial \vec{V}}{\partial t} + (\vec{V} \cdot \nabla)\vec{V} = \vec{F} - \frac{1}{\rho}\nabla p,$$

the Cartesian equations would be

$$\frac{\partial}{\partial t}(iu + jv + kw) + \left[(iu + jv + kw) \cdot \left(i\frac{\partial}{\partial x} + j\frac{\partial}{\partial y} + k\frac{\partial}{\partial z}\right)\right](iu + jv + kw)$$

$$= (iF_1 + jF_2 + kF_3) - \frac{1}{\rho}\left(i\frac{\partial p}{\partial x} + j\frac{\partial p}{\partial y} + k\frac{\partial p}{\partial z}\right)$$

or

$$\frac{\partial}{\partial t}(iu + jv + kw) + \left[\left(u\frac{\partial}{\partial x} + v\frac{\partial}{\partial y} + w\frac{\partial}{\partial z}\right)\right](iu + jv + kw)$$

$$= (iF_1 + jF_2 + kF_3) - \frac{1}{\rho}\left(i\frac{\partial p}{\partial x} + j\frac{\partial p}{\partial y} + k\frac{\partial p}{\partial z}\right)$$

Equating the like terms, we have

$$\left.\begin{aligned} \frac{\partial u}{\partial t} + u\frac{\partial u}{\partial x} + v\frac{\partial u}{\partial y} + w\frac{\partial u}{\partial z} &= F_1 - \frac{1}{\rho}\frac{\partial p}{\partial x} \\ \frac{\partial v}{\partial t} + u\frac{\partial v}{\partial x} + v\frac{\partial v}{\partial y} + w\frac{\partial v}{\partial z} &= F_2 - \frac{1}{\rho}\frac{\partial p}{\partial y} \\ \frac{\partial w}{\partial t} + u\frac{\partial w}{\partial x} + v\frac{\partial w}{\partial y} + w\frac{\partial w}{\partial z} &= F_3 - \frac{1}{\rho}\frac{\partial p}{\partial z} \end{aligned}\right\} \quad \text{.....(12.11c)}$$

12.2 Absolute and Relative Velocity

Suppose a particle P moves with a velocity $\vec{V}_r$ relative to a point M (some geometrical point).

Suppose the point M rotates about axis with a constant angular velocity $\vec{\Omega}$.

Then $$\vec{V}_M = \vec{\Omega} \times \vec{R} \quad .....(12.12)$$

as described in the rotating motion (see fig12.1)

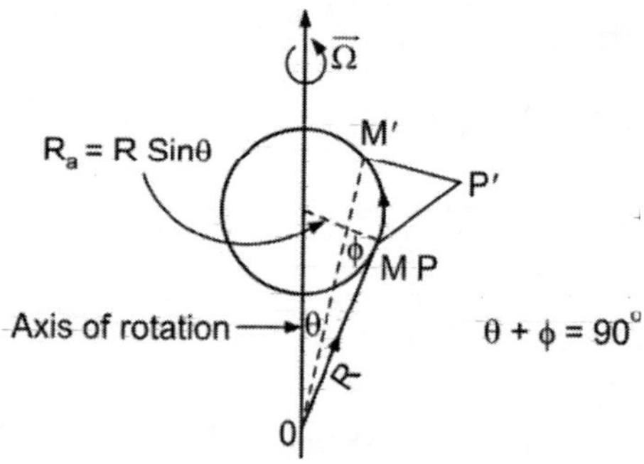

Fig. 12.1

Suppose during a small interval of time Δt, M moves to M',

so that $$MM' = \vec{V}_M \Delta t$$

During this same interval time Δt, suppose P has moved to P', so that $PP' = \vec{V}_r \Delta t$

∴ The total or absolute displacement

$$\vec{MP'} = (\vec{V}_M + \vec{V}_r)\Delta t \quad .....(12.13)$$

or $$\frac{\vec{MP'}}{\Delta t} = \vec{V}_M + \vec{V}_r$$

In the limit as $\Delta t \to 0$, we have

$$\vec{V}_a = \vec{V}_r + \vec{V}_M$$

or $$\vec{V}_a = \vec{V}_r + \vec{\Omega} \times \vec{R} \text{ [using (12.12) eq.]} \quad .....(12.14)$$

where $\vec{V}_a = \left(\frac{d\vec{R}}{dt}\right)_a$ Absolute velocity

$$\vec{V}_r = \left(\frac{d\vec{R}}{dt}\right)_r \text{ Relative velocity}$$

Suffixes a, r, stand for absolute and relative axis.

The above rotation, eq (12.14) holds for any arbitrary $\vec{R}$ say $\vec{Q}$.

The above relation (12.14)

$\vec{V}_a = \vec{V}_r + \vec{\Omega}\times\vec{R}$ (is of absolute and relative velocity) can be put as

$$\left(\frac{d\vec{Q}}{dt}\right)_a = \left(\frac{d\vec{Q}}{dt}\right)_r + \vec{\Omega}\times\vec{Q} \quad .....(12.15)$$

where $\vec{Q}$ is any arbitrary vector

In this equation (12.15) put $\vec{Q} = \vec{V}_a$, then we have

$$\left(\frac{d\vec{V}_a}{dt}\right)_a = \left(\frac{d\vec{V}_a}{dt}\right)_r + \vec{\Omega}\times\vec{V}_a \quad .....(12.16)$$

Substituting $\vec{V}_a = \vec{V}_r + \vec{\Omega}\times\vec{R}$ on RHS of eq. (12.16), we have

$$\left(\frac{d\vec{v}_a}{dt}\right)_a = \left[\frac{d}{dt}(\vec{V}_r + \vec{\Omega}\times\vec{R})\right]_r + \vec{\Omega}\times[\vec{V}_r + \vec{\Omega}\times\vec{R}]$$

$$= \frac{d}{dt}\vec{V}_r + \vec{\Omega}\times\left(\frac{d\vec{R}}{dt}\right)_r + \vec{\Omega}\times\vec{V}_r + \vec{\Omega}\times\vec{\Omega}\times\vec{R}$$

or
$$\left(\frac{d\vec{V}_a}{dt}\right)_a = \left(\frac{d\vec{V}_r}{dt}\right)_r + 2\vec{\Omega}\times\vec{V}_r + \vec{\Omega}\times\vec{\Omega}\times\vec{R} \quad .....(12.17)$$

since $\vec{\Omega}$ = constant $\left(\frac{\cdot d\vec{\Omega}}{dt} = 0\right)$ and $\left(\frac{d\vec{R}}{dt}\right)_r = \vec{V}_r$

In eq. (12.17), $\left(\frac{d\vec{V}_a}{dt}\right)_a$ = Absolute acceleration

$\left(\frac{d\vec{V}_r}{dt}\right)_r$ = Relative acceleration

$2\vec{\Omega}\times\vec{V}_r$ = coriolis acceleration, this is perpendicular to the relative velocity and exists only when the relative motion is non-zero

$\vec{\Omega}\times\vec{\Omega}\times\vec{R}$ = centripetal acceleration $(\vec{a}_c)$. This depends on position.

Note:

1. $\vec{a}_c$ depends on position $= \Omega^2 R \sin\theta$ (in kinematic rotation, $\theta + \phi = 90^\circ$)

$$= \Omega^2 R \cos\phi$$

$$\Omega = 7.29 \times 10^{-5}\ \text{s}^{-1},\ R = \frac{2}{\pi} \times 10^7\ \text{m (radius of the earth)}$$

$$\Omega = \frac{2\pi^c}{T} = \frac{2}{86164} \times \frac{22}{7} = 7.29 \times 10^{-5}\ \text{c/s} \quad \begin{pmatrix} c = \text{radiance} \\ s = \text{seconds} \end{pmatrix}$$

2. The vector equation of motion (which will be derived)

$$\frac{d\vec{V}}{dt} = -\frac{\nabla p}{\rho} - 2\vec{\Omega} \times \vec{V} + \vec{g} + \vec{F}$$

where $\vec{V} = iu + jv + kw$ — i, j, k unit vectors in x, y, z directions, $|\vec{\Omega}| = \Omega$

$$\nabla p = i\frac{\partial p}{\partial x} + j\frac{\partial p}{\partial y} + k\frac{\partial p}{\partial z}$$

$$\vec{\Omega} = i\Omega_x + j\Omega_y + k\Omega_z$$

$$-2\vec{\Omega} \times \vec{v} = -2i(w\cos\phi - v\Omega\sin\phi)$$

$$+2j(0 - u\Omega\sin\phi) - 2k(0 - u\Omega\cos\phi)$$

$$\vec{F} = iF_x + jF_y + kF_z$$

$$\vec{g} = kg \qquad |\vec{g}| = g$$

$$f = 2\Omega\sin\phi,\quad e = 2\Omega\cos\phi$$

12.3 Momentum Equation

A fluid is continuous, the mass of a single particle (element/parcel) is replaced by the mass per unit volume (i.e., density = ρ) and forces are also expressed per unit volume.

Thus, Force per unit volume = ρ (acceleration + advection)

$$\frac{dx}{dt}\ (\text{in x - direction}) = \rho\left(\frac{\partial u}{\partial t} + u\frac{\partial u}{\partial x} + v\frac{\partial u}{\partial y} + w\frac{\partial u}{\partial z}\right) \quad \text{.....(12.18)}$$

$$\frac{dy}{dt}\ (\text{in y - direction}) = \rho\left(\frac{\partial v}{\partial t} + u\frac{\partial v}{\partial x} + v\frac{\partial v}{\partial y} + w\frac{\partial v}{\partial z}\right) \quad .....(12.19)$$

$$\frac{dz}{dt}\ (\text{in z - direction}) = \rho\left(\frac{\partial w}{\partial t} + u\frac{\partial w}{\partial x} + v\frac{\partial w}{\partial y} + w\frac{\partial w}{\partial z}\right) \quad .....(12.20)$$

$$\left.\begin{array}{l}\text{Component of acceleration in x - direction} = \dfrac{\partial u}{\partial t} \\ \text{Similarly acceleration components in} \\ \text{y, z, directions } \dfrac{\partial v}{\partial t}, \dfrac{\partial w}{\partial t}\end{array}\right\} \quad .....(12.21)$$

Adding eq. (12.18), (12.19), (12.20), & (12.21) gives, total forces per unit volume

$$= \rho\left(\frac{dx}{dt} + \frac{dy}{dt} + \frac{dz}{dt}\right)$$

Also, Force per unit volume

= Pressure gradient force + gravity + friction.

Note: The acceleration of a fluid particle (element/parcel) = sum of the forces that act per unit mass of the body

The forces in Geophysical Fluid Dynamics (GFD) are – gravity, coriolis and friction.

Gravity includes pressure gradient, buoyancy and acceleration due to gravity.

Newton's second law: The rate of change of momentum = sum of applied forces.

$$\frac{dm\vec{v}}{dt} = \text{Force},$$

where m = mass;

$\vec{V}$ = velocity;

$\vec{F}$ = Force

or $$\frac{d\vec{V}}{dt} = \frac{\vec{F}}{m} = \vec{F}_m$$

where $\vec{F}_m$ = Force per unit mass

or Sum of Applied forces

= pressure gradient + coriolis force + gravity + friction.

Equating, total forces acting on unit volume to sum of applied forces, we have

$$\frac{d\vec{V}}{dt} = -\frac{\nabla p}{\rho} - 2\vec{\Omega}\times\vec{V} + \vec{g} + \vec{F} \qquad .....(12.22)$$

where

∇p = Pressure gradient

$\vec{\Omega}$ = rate of rotation of earth

$|\vec{\Omega}| = 7.292 \times 10^{-5}$radiance/second.

$\vec{g}$ = Acceleration gravity

$\vec{F}$ = Frictional force.

Fluid is continuous. In mathematical treatment, mass of a tiny volume of fluid is treated as (a single object), having mass per unit volume

$$\text{Density (acceleration + advection)} = \rho\left(\frac{\partial\vec{V}}{\partial t} + \nabla\cdot\vec{V}\right)$$

$$= \text{force per unit volume}$$

Forces per unit volume = Pressure gradient force + gravity + friction

$$= (\nabla p) + g + \vec{F}$$

Thus,

$$\rho\left(\frac{\partial\vec{V}}{\partial t} + \nabla\cdot\vec{V}\right) = \nabla p + \vec{g} + \vec{F}$$

or

$$\rho\left(\frac{\partial u}{\partial t} + u\frac{\partial u}{\partial u} + v\frac{\partial u}{\partial y} + w\frac{\partial u}{\partial z}\right) = -\frac{\partial p}{\partial x} + g_x + F_x$$

$$= -\frac{\partial p}{\partial x} + F_x - vf \qquad .....(12.23)$$

[here $g_x = 0$, $g_y = 0$, $g_z = g$; components of $\vec{g}$]

and

$$\rho\left(\frac{\partial v}{\partial t} + u\frac{\partial v}{\partial x} + v\frac{\partial u}{\partial y} + w\frac{\partial v}{\partial z}\right) = -\frac{\partial p}{\partial y} + F_y + uf \qquad .....(12.24)$$

where $\frac{\partial u}{\partial t}$ = acceleration component in x – direction

$$u\frac{\partial u}{\partial x}+v\frac{\partial u}{\partial y}+w\frac{\partial u}{\partial y} = \text{Advection component in x – direction}$$

$$\frac{\partial p}{\partial x} = \text{pressure gradient force component in x - direction}$$

$g_x = 0$, gravitational force component in x - direction

$$g_y = 0,\ g_z = \rho g$$

In inertial reference frame coriolis force and centrifugal force.

12.4 Derivation of Pressure Gradient

Pressure gradient generally means change of pressure across a horizontal distance.

Consider a parcel of volume δv with sides $\delta x, \delta y, \delta z$ (in the form of a cuboid) i.e.,

$$\delta v = \delta x\ \delta y\ \delta z \qquad .....(12.25)$$

Let δF_x be the component of force in x – direction.

Let p be the pressure force acting on the left face (A) and $(P+\delta P)$ be the pressure force acting on the right face (B) of the cube. Then $\delta F_x = P\ \delta y\ \delta z - (P+\delta P)\ \delta y\ \delta z$

force = Area × Pressure

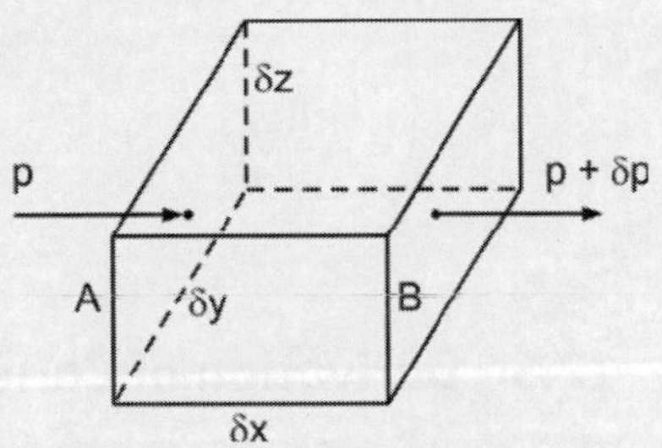

or $\qquad \delta F_x = -\delta p\ \delta y\ \delta z$

since $\qquad \delta p = \frac{\partial p}{\partial x}\delta x,$

the above equation can be written as

$$\delta F_x = \frac{-\partial p}{\partial x}\delta x\ \delta y\ \delta z$$

$$\delta F_x = -\frac{\partial p}{\partial x}\delta v \qquad .....(12.26)$$

The acceleration component of the fluid in x – direction (a_x)

We know F = m a

Force = mass × acceleration;

$$\text{density} = \frac{\text{mass}}{\text{volume}}$$

$$\rho = \frac{m}{v}$$

$$a_x = \frac{\delta F_x}{\delta m}$$

$$a_x = -\frac{1}{\delta m}\frac{\partial p}{\partial x}\delta v \qquad \text{using eq. (12.26)}$$

$$a_x = -\frac{1}{\rho}\frac{\partial p}{\partial x} \qquad \text{.....(12.27)}$$

Similary we have components of acceleration in y, z directions as

$$\left.\begin{aligned} a_y &= -\frac{1}{p}\frac{\partial p}{\partial y} \\ a_z &= -\frac{1}{p}\frac{\partial p}{\partial z} \end{aligned}\right\} \qquad \text{..... (12.28)}$$

Let $\vec{a} = ia_x + ja_y + ka_z \;\;\&\;\; \nabla P = i\frac{\partial p}{\partial x} + j\frac{\partial p}{\partial y} + k\frac{\partial p}{\partial z}$

$$\therefore \qquad \vec{a} = -\frac{1}{\rho}\nabla P \qquad \text{..... (12.29)}$$

12.5 Derivation of Coriolis Term

The rotation of the earth causes drift in a moving body, that is the coriolis effect.

The components of coriolis acceleration effect in x, y, z directions denoted as f_x, f_y, f_z

i.e., $f = if_x + jf_y + kf_z$

where

$$\left.\begin{aligned} f_x &= 2\Omega \sin\phi \, v \\ f_y &= -2\Omega \sin\phi \, u \\ f_z &= 2\Omega \cos\phi \, v_E \end{aligned}\right\} \qquad \text{..... (12.30)}$$

Ω = Angular velocity of the earth = $\frac{2\pi}{86164} = 0.729 \times 10^{-4}$ / sec

ϕ = Latitude

V_E = component toward the east of the velocity

The gravitational force per unit volume in z - direction = – ρg .....(12.31)

where g = 9.780318 m^2/sec

The gravitational force effect in x and y directions is zero

Let the components of friction in x, y, z directions be F_x, F_y, F_z

i.e., $\vec{F} = iF_x + jF_y + kF_z$ .....(12.32)

Substituting the values eq. (12.27), (12.30), (12.31), & (12.32) in eq (12.22) we have the equations of momentum as

$$\left.\begin{aligned} \frac{du}{dt} &= -\frac{1}{\rho}\frac{\partial p}{\partial x} + 2fv - e\omega + F_x \\ \frac{dv}{dt} &= -\frac{1}{\rho}\frac{\partial p}{\partial y} - 2fu + F_y \\ \frac{dw}{dt} &= -\frac{1}{\rho}\frac{\partial p}{\partial z} + 2eu - g + F_z \end{aligned}\right\} \quad .....(12.33)$$

where $f = 2\Omega \sin\phi$, $e = 2\Omega \cos\phi$

or

$$\left.\begin{aligned} \frac{\partial u}{\partial t} + u\frac{\partial u}{\partial x} + v\frac{\partial u}{\partial y} + w\frac{\partial u}{\partial z} &= -\frac{1}{\rho}\frac{\partial p}{\partial x} + fv - e\omega + F_x \\ \frac{\partial v}{\partial t} + u\frac{\partial v}{\partial x} + v\frac{\partial v}{\partial y} + w\frac{\partial x}{\partial z} &= -\frac{1}{\rho}\frac{\partial p}{\partial y} - fu + F_y \\ \frac{\partial w}{\partial t} + u\frac{\partial w}{\partial x} + v\frac{\partial w}{\partial y} + w\frac{\partial w}{\partial z} &= -\frac{1}{\rho}\frac{\partial p}{\partial z} + eu - g + F_z \end{aligned}\right\} \quad ..... (12.34)$$

The eq. (12.34) is called Eulers equations of motion or simply Eulers equations.

They are also called acceleration equations or Navier – Stokes equations.

Vector equation of fluid motion is given by

$$\frac{d\vec{V}}{dt} = -\frac{1}{\rho}\nabla p - 2\vec{\Omega} \times \vec{V} - \vec{\Omega} \times \vec{\Omega} \times \vec{r} + \vec{F} + kg \quad .....(12.35)$$

where k = unit vector in z direction

i.e., Acceleration = pressure gradient + coriolis acceleration + centrifugal acceleration + Frictional force + gravitational force

Here $\vec{\Omega}\times\vec{\Omega}\times\vec{r} = \nabla\frac{\Omega^2 r^2}{2}, |\vec{r}| = r$ distance normal to the rotating axis

$$g + \nabla\left(\frac{\Omega^2 r^2}{2}\right) = \nabla(gz)$$

= gradient of gravitational potential gz

Equation 12.35 can be put in the following form

$$\frac{d\vec{V}}{dt} = -\frac{1}{\rho}\nabla P - \nabla\phi - 2\vec{\Omega}\times\vec{V} + \vec{F} \qquad \text{.....(12.36)}$$

where $\phi = gz - \frac{\Omega^2 r^2}{2}$

= gravitational potential measured in rotating frame

$\vec{\Omega}^2 \vec{r}$ = centrifugal acceleration

12.6 Equations of Motion Applied to the Oceans

The equations of motion applied to oceans require certain simplifications.

Vertical acceleration $\frac{dw}{dt}$, frictional term $\vec{F}$ are neglected. The vertical component of coriolis force (deflection force) also neglected. Under these assumptions the approximate equations of motion for oceans are. (from 12.33)

$$\frac{du}{dt} = -\frac{1}{\rho}\frac{\partial p}{\partial x} + fv + \frac{1}{\rho}F_x$$

$$\frac{dv}{dt} = -\frac{1}{\rho}\frac{\partial p}{\partial y} - fu + \frac{1}{\rho}F_y$$

$$g\rho\, dz = dp \text{ or } \delta p = g\rho\, \delta z$$

where F_x, F_y, components of external forces in the x, y directions.

At any given time an isobaric surface is defined by

$$\frac{\partial p}{\partial x}dx + \frac{\partial p}{\partial y}dy + \frac{\partial p}{\partial z}dz = 0$$

(Mathematical definition of equi-scalar surface is

$$ds = \frac{\partial s}{\partial x}dx + \frac{\partial s}{\partial y}dy + \frac{\partial s}{\partial z}dz = 0 \text{ where s is a scalar quantity)}$$

Components of the horizontal pressure gradients $\left(\frac{\partial p}{\partial x}, \frac{\partial p}{\partial y}\right)$ are identical with components of gravity acting along the isobaric surfaces

$$g\left(\frac{dz}{dx}\right)_p = g_i p_x = \frac{-g\alpha\frac{\partial p}{\partial x}}{\alpha\frac{\partial p}{\partial z}} = -\frac{\partial p}{\partial x}$$

$$g\left(\frac{dz}{dy}\right)_p = g_i p_y = \frac{-g\alpha\frac{\partial p}{\partial y}}{\alpha\frac{\partial p}{\partial z}} = -\frac{\partial p}{\partial y}$$

The slope is positive in the direction in which the surface slopes downward.

Geopotential slope: The geopotential slope is defined

as $iD_x = \frac{dD}{dx}$, $iD_y = \frac{dD}{dy}$ and measured in dynamic meters,

$$dD = \frac{g}{10}dz \text{ where } D = \frac{1}{10}\int_0^z g dz$$

or $$z = 10\int_0^D \frac{1}{g} dD$$

the geopotential of a surface level at the geometrical depth z

or $$iD, x = -\alpha\frac{\partial p}{\partial x}$$

$$iD, y = -\alpha\frac{\partial p}{\partial y}$$

where iD, x = inclination of slope of geopotential in x-direction

The geopotential slope in dynamic metres per meter = specific volume × pressure gradient in deco–bars per meter.

12.7 Alternate Method of Equations of Motion

Euler's Method: This method uses external forces acting on unit mass and total pressure gradients per unit mass.

Let X, Y, Z be the external forces per unit mass in the direction of x, y, z respectively. Then,

$$\left.\begin{aligned} \frac{du}{dt} &= X - \alpha\frac{\partial p}{\partial x} \\ \frac{dv}{dt} &= Y - \alpha\frac{\partial p}{\partial y} \\ \frac{dw}{dt} &= Z - \alpha\frac{\partial p}{\partial z} \end{aligned}\right\} \text{ where } \alpha = \frac{1}{\rho} \qquad \text{.....(12.37)}$$

If

$$\left.\begin{aligned} u &= f_x(x, y, z, t) \\ v &= f_y(x, y, z, t) \\ w &= f_z(x, y, z, t) \end{aligned}\right\}$$

be the velocity components in x, y, z directions in the beginning time t.

Let after time $t + \delta t$, the change of particle (parcel/element) the velocity components move to x + dx, y + dy, z + dz.

The velocity components at $t + \Delta t$ time

become $= f_x(x + dx, y + dy, z + dz; t + \Delta t)$ etc

Using Taylors expansion (series) we have

$$f(x + dx, y + dy, z + dz, t + \delta t)$$

$$= f_x(x, y, z, t) + \frac{\partial u}{\partial x}dx + \frac{\partial u}{\partial y}dy + \frac{\partial u}{\partial z}dz + \frac{\delta a}{\partial t}dt$$

$$\underset{\delta t \to 0}{\text{Lt}} \frac{f(x + dx, y + dy, z + dz, t + \delta t) - f_x(x, y, z, t)}{\delta t}$$

$$= \frac{\partial u}{\partial x}\frac{dx}{dt} + \frac{\partial u}{\partial y}\frac{dy}{dt} + \frac{\partial u}{\partial z}\frac{dz}{dt} + \frac{\partial u}{\partial t} \quad \text{neglecting second order terms}$$

$$= u\frac{\partial u}{\partial x} + v\frac{du}{\partial y} + w\frac{\partial u}{\partial z} + \frac{\partial u}{\partial t}$$

∴ The acceleration of individual particle (parcel) will have components.

$$\left.\begin{aligned}\frac{du}{dt}&=\frac{\partial u}{\partial t}+u\frac{\partial u}{\partial x}+v\frac{\partial u}{\partial y}+w\frac{\partial u}{\partial z}\\ \frac{dv}{dt}&=\frac{\partial v}{\partial t}+u\frac{\partial v}{\partial x}+v\frac{\partial v}{\partial y}+w\frac{\partial v}{\partial z}\\ \frac{dw}{dt}&=\frac{\partial w}{\partial t}+u\frac{\partial w}{\partial x}+v\frac{\partial w}{\partial y}+w\frac{\partial w}{\partial z}\end{aligned}\right\} \quad \text{..... (12.38)}$$

if we consider level surface (x-y plane), then X = Y =0 and Z = g, then

eq (12.37) becomes

$$\left.\begin{aligned}\frac{du}{dt}&=-\frac{1}{\rho}\frac{\partial p}{\partial x}+\frac{1}{\rho}F_x\\ \frac{dv}{dt}&=-\frac{1}{\rho}\frac{\partial p}{\partial y}+\frac{1}{\rho}F_y\\ \frac{dw}{dt}&=-\frac{1}{\rho}\frac{\partial p}{\partial z}+g+\frac{1}{\rho}F_z\end{aligned}\right\} \quad \text{..... (12.39)}$$

where $\vec{F} = iF_x + jF_y + kF_z$ frictional force

From Eq. (12.38) & (12.39) we have

$$\left.\begin{aligned}\frac{\partial u}{\partial t}+u\frac{\partial u}{\partial x}+v\frac{\partial u}{\partial y}+w\frac{\partial u}{\partial z}&=-\alpha\frac{\partial p}{\partial x}+\alpha F_x\\ \frac{\partial v}{\partial t}+u\frac{\partial v}{\partial x}+v\frac{\partial v}{\partial y}+w\frac{\partial v}{\partial z}&=-\alpha\frac{\partial p}{\partial x}+\alpha F_y\\ \frac{\partial w}{\partial t}+u\frac{\partial w}{\partial x}+v\frac{\partial w}{\partial y}+w\frac{\partial w}{\partial z}&=-\alpha\frac{\partial p}{\partial z}+g+\alpha F_z\end{aligned}\right\} \quad \text{..... (12.40)}$$

The above equations of motion of fluid are in relative coordinate system (x, y, z) on the earth.

In above equations (12.40) of motion in rotating coordinates (absolute coordinates) with earth, the RHS have to add the accelerations

$$\left.\begin{aligned}f_x &= 2\Omega, \sin\phi v = fv\\ f_y &= 2\Omega\sin\phi u = -fu\\ f_z &= 2\Omega\cos\phi V_R = eV_E\end{aligned}\right\} \quad \text{..... (12.41)}$$

where $f = 2\Omega\sin\phi$

$e = 2\cdot\Omega\cos\phi$

and Ω = angular velocity of the earth

$$= \frac{2\pi}{86164} \simeq 0.729 \times 10^{-4}/\text{sec},$$

ϕ = Latitude

V_E = component toward the east of velocity

$\therefore$ The equations of motion (12.40) with (12.41) we have

$$\left.\begin{aligned} \frac{du}{dt} &= fv - \alpha\frac{\partial p}{\partial x} + \alpha F_x \\ \frac{dv}{dt} &= -fu - \alpha\frac{\partial p}{\partial y} + \alpha F_y \\ \frac{dw}{dt} &= -eV_E + g - \alpha\frac{\partial p}{\partial z} + \alpha F_z \end{aligned}\right\} \quad .....(12.42)$$

where F_x, F_y, F_z friction components per unit volume in x, y, z directions eq. (12.42) are the required equations of motion.

12.8 The Dynamic Energy Equation

The equations of motion are:

$$\frac{du}{dt} = fv - \alpha\frac{\partial p}{\partial x} + \alpha F_x \quad .....(12.43)$$

$$\frac{dv}{dt} = -fu - \alpha\frac{\partial p}{\partial y} + \alpha F_y \quad .....(12.44)$$

$$\frac{dw}{dt} = -eV_E + g - \alpha\frac{\partial p}{\partial z} + \alpha F_z \quad .....(12.45)$$

In horizontal motion $\frac{dw}{dt} = 0,\ V_E = 0,\ F_z = 0$

and $\alpha\frac{\partial p}{\partial z} = g$ [from eq(12.45)]

or $\delta p = g\rho\ \delta z$ which is Hydrostatic equation .....(12.46)

Multiplying eq. (12.43) by u and eq. (12.44) by v we get

$$\frac{1}{2}\frac{du^2}{dt^2} = -fuv - \alpha u\frac{\partial p}{\partial x} + \alpha uF_x$$

$$\frac{1}{2}\frac{d\overset{2}{v}}{dt^2} = -fuv - \alpha v\frac{\partial p}{\partial y} + \alpha v F_y$$

Adding we get

$$\frac{1}{2}\frac{\overset{2}{d}\left(\overset{2}{u}+\overset{2}{v}\right)}{dt^2} = -\alpha\left(u\frac{\partial p}{\partial x} + v\frac{\partial p}{\partial y}\right) + \alpha\left(uF_x + vF_y\right)$$

or
$$\frac{d}{dt}(\rho\vec{V}^2) = -u\frac{\partial p}{\partial x} - v\frac{\partial p}{\partial y} + uF_x + vF_y \qquad .....(12.47)$$

LHS of eq. (12.47) $\frac{d}{dt}\left(\frac{\rho}{2}\overset{2}{\vec{V}}\right)$ = increase in kinetic energy per unit volume

RHS of eq. (12.47) $-u\frac{\partial p}{\partial x} - v\frac{\partial p}{\partial y} + uF_x + vF_y$ = sum of the (activity) power per unit of volume of the forces due to pressure distribution and friction.

∴ Eq. (12.47) gives increase in KE per unit volume = work done per unit volume by acting forces.

Combining with thermodynamic energy equation, we have

$$\frac{dW}{dt} = \frac{d}{dt}(T + \phi + E) + \iiint (uF_x + vF_y)dx\, dy\, dz \qquad .....(12.48)$$

where

W = Amount of heat added to the system

T = Total KE of the system

ϕ = Potential energy of the system

E = Internal energy of the system

The above equation is,

If the total energy of the system remains constant, the amount of heat added per unit of time = the work done per unit time of the frictional forces. If heat added is zero, then

The work of the frictional forces = a change of the total energy of the system.

Hydrostatic equation is given by

$$dp = K\ \rho_{s\ \theta\ p}\ gdz$$

where

K = numerical factor depends on the units

Notation: Density of water = $\rho_{s,\ \theta,\ p}$, that is density is a function of$[\rho = \rho(s, \theta, p)]$

s = salinity

θ = Potential temperature

P = Pressure

CHAPTER 13
FIELDS

13.1 Field Variables

Atmosphere and oceans are continuous fluid mediums. The physical quantities that specify the atmosphere or ocean such as temperature, pressure, density, salinity, wind velocity etc., are continuous function of space and time. They are called field variables or simply fields. Field variables are differentiable. A field (D) is a part of space where every point of space has a value for any measurable quantity such as temperature (T), pressure (P), density (ρ), salinity (S) etc., of fluid. The plot of these variables provide us temperature field, pressure field, density field, salinity field etc.

13.2 Scalar Fields

Scalar quantities, like temperature, pressure, density, speed, mass etc., can be fully specified if its magnitude known (degrees celsius, pascals, etc). A scalar quantity has only magnitude but has no direction (i.e., not associated with direction).

Temperature is a scalar quantity. The measurements of temperature at points in the atmosphere is required for the studies of atmosphere.

Scientists use the word Field to describe the conditions which exists within a specified (given) space.

A field is a portion of space where there is a value for a measurable quantity at every point in that space. Thus we have temperature field, pressure field, density field etc. The measurements may be plotted on maps, which represent various surfaces cutting through the atmosphere. Plotting of the temperature measurements in vertical section or horizontally on maps gives temperature field of the atmosphere. Lines passing through all points having same temperature is called isotherms. In Greek, 'iso' means equal and 'Therme' means heat. An isotherm is a line in space along which temperatures are equal. The temperature field is a scalar quantity. Similarly we have pressure field, density field, mass field etc., in general we call isopleths (having same quantity) like isopycnals, isotachs etc.

Consider a three dimensional space (x, y, z). Let D be a part of this space in which $\phi = \phi(x, y, z)$ defined a scalar field. If $\phi(x, y, z)$ denotes temperature at the point M(x, y, z), then we say that scalar field of temperature defined. If D occupies some fluid (liquid or gas) and $\phi(x, y, z)$ denotes T, P, ρ etc. We have scalar fields of T, P, ρ etc.

Consider the points of the region D in which the function $\phi(x, y, z)$ has a fixed value, say C

$$\phi(x, y, z) = C$$

The totality of these points form a certain surface. For different values of C we get different surfaces. These surfaces are called level surfaces.

For example $\phi(x,y,z) = x^2 + y^2 + z^2$ or $\frac{x^2}{4} + \frac{y^2}{16} + \frac{z^2}{25}$,

then $x^2 + y^2 + z^2 = r^2$ or $\frac{x^2}{4} + \frac{y^2}{16} + \frac{z^2}{25} = 1$ present level surface of sphere or ellipsoid respectively.

If $\phi = \phi(x, y)$ is a function of two variables x and y, then the level surfaces are lines on the xy-plane $\phi(x, y) = C$ which are called level lines.

13.3 Vector Field

A vector quantity has both magnitude and direction. Velocity, acceleration, force etc., are vectors, because they are specified by both magnitude and direction.

A vector field is a portion (part) of space, where there is a value for that vector quantity at every point in that space. For example earth exerts a gravitational force of attraction on objects located in the space around it and which determine the gravitational field of the earth. Similarly earth has magnetic field.

13.4 Equi-scalar Surface(s)

An equi-scalar surface is a function of coordinates.

For an equi-scalar surface (s), the differential equation is given by

$$ds = \frac{\partial s}{\partial x}dx + \frac{\partial s}{\partial y}dy + \frac{\partial s}{\partial z}dz = 0$$

i.e., s = constant, like isobar, isotherm, isopyanal etc.

In a vertical plane (x-z plane) the equi-scalar curves similarly defined by

$$\frac{\partial s}{\partial x}dx + \frac{\partial s}{\partial z}dz = 0$$

or surface slope x- direction $\frac{dz}{dx} = \frac{-\left(\frac{\partial s}{\partial x}\right)}{\left(\frac{\partial s}{\partial z}\right)}$

slope in y – direction $\frac{dz}{dy} = \frac{-\left(\frac{\partial s}{\partial y}\right)}{\left(\frac{\partial s}{\partial z}\right)}$

At a perfect hydrostatic equilibrium, the isobaric surfaces coincide with level surfaces but this is not valid if motion exists.

At any given time an isobaric surface is defined by

$$\frac{\partial p}{\partial x}dx + \frac{\partial p}{\partial y}dy + \frac{\partial p}{\partial z}dz = 0$$

Note: A surface level is considered as a level surface.

The work required (or gained) in moving a unit mass from sea level to a point above (or below) sea level is called the **gravity potential**.

In m, t, s, (meters, tonnes, & seconds) system, the unit of gravity potential is one dynamic meter.

13.5 The Pressure Field

The distribution of pressure in the sea is given by the equation of static equilibrium

$$dp = k\,\rho\,g\,d\,z \qquad .....(i)$$

where dp = pressure change

k = numerical factor, depend on the units

$\rho = \rho(s, \theta, p)$, density of sea water which is a function of salinity (S), potential temperature (θ) and pressure (P)

The geopotential expressed in dynamic meters (D) as vertical coordinate
10 dD = g dz

where $$D = \frac{1}{10}\int_0^z g\,dz \qquad \text{....(ii)}$$

If pressure p is measured in decibars.

$$1 \text{ bar} = 10^5 \text{ h Pa or } 10^6 \text{ dynes/cm}^2\text{, the factor K is } \frac{1}{10}.$$

The equation (i) reduces to

$$dp = \rho_{s,\theta,p}\, dD \quad \text{or} \quad dD = \propto_{s,\theta,p}\, dp \qquad \text{.....(iii)}$$

where $$\propto_{s,\theta,p} = \frac{1}{\rho_{s,\theta,p}} = \text{specific volume}$$

$$\int_{p_1}^{p_2} dp = \rho_{s,\theta,p} \int_{D_1}^{D_2} dD$$

or $$p_2 - p_1 = \rho_{s,\theta,p}(D_2 - D_1)$$

$$= Z_2 - Z_1$$

Thus the pressure field is completely described by means of a system isobaric surfaces.

The field of mass: The field of mass in the ocean is described by specific volume.

$$\propto_{s,\theta,p} = \propto_{-35,0,p} + \delta \quad \text{or} \quad \delta = \propto_{s,\theta,p} - \propto_{35,0,p}$$

where $$\propto_{s,\theta,p} = \frac{1}{\rho_{s,\theta,p}} \text{specific volume}$$

$\propto_{35,0,p}$ = specific volume of water of salinity 35‰ at

temperature 0°C and pressure = P.

δ = anomaly = $\propto_{s,\theta,p} - \propto_{35,0,p}$

where $\propto_{s,\theta,p}$ = Specific volume in situ

$\propto_{35,\theta,p}$ = Standard specific volume

13.6 Dynamic Meter (D)

Dynamic meter (D) is defined as

$$D = \frac{1}{10}\int_{o}^{z} g\, d\, z$$

where the geopotential of a level surface at the geometrical depth z.

D = Dynamic meter, the geopotential of a level surface.

The geopotential of a level surface at the geometrical depth z is in dynamic meters.

$$D = \frac{1}{10}\int_{o}^{z} g\, d\, z$$

On the other hand, the geometrical depth in meters of a level surface is

$$z = 10\int_{o}^{D} \frac{1}{g}\, d\, D$$

At the north pole, the geometrical depth of the

D = 1000 (dynamic meters) surface is Z = 1017 m

At the equator, the depth (z) is 1022.3 m for the D = 1000 (dynamic meters), because 'g' is greatest at the poles as compared at equator.

Normal values of 'g' at the pole 9.83205

'g' at the equator 9.78027

At pole: $z = 10\int_{o}^{D} \frac{1}{g} dD$

$$= \frac{10}{g}(D-0) = \frac{10}{9.8320}\times 1000$$

$$= 1017.08 \text{ m}$$

At equator: $z = 10 \int_{o}^{D} \frac{1}{g} dD$

$$= \frac{10}{g}[D-0] = \frac{10(1000-0)}{9.78023} = \frac{10}{9.78023}\times 1000$$

$$= 1022.3 \text{ m}$$

Note: Level surfaces (depths) and surfaces of equal geometrical depths do not coincide. Therefore components of the acceleration of gravity (g) acts along surfaces of equal geometrical depth.

The topography of the sea bottom denoted by isobaths (lines of equal geometrical depth) could be shown by lines of equal geopotential.

Contour lines represent the lines of intersection of level surfaces with that of irregular surface of the bottom.

Note:

(i) An isobaric surface (an imaginary surface of equal pressure) along which fluid pressure is constant.

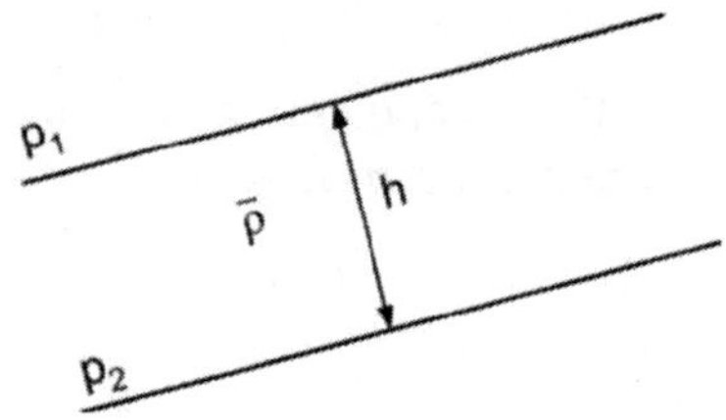

(ii) A level surface (an imaginary surface) along which component of gravity is zero.

(iii) The vertical distance (h) between two isobaric surfaces is given by

$$h = \frac{P_1 - P_2}{g\bar{\rho}}$$

where g = acceleration due to gravity

$\bar{\rho}$ = average density between the isobars P_1, P_2

13.7 Some Mathematical Definitions

1. Hamilton operator $\nabla = i\frac{\partial}{\partial x} + j\frac{\partial}{\partial y} + k\frac{\partial}{\partial z}$

   ∇ is a vector differential

   If U is a scalar quantity, then

2. $\text{Grad } U = \nabla U = i\frac{\partial U}{\partial x} + j\frac{\partial U}{\partial y} + k\frac{\partial U}{\partial z}$ or $\left(i\frac{\partial}{\partial x} + j\frac{\partial}{\partial y} + k\frac{\partial}{\partial z}\right)U$

   Grad U = gradient of U.

3. $\text{Div } \vec{U} = \nabla \cdot \vec{U} = \left(i\frac{\partial}{\partial x} + j\frac{\partial}{\partial y} + k\frac{\partial}{\partial z}\right) \cdot (iu + jv + kw)$

   $$= \frac{\partial u}{\partial x} + \frac{\partial v}{\partial y} + \frac{\partial w}{\partial z}$$

   where $\vec{U} = iu + jv + kw$

   Div $\vec{U}$ = divergent of $\vec{U}$

4. $\text{rot } \vec{U} = \text{curl } \vec{U}$

$$= \nabla \times \vec{U}$$

$$= \left(i\frac{\partial}{\partial x} + j\frac{\partial}{\partial y} + k\frac{\partial}{\partial z} \right) \times (iu + jv + kw)$$

$$= k\frac{\partial v}{\partial x} - j\frac{\partial w}{\partial x} - k\frac{\partial u}{\partial y} + i\frac{\partial w}{\partial y} + j\frac{\partial u}{\partial z} - i\frac{\partial v}{\partial z}$$

$$= i\left(\frac{\partial w}{\partial y} - \frac{\partial v}{\partial z} \right) + j\left(\frac{\partial u}{\partial z} - \frac{\partial w}{\partial x} \right) + k\left(\frac{\partial v}{\partial x} - \frac{\partial u}{\partial y} \right)$$

5. **Gauss or Ostrogradsky formula**

The flux of a vector field through a closed surface (s) is defined as

$$\oint_s \vec{A} \cdot d\vec{s} = \oint_s A_n ds$$

or $$\oint_s \vec{A} \cdot d\vec{s} = \oint_\Omega \text{div}\vec{A} d\Omega$$

$$= \oint_\Omega \nabla \cdot \vec{A} \, d\Omega$$

where Ω is the (domain) region bounded by the surface s and A_n = projection of $\vec{A}$ on the outward normal to s.

6. **Stoke's formula:** The circulation of a vector field $\vec{A}$ over a closed oriented contour (L) is defined as

$$\oint_L \vec{A} \cdot d\vec{r} = \oint_L A_T dL$$

or $$\oint_L \vec{A} \cdot d\vec{r} = \oint_s \text{rot}\vec{A}.ds$$

$$= \int_s \left(\nabla \times \vec{A} \right) \cdot ds$$

where $\vec{r}$ = ix + jy + kz, radius vector

s = Arbitrary surface bounded by L.

7. rot grad U = $\nabla \times (\nabla U)$

$$= \left(i\frac{\partial}{\partial x} + j\frac{\partial}{\partial y} + k\frac{\partial}{\partial z} \right) \times \left(i\frac{\partial u}{\partial x} + j\frac{\partial u}{\partial y} + k\frac{\partial u}{\partial z} \right)$$

$$= k\frac{\partial^2 u}{\partial x \partial y} - j\frac{\partial^2 u}{\partial x \partial z} - k\frac{\partial^2 u}{\partial x \partial y} + i\frac{\partial^2 u}{\partial y \partial z} + j\frac{\partial^2 u}{\partial x \partial z} - i\frac{\partial^2 u}{\partial y \partial z}$$

$$= 0$$

8. div rot $\vec{U} = \nabla \cdot (\nabla \times \vec{U})$

$$= \nabla \cdot \left[\left(i\frac{\partial}{\partial x} + j\frac{\partial}{\partial y} + k\frac{\partial}{\partial z} \right) \times \left(iu + jv + kw \right) \right]$$

$$= \nabla \cdot \left[k\frac{\partial v}{\partial x} - j\frac{\partial w}{\partial x} - k\frac{\partial u}{\partial y} + i\frac{\partial w}{\partial y} + j\frac{\partial u}{\partial z} - i\frac{\partial v}{\partial z} \right]$$

$$= \left(i\frac{\partial}{\partial x} + j\frac{\partial}{\partial y} + k\frac{\partial}{\partial z} \right) \cdot \left(k\left(\frac{\partial v}{\partial x} - \frac{\partial u}{\partial y} \right) + j\left(\frac{\partial u}{\partial z} - \frac{\partial w}{\partial x} \right) + i\left(\frac{\partial w}{\partial y} - \frac{\partial v}{\partial z} \right) \right)$$

$$= \frac{\partial}{\partial x}\left(\frac{\partial w}{\partial y} - \frac{\partial v}{\partial z} \right) + \frac{\partial}{\partial y}\left(\frac{\partial u}{\partial z} - \frac{\partial w}{\partial x} \right) + \frac{\partial}{\partial z}\left(\frac{\partial v}{\partial x} - \frac{\partial u}{\partial y} \right)$$

$$= 0$$

9. $\vec{U} \cdot \nabla = (iu + jv + kw) \cdot \left(i\frac{\partial}{\partial x} + j\frac{\partial}{\partial y} + k\frac{\partial}{\partial z} \right)$

$$= u\frac{\partial}{\partial x} + v\frac{\partial}{\partial y} + w\frac{\partial}{\partial z}$$

10. $\nabla \cdot \left[\phi\vec{A} \right] = \phi(\nabla \cdot \vec{A}) + \vec{A} \cdot (\nabla\phi)$, where ϕ is a scalar

11. $\nabla \times \left[\phi\vec{A} \right] - \phi(\nabla \times \vec{A}) + (\nabla\phi) \times \vec{A}$

12. $\nabla \cdot \left[\vec{A} \times \vec{B} \right] = \vec{B} \cdot (\nabla \times \vec{A}) - \vec{A} \cdot (\nabla \times \vec{B})$

13. $\nabla \left[\vec{A} \cdot \vec{B} \right] = (\vec{A} \cdot \nabla)\vec{B} + (\vec{B} \cdot \nabla)\vec{A} + \vec{A} \times (\nabla \times \vec{B}) + \vec{B} \times (\nabla \times \vec{A})$

14. $\nabla \times (\nabla \times \vec{A})$ or rot of rot $\vec{A}$

$$= \nabla(\nabla \cdot \vec{A}) - \nabla^2\vec{A}$$

15. $\nabla \cdot (\nabla\phi) = \nabla^2\phi$ div of grad ϕ = Laplacian ϕ, where ϕ is a scalar

$$\left(i\frac{\partial}{\partial x} + j\frac{\partial}{\partial y} + k\frac{\partial}{\partial z} \right) \cdot \left(i\frac{\partial\phi}{\partial x} + j\frac{\partial\phi}{\partial y} + k\frac{\partial\phi}{\partial z} \right)$$

$$= \frac{\partial^2\phi}{\partial x^2} + \frac{\partial^2\phi}{\partial y^2} + \frac{\partial^2\phi}{\partial z^2} = \nabla^2\phi$$

CHAPTER 14

EKMAN LAYER – INTRODUCTION

The wind driven ocean friction layer is called the Ekman layer. Ekman layer accounts only eddy viscosity (friction) and coriolis acceleration. Changes in the wind stress develop transcient oscillations in the ocean which are called inertial currents. The inertial currents have the period $2\pi/f$ (where f is coriolis parameter). Steady winds produce a thin boundary layer. Ekman layer at the top of the ocean have the following properties.

Direction of the fluid flow is 45° to the right of the wind looking down wind in the Northern hemisphere (and to the left in SH). Surface speed of ocean water is 1-2.5% of wind speed. The depth of wind effect on ocean water 50-300 meters depending on latitude and wind velocity. Surface ocean current measurements show that inertial oscillations are the main component of the current in the mixed top layer. The flow is roughly independent of depth within mixed layer for inertial period $2\pi/f$ Mixed layer moves like slab at the inertial period. Surface drifters tend to drift parallel to the isobaric lines (of constant atmospheric pressure) at the sea surface. As said earlier the physical process in Ekman layer considers eddy viscosity and coriolis acceleration. In Ekman layer current velocity is strongest at the ocean surface and reduces exponentially downwards and becomes zero at about 50 m depth.

14.1 Ocean Surface Ekman Layer

Steady winds blowing on the sea surface develop a thin horizontal boundary layer (a few hundred meters thickness), in which friction plays a part. This layer is called Ekman layer. Ekman layer accounts only friction (eddy viscosity) and Coriolis acceleration. A similar boundary layer exists at the bottom of the ocean, which is termed bottom Ekman layer. Another boundary layer that exists at the bottom of the atmosphere and just above sea level (also called planetary boundary layer) also called Ekman layer. Ekman expanded the layer to study the influence of density differences of water with the following assumptions.

State of fluid is steady, homogeneous, flow is horizontal with the effect of friction on a rotating earth.

The horizontal and temporal derivatives are zero i.e.,

$$\frac{\partial}{\partial t} = 0 = \frac{\partial}{\partial x} = \frac{\partial}{\partial y}$$

There is a balance between frictional and coriolis forces.

A constant vertical eddy viscosity acts in the fluid in the form

$$\left.\begin{aligned} \tau_{xz} &= \rho A_z \frac{\partial u}{\partial z} \\ \tau_{yz} &= \rho A_z \frac{\partial v}{\partial z} \end{aligned}\right\}$$

where A_z = eddy viscosity or eddy diffusing: τ_{xz}, τ_{yz} are components of wind stress in x, y directions (i.e., zonal and meridional directions).

The ocean circulation in the top friction layer is driven by: tangential stresses due to prevailing wind at the ocean surface which give (impact) momentum to the ocean and convection by the loss of buoyancy in polar latitudes because of cooling and/or salt input that causes the water to sink to depth.

The effect of wind stress over ocean surface induce Ekman pumping or suction (a pattern of vertical motion or curl). Pumping down of buoyanent surface water in sub–tropics and sucking up of salty heavier interior fluid at the poles and equator. Wind stress pattern varies in space and results in vertical motion across the Ekman layer base. The flow within the Ekman layer is convergent in association with anticyclonic flow and divergent in association with cyclonic flow. Convergent flow causes (or drives) downward vertical motion which is termed Ekman pumping, whereas divergent flow causes (or drives) upward vertical motion beneath, which is called Ekman suction. The (Figure 14.1) illustrates Ekman pumping and suction.

As stated earlier, the wind driven friction layer is called the Ekman layer. In this layer there is a balance between friction (or eddy viscosity) and coriolis acceleration.

The momentum equations for homogeneous steady state, turbulent boundary layer above or below horizontal surface are:

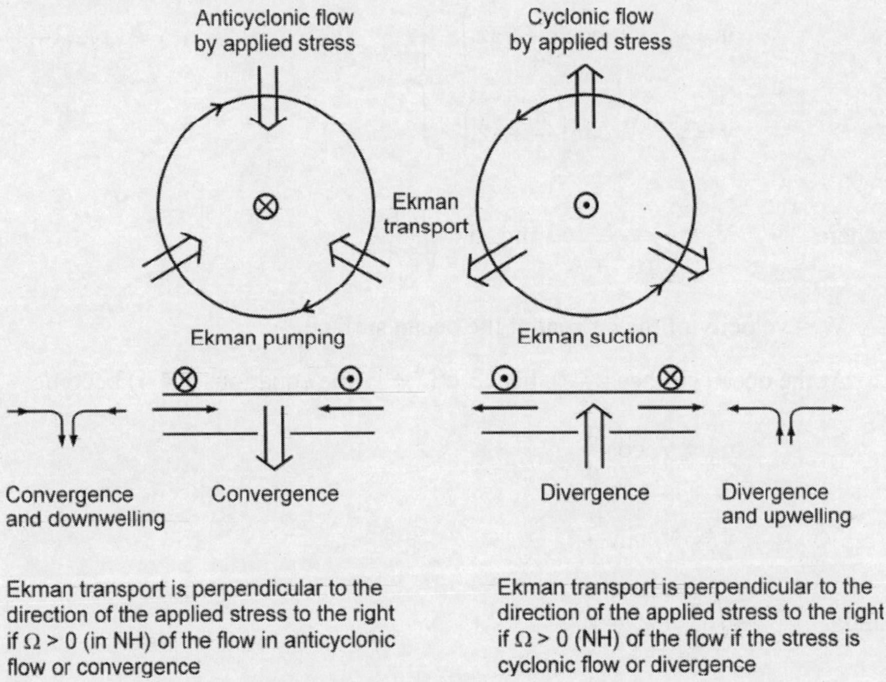

Fig. 14.1

$$\left.\begin{aligned} \rho fv + \frac{\partial \tau_{xz}}{\partial z} = 0 \\ \rho fu - \frac{\partial \tau_{yz}}{\partial z} = 0 \end{aligned}\right\} \quad \ldots..(14.1)$$

where $f = 2\Omega \sin\phi$, τ_{xz}, τ_{yz} are components of wind stress in x, y directions

$$\text{Since} \quad \left.\begin{aligned} \tau_{xz} = \rho A_z \frac{\partial u}{\partial z} \\ \tau_{yz} = PA_z \frac{\partial v}{\partial z} \end{aligned}\right\} \quad \ldots..(14.2)$$

Ekman vertical eddy viscosity components assumed to be constant

Using eq. (14.2), the above equations of momentum become

$$\left.\begin{aligned} \rho fv + \frac{\partial}{\partial z}\left(\rho A_z \frac{\partial u}{\partial z}\right) = 0 \text{ or } fv + A_z \frac{\partial^2 u}{\partial z^2} = 0 \\ \rho fu - \frac{\partial}{\partial z}\left(\rho A_z \frac{\partial v}{\partial z}\right) = 0 \text{ or } fu - A_z \frac{\partial^2 v}{\partial z^2} = 0 \end{aligned}\right\} \quad \ldots..(14.3)$$

The differential equations (14.3) have the solution of the form (when the wind blowing to the north $\tau = \tau_{yz}$)

$$\left.\begin{aligned} u &= V_o\, e^{az} \cos\left(\frac{\pi}{4} + az\right) \\ v &= V_o\, e^{az} \sin\left(\frac{\pi}{4} + az\right) \end{aligned}\right\} \quad \text{.....(14.4)}$$

where $V_o = \dfrac{\tau}{\sqrt{\rho_\omega{}^2 + A_z}}$ and $a = \sqrt{\dfrac{f}{2A_z}}$

V_o = velocity of the current at the ocean surface.

At the ocean surface z = 0, hence $e^{z=0} = 1$, the equations (14.4) become

$$u_o = V_o \cos\frac{\pi}{4}$$

$$v_o = V_o \sin\frac{\pi}{4}$$

and $\sqrt{u_o^2 + v_o^2} = v_o e^{az}$

14.2 Ekman Mass Transport (M_{EK})

Ocean flow in the top (surface) Ekman layer carries mass.

The Ekman mass transport M_{EK} is defined as below.

Let $M_{EK\,X}$, $M_{EK\,Y}$ be x, y components of the transport, then

$$M_{EK\,x} = \int_{-d}^{0} \rho U_{EK}\, dz \quad \text{.....(14.5)}$$

$$M_{EK\,y} = \int_{-d}^{0} \rho V_{EK}\, dz \quad \text{.....(14.6)}$$

where $\vec{U} = iU_{EK} + jV_{EK}$,

The integral limits are z = – d to z = 0 (from depth z = – d to the surface z = 0)

Note: ρ dz = mass of unit volume

$M_{EK\,x}$ is the mass of water passing through a vertical plane yz = 1 meter wide and is perpendicular to the transport extending from surface z = 0 to depth z = – d.

U_{EK}, V_{EK} components of Ekman velocity in x, y directions (zonal & meridional components)

For homogeneous steady, turbulent incompressible flow in boundary layer below or above horizontal surface, the momentum equations (x, y components) are:

$$f\rho v + \frac{\partial \tau_{xz}}{\partial z} = 0 \qquad .....(14.7)$$

$$f\rho u - \frac{\partial \tau_{yz}}{\partial z} = 0 \qquad .....(14.8)$$

where τ_{xz}, τ_{yz} are components of wind stress in x, y directions.

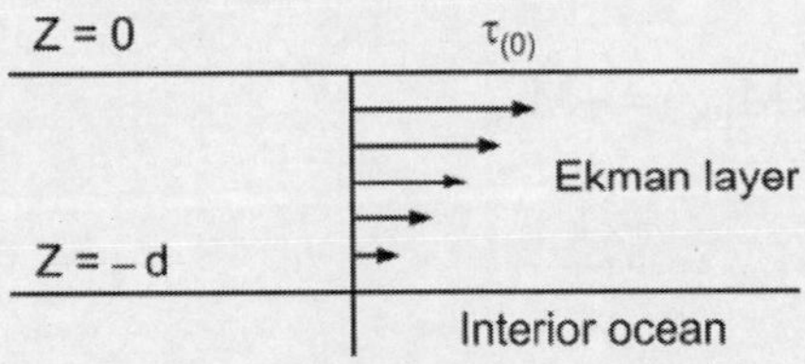

Fig. 14.2

Integrating (14.7) and (14.8) between limits z = – d to z = 0, we have

$$f\int_{-d}^{o} \rho V_{EK} dz = -\int_{-d}^{o} \frac{\partial \tau_{xz}}{\partial z} dz = -\int_{-d}^{o} d\,\tau_{xz}$$

$$f\int_{-d}^{o} \rho U_{EK} dz = +\int_{-d}^{o} \frac{\partial \tau_{yz}}{\partial z} dz = \int_{-d}^{o} d\,\tau_{yz}$$

Using (14.5) & (14.6), the above integrals become

$$fM_{Eky} = -\int_{-d}^{o} d\tau_{xz} = -\left[\tau_{xz}\right]_{-d}^{o} = -\tau_{xz}\big|_{z=0} + \tau_{xz}\big|_{z=-d}$$

$$fM_{Eky} = -\tau_{xz}(0) \text{ since } \tau_{xz} \text{ at depth } (-d) = 0$$

and $$fM_{Ekx} = \int_{-d}^{o} d\tau_{yz} = \left[\tau_{yz}\right]_{-d}^{o} = \tau_{yz}\big|_{z=0}, \text{ since } T_{yz}\big|_{z=-d} = 0$$

$$\therefore \left.\begin{aligned} fM_{EKY} &= -\tau_{xz}(0) \\ fM_{EKx} &= +\tau_{yz}(0) \end{aligned}\right\} \qquad .....(14.9)$$

We know $M_{EK} = \int_{-d}^{o} \rho \vec{U} dz$

where $\vec{U} = i\, U_{EKX} + j v_{EKY}$ are x, y components

$$M_{EK} = \frac{\tau \times \hat{k}}{f}$$

where $\tau_{xz}(0)$, $\tau_{yz}(0)$ are x, y components stresses at sea surface ($z = 0$)

Note 1. $\hat{k} \times \left[\hat{k} \times M_{EK}\right] = -M_{EK}$

Note 2. Friction is confined to Ekman layer and below that friction is zero

$$M_{EK} = \int_{-d}^{o} \rho \vec{U} dz$$

$$f\hat{k} \times M_{EK} = \tau_{(z=0)} = \tau$$

Since $\hat{k} \times (\hat{k} \times M_{EK}) = -M_{EK,}$

By rearranging $M_{EK} = \dfrac{\tau \times \hat{k}}{f}$

where $\hat{k}$ = unit vector in z – direction

$\vec{U} = i\, U_{ekx} + j V_{eky}$ horizontal velocity.

Note: The Ekman transport is perpendicular to the right of the wind in NH (and to the left of the wind in SH)

14.3 Ekman's Constants

Wind stress bulk formula is $\tau_{yz} = \tau = \rho_{air} C_D U_{10}^2$

where τ_{yz} = wind stress component in horizontal y – direction (meridional)

ρ_{air} = density of air

C_D = drag coefficient

U_{10} = wind speed at 10m above sea.

V_o is a function of wind speed, given by

$$V_o = \frac{0.0127}{\sqrt{\sin|\phi|}} U_{10}, \quad |\phi| \geq 10$$

14.4 Ekman Layer Depth (D_{EK})

Ekman layer depth is arbitrary, since current speed decreases exponentially with depth.

However Ekman proposed thickness depth D_{EK}, at which the current velocity is opposite of the velocity at surface (which occurs at depth $D_{EK} = \frac{\pi}{a}$,)

$$\text{Ekman layer depth } D_{EK} = \sqrt{\frac{2\pi^2 A_K}{f}} \text{ or } \frac{7.6}{\sqrt{\sin H}} U_{10}$$

$$\text{Ekman Number} \quad E_z = \frac{\text{friction force}}{\text{coriolis force}} = \frac{A_z \frac{u}{d^2}}{fu}$$

$$= \frac{A_z}{fd^2} \text{ (non – dimensional)}$$

14.4.1 Ekman Volume Transport

$$\text{volume} = \frac{\text{mass}}{\text{density}}$$

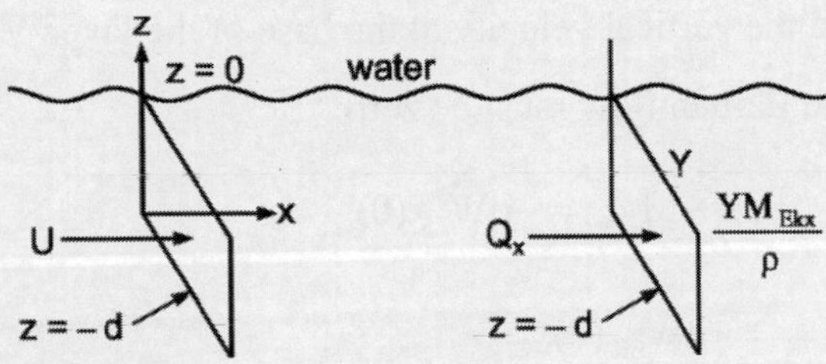

Fig. 14.3

Volume transport Q [see fig.14.3]

$$Q = \frac{\text{mass transport}}{\text{density of water}} \times [\text{width perpendicular to the transport}]$$

$$Q_x = \frac{YM_{EK\,x}}{\rho}, \qquad Q_y = \frac{X\,M_{EK\,Y}}{\rho}$$

where Y = North south distance; X = East west distance

14.5 Ekman Pumping

Horizontal variability of Ekman transport depends on the horizontal wind variability blowing on the sea surface.

Since mass is conserved, the spatial variability of the transports lead to vertical velocities at the top of the Ekman layer.

Integrating the continuity equation in the vertical between z = 0 to z = – d

$$\rho \int_{-d}^{0} \left(\frac{\partial u}{\partial x} + \frac{\partial v}{\partial y} + \frac{\partial w}{\partial z} \right) dz = 0 \qquad \text{.....(14.10)}$$

Rewriting as

$$\frac{\partial}{\partial x} \int_{-d}^{0} \rho u dz + \frac{\partial}{\partial y} \int_{-d}^{0} \rho v dz = -\rho \int_{-d}^{0} \frac{\partial w}{\partial z} dz$$

Using Ekman transport definition, the above equation becomes

$$\frac{\partial}{\partial x}(M_{EKx}) + \frac{\partial}{\partial y}(M_{EXY}) = -[w(0) - w(-d)]$$

By definition, the Ekman velocities approach zero at the base of the Ekman layer and the vertical velocity at the base of the layer $W_{EK}(-d)$ due to divergence of the Ekman flow must be zero.

$$\therefore \qquad \frac{\partial}{\partial x} M_{EKx} + \frac{\partial}{\partial y} M_{EXy} = -\rho W_{EK}(0) \qquad \text{.....(14.11)}$$

or $$\nabla_H M_{EK} = -\rho W_{EK}(0) \qquad \text{.....(14.12)}$$

where $M_{EK} = iM_{EK\,x} + j\,M_{EK\,y}$

= vector mass transport due to Ekman flow in the upper boundary of the ocean.

∇_H is the horizontal divergence operator

eq., (14.12) states that the horizontal divergence of the Ekman transport lead to a vertical velocity in the upper boundary layer (of the friction) of the ocean. This process is called Ekman pumping.

We have proved

$$\left.\begin{aligned} & f\, M_{EK\,x} = +\tau_{yz(0)} \\ \text{and} \quad & f\, M_{EK\,y} = -\tau_{xz(0)} \end{aligned}\right\} \quad \text{Ekman transport eq. (14.9) components.}$$

Substituting these values in the eq(14.11), we have

$$\frac{\partial}{\partial x}\left(\frac{\tau_{yz(0)}}{f}\right) - \frac{\partial}{\partial y}\left(\frac{\tau_{xz(0)}}{f}\right) = -\rho\, W_{EK}(0)$$

$$\text{or} \qquad W_{EK}(0) = -\frac{1}{\rho}\left[\frac{\partial}{\partial x}\left(\frac{\tau_{yz}(0)}{f}\right) - \frac{\partial}{\partial y}\left(\frac{\tau_{xz(0)}}{f}\right)\right]$$

$$= -\left[\frac{\partial}{\partial x}\left(\frac{\tau_{yz}(0)}{\rho f}\right) - \frac{\partial}{\partial y}\left(\frac{\tau_{xz(0)}}{\rho f}\right)\right]$$

$$\left.\begin{aligned} \text{or} \qquad W_{EK}(0) &= -\,\text{curl}\left(\frac{\tau}{\rho f}\right) \\ &= -\hat{k}\cdot\nabla\times\left(\frac{\tau}{\rho f}\right) \end{aligned}\right\} \qquad \text{.....(14.13)}$$

where $\tau = i\,\tau_{xz}(0) + j\,\tau_{yz}(0)$

τ_{xz} – zonal wind stress

τ_{yz} – Meridional wind stress

$W_{EK}(0)$ = vertical velocity at sea surface

Since the vertical velocity at sea surface $W_{EK}(0) = 0$ (because surface cannot rise) which must be balanced by another velocity viz., a vertical geostrophic velocity $W_G(0)$.

$$\text{i.e.,} \qquad W_{EK}(0) = -\,W_G(0) = -\,\text{curl}\left(\frac{\tau}{\rho f}\right)$$

Ekman pumping $W_{EK}(0)$ drives a vertical geostrophic current $-W_G(0)$ in the ocean's interior.

14.6 Ekman Transport Convergence

With position changes there are changes in wind stress. Corresponding to this wind stress changes there are changes in Ekman transport with position. Because of this there can be a convergence or divergence of water out of Ekman layer. Convergence results in down welling and divergence results in upwelling into the Ekman layer.

With the changes in zonal wind the Ekman transport is to the right of wind direction in NH and is convergent.

The Ekman transport divergence to the wind stress is

$$\nabla \cdot \vec{U}_{EK} = \frac{\partial}{\partial x}\left(\frac{\tau_{yz}}{\rho f}\right) - \frac{\partial}{\partial y}\left(\frac{\tau_{xz}}{\rho f}\right)$$

$$= \hat{k} \cdot \nabla \times \left(\frac{\tau}{\rho f}\right) \quad \text{.....(14.14)}$$

Where $\tau = i\,\tau_{xz} + j\,\tau_{yz}$ wind stress vector

[since $\vec{U}_{EK} = i\,u + j\,v$ (two dimensional)

$$\nabla \cdot \vec{U}_{EK} = \frac{\partial u}{\partial x} + \frac{\partial v}{\partial y}$$

Integrating $z = 0$ to $z = -d$

$$\int_{-d}^{0} \nabla_H \cdot \vec{U}_{EK} dz = \frac{\partial}{\partial x}\int_{-d}^{0} u dz + \frac{\partial}{\partial y}\int_{-d}^{0} v dz$$

$$= \frac{\partial}{\partial x}\left(\frac{M_{EK\,x}}{\rho}\right) + \frac{\partial}{\partial y}\left(\frac{M_{EK\,y}}{\rho}\right) \quad \text{using eq. (14.5) and eq. (14.6)}$$

$$= \frac{\partial}{\partial x}\left(\frac{\tau_{yz}(0)}{\rho f}\right) + \frac{\partial}{\partial y}\left(\frac{-\tau_{xz}(0)}{\rho f}\right) \quad \text{using eq. (14.9)}$$

$$= \hat{k} \cdot \nabla \times \left(\frac{\tau}{\rho f}\right)]$$

$\therefore$ In NH, $f > 0$, upwelling into the Ekman layer results from positive (+) wind stress curl and down welling results from negative (–) wind stress curl.

Down welling is called Ekman pumping, while upwelling is called Ekman suction.

Equatorial upwelling due to Ekman transport results from the westward wind stress (i.e., trade winds). These cause northward Ekman transport (northward of equator) and southward transport south of the equator.

In general sub-polar regions are subjected to upwelling, while sub-tropical regions to down welling and Tropical regions to upwelling.

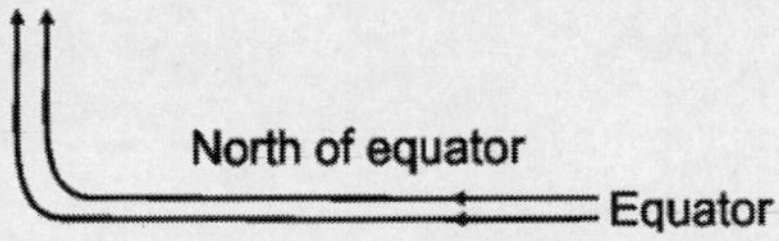

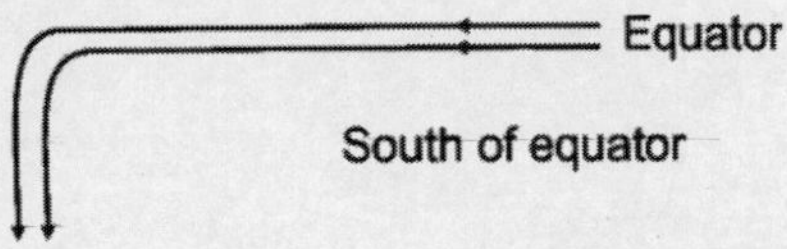

CHAPTER 15

WIND DRIVEN OCEAN CIRCULATION

Modern theory of ocean circulation is the sum of: (a) Harald Sverdrup's ocean circulation (b) Henry Stommel's circulation in ocean gyres and (c) Walter Munk's eddy viscosity related circulation in the upper layer of the Pacific.

15.1 Sverdrup's Theory of the Ocean Circulation

Sverdrup's theory is based on the following assumptions:

(i) The flow is stationary.
(ii) Lateral friction and molecular viscosity are small.
(iii) Non-linear terms (like $u\frac{\partial u}{\partial x}$....) are small.
(iv) Turbulence near the sea surface can be described by the use of vertical eddy viscosity.
(v) Wind driven circulation becomes zero at some depth of no motion.

Sverdrup's theory is also called depth integrated circulation.

Under the above assumptions, the horizontal components of momentum equations are:

$$\frac{\partial p}{\partial x} - f\rho v = \frac{\partial \tau_{xz}}{\partial z} \quad .....(15.1)$$

$$\frac{\partial p}{\partial y} + f\rho u = \frac{\partial \tau_{yz}}{\partial z} \quad .....(15.2)$$

where horizontal wind stress vector $\tau = i\ \tau_{xz} + j\ \tau_{yz}$

Note: $\tau_x(z)$, $\tau_y(z)$ are written as τ_{xz}., τ_{yz}

$\rho_{ref} = \rho_0$ is replaced by ρ

$$\begin{array}{ll} z = 0 & \tau_{(0)} \\ \hline & \tau_{(z)} \\ z = -d & \\ \hline \text{interior ocean} & \end{array}$$

Let
$$\left.\begin{aligned} M_{EK\,x} &= M_x = \int_{-d}^{0} \rho\, u(z)\, dz \\ M_{EK\,y} &= M_y = \int_{-d}^{0} \rho\, v(z)\, dz \end{aligned}\right\} \quad \text{.....(15.3)}$$

where M_x, M_y are mass transport in wind driven layer, extending down to a depth of 'd' of the motion.

Denote
$$\left.\begin{aligned} \frac{\partial P}{\partial x} &= \int_{-d}^{0} \frac{\partial p}{\partial x}\, dz, \\ \frac{\partial P}{\partial y} &= \int_{-d}^{0} \frac{\partial p}{\partial y}\, dz \end{aligned}\right\} \quad \text{.....(15.4)}$$

Integrating eq. (15.1) and (15.2) under the limits z = – d to z = 0 and using above notations (15.3) & (15.4) we have

$$\int_{-d}^{0} \frac{\partial p}{\partial x}\, dz - \int_{-d}^{0} f\rho v dz = \int_{-d}^{0} \frac{\partial\, \tau_{xz}}{\partial z}\, dz$$

$$\int_{-d}^{0} \frac{\partial p}{\partial y}\, dz + \int_{-d}^{0} f\rho u dz = \int_{-d}^{0} \frac{\partial\, \tau_{yz}}{\partial z}\, dz$$

$$\frac{\partial P}{\partial x} - f\, M_y = \tau_x \quad \text{.....(15.5)}$$

$$\frac{\partial P}{\partial y} + f\, M_x = \tau_y \quad \text{.....(15.6)}$$

Note: Changes; $\tau_x = \tau_{xz}$

$$\tau_y = \tau_{yz}$$

where horizontal boundary conditions at sea surface is the wind stress and the boundary at depth '– d' is zero stress (because the current become zero)

$$\tau_{xz}(0) = \tau_x\,,\ \ \tau_{xz}(-d) = 0$$

$$\tau_{yz}(0)=\tau_y,\ \tau_{yz}(-d)=0$$

Integrating the equation of continuity between z = – d to z = 0 (continuity equation $\frac{\partial \rho u}{\partial x}+\frac{\partial \rho v}{\partial y}+\frac{\partial \rho w}{\partial z}=0$)

$$\int_{-d}^{0}\frac{\partial \rho u}{\partial x}\,dz+\int_{-d}^{0}\frac{\partial \rho v}{\partial y}\,dz+\int_{-d}^{0}\frac{\partial \rho w}{\partial z}\,dz=0$$

The last term $\int_{-d}^{0}\frac{\partial \rho w}{\partial z}\,dz=0$, because the motion is horizontal.

The above equation becomes

$$\frac{\partial M_x}{\partial x}+\frac{\partial M_y}{\partial y}=0 \qquad \text{[using eq. 15.3].....(15.7)}$$

Differentiating eq. (15.5) w r t y and eq. (15.6) w r t x and subtracting, we have

$$\frac{\partial}{\partial y}\left(\frac{\partial P}{\partial x}\right)=\frac{\partial}{\partial y}\left[f\,M_y+\tau_x\right]$$

$$\frac{\partial}{\partial x}\left(\frac{\partial P}{\partial y}\right)=\frac{\partial}{\partial x}\left[-f\,M_x+\tau_y\right]$$

$$\frac{\partial P^2}{\partial x\partial y}=\frac{\partial f}{\partial y}M_y+f\frac{\partial M_y}{\partial y}+\frac{\partial \tau_x}{\partial y}$$

$$\frac{\partial P^2}{\partial y\partial x}=-\frac{\partial f}{\partial x}M_x-f\frac{\partial M_x}{\partial x}+\frac{\partial \tau_y}{\partial x}$$

Subtracting

$$0=\beta M_y+f\left[\frac{\partial M_y}{\partial x}+\frac{\partial M_x}{\partial y}\right]+\frac{\partial \tau_x}{\partial y}-\frac{\partial \tau_y}{\partial x}$$

where $\frac{\partial f}{\partial y}=\beta$ and $\frac{\partial f}{\partial x}=0$

Now using eq. (15.7) the above equation becomes

$$0=\beta M_y+0+\frac{\partial \tau_x}{\partial y}-\frac{\partial \tau_y}{\partial x}$$

or $$\beta M_y = -\left(\frac{\partial \tau_x}{\partial y} - \frac{\partial \tau_y}{\partial x}\right)$$

$$= \hat{k} \cdot \nabla \times \tau$$

since $$\hat{k} = \left[\left(i\frac{\partial}{\partial x} + j\frac{\partial}{\partial y}\right) \times \left(i\tau_x + j\tau_y\right)\right] = \frac{\partial \tau_y}{\partial x} - \frac{\partial \tau_x}{\partial y}$$

or $$\beta M_y = \text{curl}_z(\tau) \quad \text{.....(15.8)}$$

where curl_z (τ) = vertical components curl of the wind stress τ.

M_y mass transport northwards (in NH) and

$$\beta = \frac{\partial f}{\partial y},$$ change of coriolis force with latitude.

The last eq. (15.8), states that northward mass transport (i.e., meridional transport of mass) of wind driven currents is equal to the curl of the wind stress. Eq.(15.8) is the Sverdrup relation.

15.2 Some Uncertainties in Sverdrup's Relations

Sverdrup assumed internal ocean flow to be geostrophic, there is uniform depth of no motion (layer).

Ekmans transport validity. However the depth of no motion in tropical Pacific is uncertain. Observed eastward transport M_x is valid subject to negligible friction, while western boundary currents not satisfied as it observed intensification. Sverdrup's solutions do not give the vertical distribution of current. Instead of steady state the average wind data suited. His level of no motion 500 m is uncertain inspite of his theory of ocean circulation is widely globally applied.

15.3 Stommel's Theory of Western Boundary Currents

We know the Sverdrup's momentum equations are:

$$\left.\begin{aligned} \frac{\partial p}{\partial x} &= f\rho v + \frac{\partial \tau_{xz}}{\partial x} \\ \frac{\partial p}{\partial y} &= -f\rho u + \frac{\partial \tau_{yz}}{\partial z} \end{aligned}\right\} \quad \text{.....(15.9)}$$

The mass transports

$$\left.\begin{aligned} M_x &= \int_{-d}^{0} \rho u dz \\ M_y &= \int_{-d}^{0} \rho v dz \end{aligned}\right\} \quad \text{.....(15.10)}$$

Integrated pressure gradients

$$\left.\begin{aligned} \frac{\partial P}{\partial x} &= \int_{-d}^{o} \frac{\partial p}{\partial x} dz \\ \frac{\partial P}{\partial y} &= \int_{-d}^{o} \frac{\partial P}{\partial x} \end{aligned}\right\} \quad \text{.....(15.11)}$$

and horizontal boundary conditions at sea surface wind stress

$$\left.\begin{aligned} &\tau_{xz}(0) = \tau_x, \\ &\tau_{xz}(-d) = 0 \\ \text{and} \quad &\tau_{yz}(0) = \tau_y \\ &\tau_{yz}(-d) = 0 \end{aligned}\right\} \quad \text{.....(15.12)}$$

In order to justify observed western boundary currents, Stommel added a bottom stress proportional to the velocity (given below) to eq. (15.12) and retained other equations.

$$\left.\begin{aligned} \left(A_z \frac{\partial u}{\partial z}\right)_0 &= -\tau_x = -F\cos\left(\frac{\pi y}{b}\right); \left(A_z \frac{\partial u}{\partial z}\right)_{+d} = -Ru \\ \left(A_z \frac{\partial v}{\partial z}\right)_0 &= -\tau_y = 0; \left(A_z \frac{\partial v}{\partial z}\right)_{+d} = -Rv \end{aligned}\right\} \quad \text{.....(15.13)}$$

where F and R are constants, with rectangular basin $0 \le y \le b$, $0 \le x \le \lambda$

with depth 'd', constant density ρ and variation of coriolis force with latitude and constant rotation (Ω) of earth.

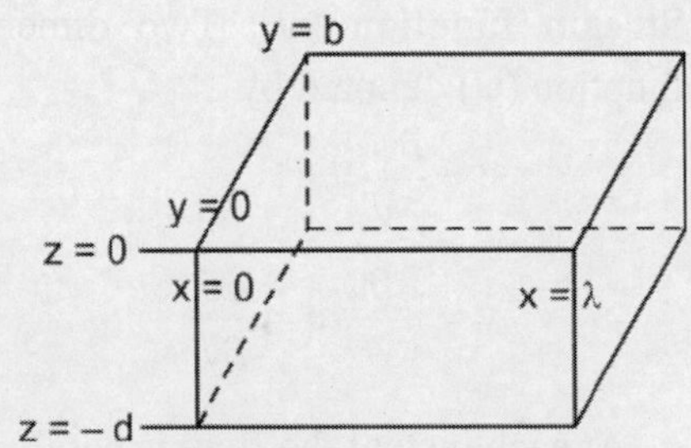

This explained the cause of crowding stream lines in the western boundary (Ex. Gulf stream explanation). Other models that explain western boundary current with the help of friction. See the following figures.

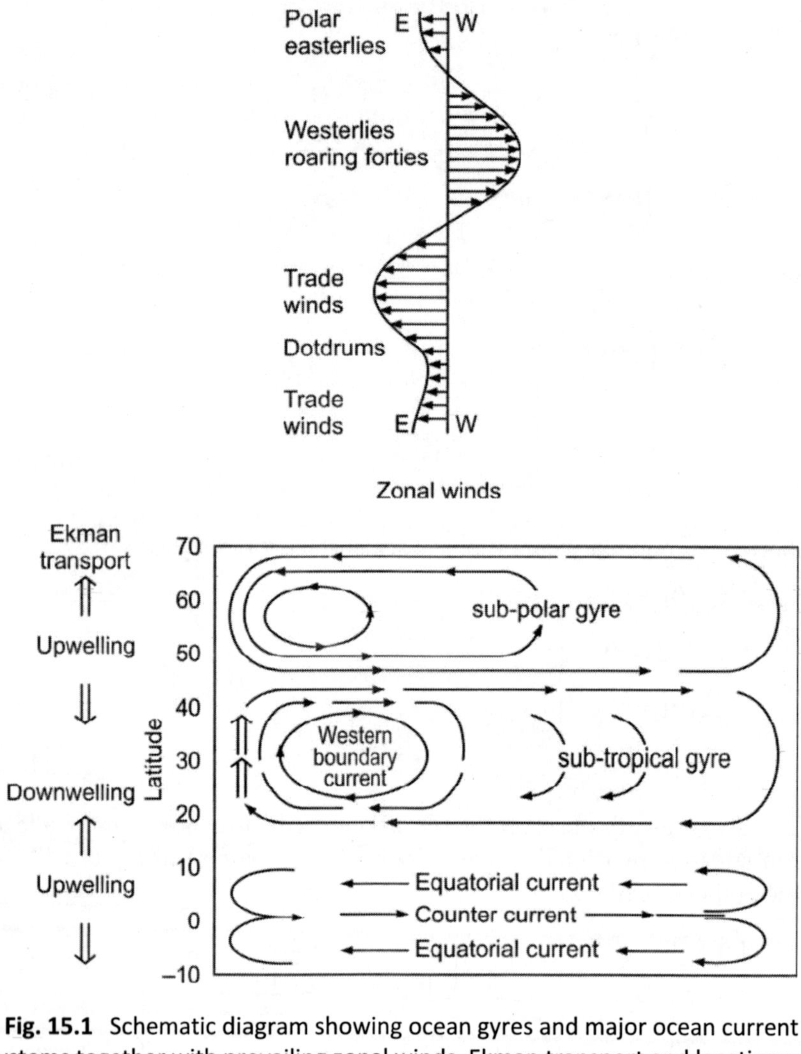

Fig. 15.1 Schematic diagram showing ocean gyres and major ocean current systems together with prevailing zonal winds. Ekman transport and locations of upwelling and downswelling

Stream function (ψ): Two dimensional incompressible flow by stream function (ψ) denoted by

$$u = \frac{\partial \psi}{\partial y}$$

$$v = -\frac{\partial \psi}{\partial x} u = \frac{\partial \psi}{\partial y}$$

At each instant the flow is parallel to lines of constant ψ

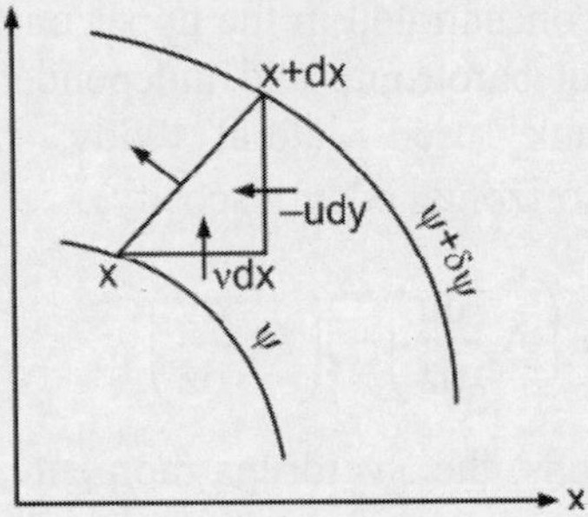

Fig. 15.2 Volume of transport between stream lines ψ, ψ + δψ in a two dimensional steady flow

The rate of flow of volume between stream lines ψ_1, $\psi_2 = \psi_1 - \psi_2$

Let an arbitrary line dx (= dx, dy) between the two stream lines

The rate of flow of volume between the stream lines is

$$vdx + (-u)dy = \frac{\partial \psi}{\partial x}dx - \frac{\partial \psi}{\partial y}dy$$

$$= -d\psi$$

$$\left.\begin{aligned} u_s &= -\frac{g}{f}\frac{\partial \zeta}{\partial y} \\ v_s &= +\frac{g}{f}\frac{\partial \zeta}{\partial x} \\ \psi &= -\frac{g}{f}\zeta \end{aligned}\right\}$$

by satellite – altimeter maps of oceanic topography

[surface geostrophic current components given by eq. 16.57]

The mass transport stream function ψ defined by

$$M_x \equiv \frac{\partial \psi}{\partial y}$$

$$M_y \equiv -\frac{\partial \psi}{\partial x}$$

15.4 Munk's Solution

In order to obtain a solution for the circulation within an ocean basin, Munk added lateral viscosity to Sverdrups solution.

Ocean currents are concentrated in the upper one kilometer of the ocean, and the currents are not barotropic and independent of depth. In order to account friction, Munk used lateral eddy friction with constant $A_H = A_x = A_y$ (A_H = Horizontal eddy viscosity).

$$\tau = i\,\tau_{xz} + j\,\tau_{yz} = i\left(A_z \frac{du}{dz}\right) + j\left(A_z \frac{dv}{dz}\right)$$

Under these conditions, the Sverdrups momentum equations (15.1,15.2) become:

$$\frac{1}{\rho}\frac{\partial p}{\partial x} = fv + \frac{\partial}{\partial z}\left(A_z \frac{\partial u}{\partial z}\right) + A_H \frac{\partial^2 u}{\partial x^2} + A_H \frac{\partial^2 u}{\partial y^2} \quad .....(15.14)$$

$$\frac{1}{\rho}\frac{\partial p}{\partial y} = -fu + \frac{\partial}{\partial z}\left(A_z \frac{\partial v}{\partial z}\right) + A_H \frac{\partial^2 v}{\partial x^2} + A_H \frac{\partial^2 v}{\partial y^2} \quad .....(15.15)$$

Munk integrated equations from z = – d to the surface $z = z_0$ (incase of Sverdrup $z_0 = 0$), further assumed at depth z = – d the currents vanish. At horizontal boundaries, at the top and bottom of the layer A_H = constant.

The above equations can be written as

$$\frac{1}{\rho}\frac{\partial p}{\partial x} = fv + \frac{\partial}{\partial z}(\tau_{xz}) + A_H\left[\frac{\partial^2 u}{\partial x^2} + \frac{\partial^2 u}{\partial y^2}\right]$$

$$\frac{1}{\rho}\frac{\partial p}{\partial y} = -fu + \frac{\partial}{\partial z}(\tau_{yz}) + A_H\left[\frac{\partial^2 v}{\partial x^2} + \frac{\partial^2 v}{\partial y^2}\right]$$

Using stream function ψ, $u = \frac{\partial \psi}{\partial y}, v = -\frac{\partial \psi}{\partial x}$ the above equations can be written as

$$\frac{1}{\rho}\frac{\partial p}{\partial x} = -f\frac{\partial \psi}{\partial x} + \frac{\partial}{\partial z}(\tau_{xz}) + A_H\left[\frac{\partial^3 \psi}{\delta^2 x \partial y} + \frac{\partial^3 \psi}{\partial y^3}\right] \quad .....(15.16)$$

$$\frac{1}{p}\frac{\partial p}{\partial y} = -f\frac{\partial \psi}{\partial y} + \frac{\partial}{\partial z}(\tau_{yz}) + A_H\left[\frac{-\partial^3 \psi}{\partial x^3} - \frac{\partial^3 \psi}{\partial y^2 \partial x}\right] \quad .....(15.17)$$

Differentiating (15.16) w r t y and (15.17) w r t x and subtracting

$$\frac{1}{\rho}\frac{\partial^2 p}{\partial x \partial y} = -f\frac{\partial^2 \psi}{\partial x \partial y} - \frac{\partial \psi}{\partial x}\frac{\partial f}{\partial z} + \frac{\partial^2}{\partial z}\left(\frac{\partial \tau_{xz}}{\partial y}\right) + A_H\left[\frac{\partial^4 \psi}{\partial x^2 \partial y^2} + \frac{\partial^4 \psi}{\partial y^4}\right]$$

$$\frac{1}{\rho}\frac{\partial^2 p}{\partial y \partial x} = -f\frac{\partial^2 \psi}{\partial y \partial x} - \frac{\partial \psi}{\partial y}\frac{\partial f}{\partial x} + \frac{\partial}{\partial z}\left(\frac{\partial \tau_{yz}}{\partial x}\right) + A_H\left[-\frac{\partial^4 \psi}{\partial x^4} - \frac{\partial^4 \psi}{\partial y^2 \partial x^2}\right]$$

Subtracting and using $\frac{\partial f}{\partial y} = \beta,\ \frac{\partial f}{\partial x} = 0$, we get

$$0 = -\frac{\partial \psi}{\partial x}\ \beta + \frac{\partial}{\partial z}\left(\frac{\partial \tau_{xz}}{\partial y} - \frac{\partial \tau_{yz}}{\partial x}\right) + A_H\left[2\frac{\partial^4 \psi}{\partial x^2 \partial y^2} + \frac{\partial^4 \psi}{\partial x^4} + \frac{\partial^4 \psi}{\partial y^4}\right]$$

or $\quad 0 = -\beta\frac{\partial \psi}{\partial x} + \frac{\partial}{\partial z}(\tau) + A_H \nabla^4 \psi$

where $\quad \nabla^4 = \frac{\partial^4}{\partial x^4} + \frac{2\partial^4}{\partial x^2 \partial y^2} + \frac{\partial^4}{\partial y^4}$ .....(15.18)

= biharmonic operator

or $\quad A_H \nabla^4 \psi - \beta\frac{\partial \psi}{\partial x} = -\text{curl}_z \tau$

$$A_H \nabla^4 \psi = \text{function}$$

$\beta = \frac{\partial \psi}{\partial x} = \text{curl}_z \tau$ Sverdrup balance.

At the lateral boundaries the friction term is large (where derivatives of horizontal velocity fields are large), while it is small in the interior of the basin.

In the interior of the basin Munk's and Sverdrups solutions are same, but they differ widely at the lateral boundaries. Munk assumed flow at a boundary is parallel to the boundary and that there is no slip at the boundary.

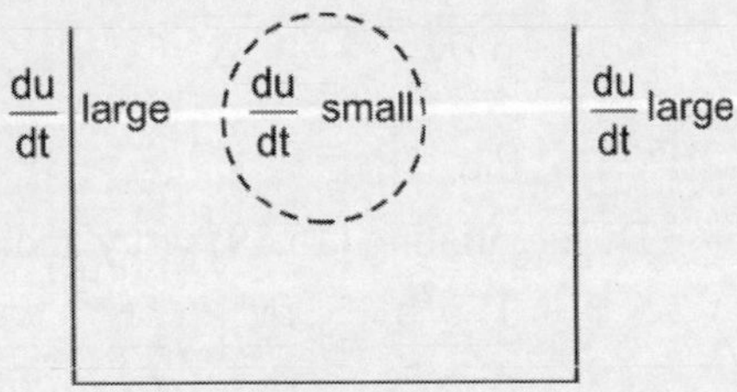

$$\psi_b = 0,\ \left(\frac{\partial \psi}{\partial n}\right)_b = 0$$

b = boundary

n = normal to the boundary

Munk assumed rectangular basin extending x = r and y = – s to y = + s

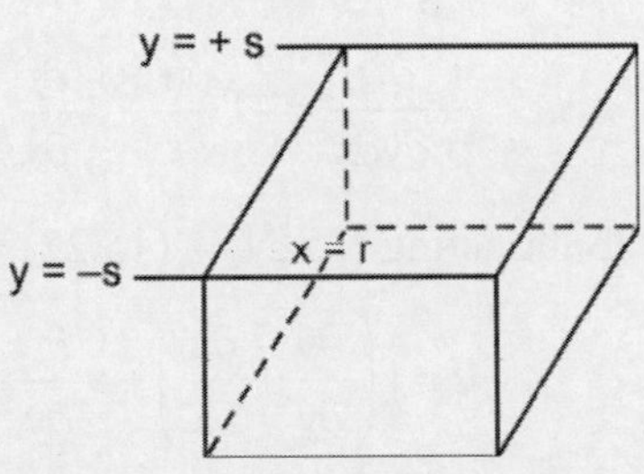

Wind stress zonal in the form

$$(\tau = a \cos ny + b \sin ny + c)$$

$$n = \frac{j\pi}{s},\ j = 1, 2$$

15.5 Ageostrophic Ocean Flow

Ocean currents generally concentrated in the upper (top) one kilometer depth of ocean because of wind force. These currents are not barotropic and they are not independent of depth. In Sverdrups theory of ocean currents it was assumed the fluid is frictionless. In ageostrophic flow, friction is accounted in the form of frictional drag.

Let $\vec{F}$ = frictional drag $= -\frac{`k\,\vec{U}}{\delta}$

where $`k$ = drag coefficient (which depends on the roughness of the underlying surface)

$\vec{U}$ = ageostrophic wind

$z = \delta$, depth of friction layer or Ekman layer

Ageostrophic balance modified momentum equations

are:

$$\frac{1}{\rho}\frac{\partial p}{\partial x} = fv - \frac{`ku}{\delta} \qquad \text{.....(15.19)}$$

$$\frac{1}{\rho}\frac{\partial p}{\partial y} = -fu - \frac{`kv}{\delta} \qquad \text{..... (15.20)}$$

where $\rho_{ref} = \rho_o = \rho$

Differentiating (15.19) wrt y and (15.20) wrt x, we have

$$\frac{1}{\rho}\frac{\partial^2 \rho}{\partial x \partial y} = f\frac{\partial v}{\partial y} + v\frac{\partial f}{\partial y} - \varepsilon\frac{\partial u}{\partial y} \qquad \text{..... (15.21)}$$

where $\varepsilon = \frac{`k}{\delta}$

$$\frac{1}{\rho}\frac{\partial^2 \rho}{\partial y \partial x} = -f\frac{\partial u}{\partial x} - u\frac{\partial f}{\partial x} - \varepsilon\frac{\partial v}{\partial x} \qquad \text{..... (15.22)}$$

Subtracting (15.21) – (15.22) we get

$$0 = f\left(\frac{\partial v}{\partial y} + \frac{\partial u}{\partial x}\right) + v\frac{\partial f}{\partial y} + \varepsilon\left(\frac{\partial v}{\partial x} - \frac{\partial u}{\partial y}\right) \qquad \because \frac{\partial f}{\partial x} = 0$$

$$0 = -f\frac{\partial w}{\partial z} + v\beta + \varepsilon\zeta \qquad .....(15.23)$$

where $\frac{\partial u}{\partial x} + \frac{\partial v}{\partial y} = -\frac{\partial w}{\partial z}$ continuity equation

$$\zeta = \frac{\partial v}{\partial x} - \frac{\partial u}{\partial y} = \text{curl}_z\vec{U}$$

$$\therefore \quad f\frac{\partial w}{\partial z} = \beta v + \varepsilon\zeta \qquad .....(15.24)$$

If we use stream function ψ, $u = \frac{\partial \psi}{\partial y}$, $v = -\frac{\partial \psi}{\partial x}$

The above equation eq.(15.24) becomes

$$f\frac{\partial w}{\partial z} + \beta\frac{\partial \psi}{\partial x} = \varepsilon\zeta \qquad .....(15.25)$$

Eq. (15.25) is similar to the Munks flow equation.

15.6 Fluid Dynamics on the Beta-Plane or Ekman Pumping

According to Taylor Proudman Theorem $\frac{\partial w}{\partial z} = 0$ is true when the flow cannot expand or contract vertically (and behaves like very rigid). Also there cannot be gradient of vertical velocity in an ocean of planetary vorticity.

Let $f = f_0$ and consider on beta plane

$F = f_0 + \beta y$

For geostrophic flow

$$fv = \frac{1}{\rho_o}\frac{\partial p}{\partial x} \text{ or } v = \frac{1}{\rho_o f}\frac{\partial p}{\partial x} \qquad .....(15.26)$$

$$fu = -\frac{1}{\rho_o}\frac{\partial p}{\partial y} \text{ or } u = -\frac{1}{\rho_o f}\frac{\partial p}{\partial y} \qquad .....(15.27)$$

and $g = -\frac{1}{\rho_o}\frac{\partial p}{\partial z}$

equation of continuity

$$\frac{\partial u}{\partial x} + \frac{\partial v}{\partial y} + \frac{\partial w}{\partial z} = 0 \qquad .....(15.28)$$

or $$\frac{\partial u}{\partial x}+\frac{\partial v}{\partial y}=-\frac{\partial w}{\partial z}$$

$$\therefore \quad \frac{\partial}{\partial x}\left(\frac{-1}{\rho_o f}\frac{\partial p}{\partial y}\right)+\frac{\partial}{\partial y}\left(\frac{1}{\rho_o f}\frac{\partial p}{\partial x}\right)=-\frac{\partial w}{\partial z} \text{ (using 15.26, 15.27)} \quad .....(15.29)$$

using eq. (15.26) & (15.27)

LHS of (15.28) is

$$\frac{\partial u}{\partial x}+\frac{\partial v}{\partial y}=\frac{-1}{\rho_o f}\frac{\partial p^2}{\partial y \partial x}-\frac{1}{\rho_o}\frac{\partial p}{\partial y}\times\frac{-1}{f^2}\frac{\partial p}{\partial x}+\frac{1}{\rho_o f}\frac{\partial p^2}{\partial x \partial y}+\frac{1}{\rho_o}\frac{\partial p}{\partial x}\times\frac{-1}{f^2}\frac{\partial f}{\partial z}$$

Since $\frac{\partial f}{\partial x}=0$, $\frac{\partial f}{\partial y}=\beta$, the above equation becomes

$$\frac{\partial u}{\partial x}-\frac{\partial v}{\partial y}=-\frac{1}{\rho_o}\frac{\partial p}{\partial x}\times\frac{1}{f^2}\frac{\partial f}{\partial y}=\frac{-1}{\rho_o f}\frac{\partial p}{\partial x}\times\frac{1}{f}\beta$$

$$=-\frac{v\beta}{f} \quad \text{Using (15.26)} \quad .....(15.30)$$

From (15.28) & (15.30) we have

$$-\frac{v\beta}{f}=\frac{-\partial w}{\partial z}$$

or $$v\beta=\frac{f\partial w}{\partial z}$$

or $$f_o\frac{\partial w_G}{\partial z}=v\beta \quad .....(15.31)$$

where W_G = geostrophic flow component.

Eq. (15.31) applies to the oceans interior geostrophic flow.

In the interior of the ocean, vertical velocity W_G leads to north-south (meridional) currents.

15.7 Langmuir Circulation

Langmuir circulation is a transcient reponse to wind forcing in which helical vortices form near the sea surface. Langmuir cells (LC) are clearly seen as numerous long parallel lines or streaks of flotsam (wind rows) that are generally aligned with the wind (which may deviate by about 20°). The streaks are formed by the convergence caused by vortices. Alternate cells rotate in opposite directions so that convergence and down welling occurs at the surface (to form streaks of flotsam) between pairs of adjacent cells, while divergence and upwelling occurs between alternate pairs. The water in the cells progress downwind so that its motion is helical.

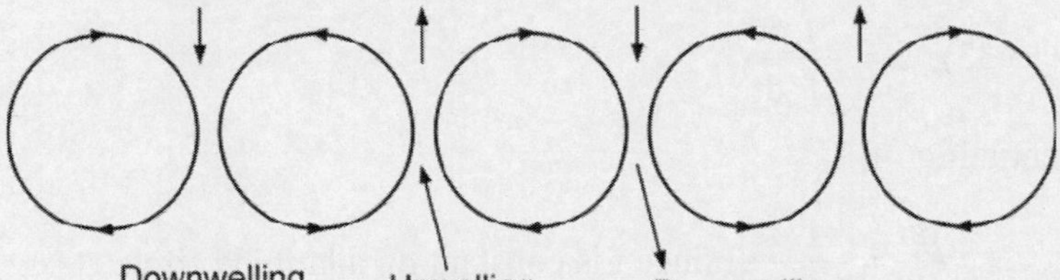

Langmuir cells have typical depth and horizontal spacing of 4 to 6 meters and 10 to 50 meters, but they can range up to several hundred meters horizontal separation and 2 to 3 times the mixed layer depth.

The cells can be many kilometers long. The horizontal flow speed at the surface in the streaks can add 10 cm/sec to the non–Longmuir currents elsewhere in the surface layer. The vertical down-welling at the convergence is about one third of the surface water speed as driven by the wind. Down welling velocities are several cm/sec and up to 20 cm/sec.

Langmuir circulations provide a mechanism for converting wave energy to turbulent energy and mixing and causing the upper layer to deepen. Thus Langmuir cells may contribute to surface mixing (besides several process that help mixing).

15.8 Inertial Currents

It is a transcient response to wind forcing on ocean. Initial windstress impulse that causes transcient motion is called inertial currents. Wind blowing over the ocean surface exerts stress. This stress causes water to move within the top 50 m depth. Initially the wind excites small capillary waves which propagate in the direction of wind. The effect of this atmospheric momentum is stress on ocean surface. The inertial currents are the result of balance of horizontal velocity and the coriolis force (earths rotation – on a scale of about a day or more). In northern hemisphere the coriolis force acts (drifts) to the right of the velocity (and to the left in SH).

Thus, we have

$$\frac{du}{dt} = fv \qquad \text{.....(15.32)}$$

$$\frac{dv}{dt} = -fu \qquad .....(15.33)$$

where f = coriolis parameter

$$f = 2\Omega \sin\phi \;\simeq 7.292 \times 10^{-5} / \sec$$

From (15.33), $u = -\frac{1}{f}\frac{dv}{dt}$

Differentiating

$$\frac{du}{dt} = -\frac{1}{f}\frac{dv^2}{dt^2} \text{ since f is constant on a latitude circle}$$

Using (15.32) in the above equation, we have

$$fv = -\frac{1}{f}\frac{dv^2}{dt^2}$$

or

$$\frac{dv^2}{dt^2} + f^2 v = 0 \qquad .....(15.34)$$

eq. (15.34) is a type of differential equation of harmonic oscillator.

The solution of (15.34) is given by

$$\left.\begin{aligned} u &= V \sin f\,t \\ v &= V \cos f\,t \\ V^2 &= u^2 + v^2 \end{aligned}\right\} \qquad .....(15.35)$$

This currents (V) is inertial current or inertial oscillation.

Eq. (15.35) represents a circle with diameter $= \frac{2V}{f}$

and period $T = \frac{2\pi}{f} = \frac{\text{sidereal day}}{f}$

Inertial currents are free motion of water parcels on a rotating plane. A few values are given below.

Latitude ϕ	Time (hours)	Diameter (km)
	For V=20 cm/sec	
90°	11.97	2.7
35°	20.87	4.8
10°	68.93	11.5

The most common currents in the ocean are inertial currents (circles). They are found at depths in the ocean and at all latitudes. The motions are short lived (transcients) and they die down (or transform) in a few days into other surface currents. The motion in the inertia circle has been shown in the figure for NH.

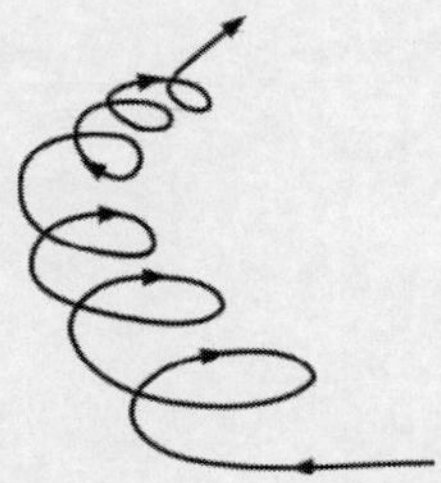

Note: The wind driven ocean friction layer (top & bottom) is called Ekman layer. Ekman layer includes only eddy viscosity (friction) and coriolis acceleration.

CHAPTER 16

GEOSTROPHIC MOTION

16.1 Geostrophic Flow

The primary source of energy is sun's radiation, while wind drives the ocean waters. Moon and tectonic process together with geothermal process also take part in ocean waves and Tidal motion. Tides create internal waves which break and create turbulence and mixing. Earthquakes and turbidity currents create random and irregular waves including Tsunamis. Geothermal process heat the water very gradually, however they cause little effect on circulation.

A fluid flow that is dominated by rotation can be geostrophic which is perpendicular to the pressure gradient force circling around the centres of High pressure or Low pressure due to coriolis force. Ocean circulations are mainly caused by wind-stress blowing over the surface and buoyancy governed theromohaline.

Wind blowing on the ocean surface initially cause capillary waves and then into a spectrum of waves and swell in the ocean. Impulsive changes in wind cause small time scale internal currents and Langmuir cells. Steady or gradual strengthening wind or change of wind direction create ocean top (surface) frictional Ekman-layer which include the coriolis effect. As the wind momentum transfer persists the geostrophic wind driven circulation develops. Over the entire width of Pacific or Atlantic ocean, Anticyclonic gyres the total difference of sea surface level height is about one meter. In the ocean, High pressure is caused by a higher mass of water lying above the observational ocean depth.

Geostrophic balance equation is

1. Horizontal pressure gradient $\left(\frac{\partial p}{\partial x} \text{ or } \frac{\partial p}{\partial y}\right)$ force

   = Horizontal coriolis acceleration ($f\rho v$ or $-f\rho u$ respectively).

i.e.,
$$\frac{\partial p}{\partial x} = f\rho v$$
$$\frac{\partial p}{\partial y} = -f\rho u$$

2. In hydrostatic balance
The vertical pressure gradient force (which points upward from High to Low) $\left(+\frac{1}{\rho}\frac{\partial p}{\partial z}\right)$
= Gravitational force which points downward (–g)

i.e.,
$$\frac{1}{\rho}\frac{\partial p}{\partial z} = -g \quad \text{or} \quad \delta p = -g\rho\delta z$$

From above statement (1) we have geostrophic balance
$$\left.\begin{array}{ll} f\,v_g = \frac{1}{\rho}\frac{\partial p}{\partial x}, & f\,u_g = -\frac{1}{\rho}\frac{\partial p}{\partial y} \\ \text{or} \quad \vec{U}_g = \frac{1}{f\,\rho}K \times \nabla p & \end{array}\right\} \quad \text{.....(A)}$$

16.1.1 Boussinesq Approximation

Taking f and ρ constant, the derivative of above equations (A)

w.r.t y and x we have
$$f\frac{\partial v_g}{\partial y} = \frac{1}{\rho}\frac{\partial^2 p}{\partial x \partial y}$$
and
$$f\frac{\partial u_g}{\partial x} = -\frac{1}{\rho}\frac{\partial^2 p}{\partial y \partial x}$$

Adding these two equations we get
$$\frac{\partial u_g}{\partial x} + \frac{\partial v_g}{\partial y} = 0$$

This equation is Boussinesq approximation of geostrophic equations which represents non-divergent geostrophic velocities.

Stream function

In fluid dynamics, a non-divergent velocity field is written in terms of stream function ψ as:
$$u = -\frac{\partial \psi}{\partial y}, \text{ and } v = +\frac{\partial \psi}{\partial x}$$

In geostrophic flow $\quad \nu_g = \frac{1}{f\rho}\frac{\partial p}{\partial x}$

$$\therefore \quad \frac{\partial \psi}{\partial x} = \frac{1}{f\rho}\frac{\partial p}{\partial x}$$

Integrating $\quad \psi = \frac{1}{f_o \rho_o}$

This shows the pressure distribution maps geostrophic stream function.

Note: Quasi-geostrophic means as if or almost geostrophic.

16.1.2 Geostrophic Equation and Thermal Wind Relation

In geostrophic balance, the horizontal pressure gradient is balalnced by Coriolis force. For the balance, the equations of motion we have no acceleration, that is

$$\frac{du}{dt} = \frac{d\nu}{dt} = \frac{d\omega}{dt} = 0 \quad \text{and} \quad \omega << u, \nu$$

Frictional force is small and the external force is only g (gravity).

With these assumptions, the geostrophic flow equations become

$$\frac{\partial p}{\partial x} = \rho f \ \nu_g, \ \frac{\partial p}{\partial y} = -\rho f \ u_g, \ \frac{\partial p}{\partial z} = -\rho g \qquad \text{......(16.1)}$$

where u_g, ν_g geostrophic velocity components in x, y, directions.

f = Coriolis parameters

ρ = Density of water

σ = Density anomaly = $\rho - \rho_o$

or $\quad \rho = \sigma + \rho_o$

where $\rho_o = 1000$ kg/m^3 = Reference density

The above equations 16.1 can be written in vector equation as:

$$i\,u_g + j\,\nu_g + k\omega_g = \frac{1}{\rho_f} k \times \left(i\frac{\partial p}{\partial x} + j\frac{\partial p}{\partial y} + k\frac{\partial p}{\partial z} \right)$$

$$= \frac{1}{\rho_f}\left(j\frac{\partial p}{\partial x} - i\frac{\partial p}{\partial y} \right)$$

or

$$\vec{U}_g = \frac{1}{\rho_f} k \times \nabla p \quad \text{geostrophic equation} \qquad \text{.....(16.2)}$$

The hydrostatic equation $\delta p = (-g\rho\delta z)$ .....(16.3)

Integrating eq. 16.3 with the boundary conditions

$$z = -h, p = p_o$$

and $$z = \zeta, p = p$$

$$\int_{p_o}^{p} \frac{\partial p}{\partial z} dz = -\int_{-h}^{\zeta} g \, \rho \, dz$$

where $$g = g(\phi, z)$$

and $$\rho = \rho(z)$$

$$p - p_o = \int_{-h}^{\zeta} g(\phi, z) \, \rho_{(z)} \, dz \qquad \text{..... (16.4)}$$

Differentiating the components of (16.1) viz

$$u_g = -\frac{1}{f \rho_o} \frac{\partial \rho}{\partial y}, \; v_g = +\frac{1}{f\rho_o} \frac{\partial t}{\partial x}$$

w.r.t z.

$$\frac{\partial u_g}{\partial z} = -\frac{1}{f \rho_o} \frac{\partial}{\partial z}\left(\frac{\partial \rho}{\partial y}\right) = -\frac{1}{f \rho_o} \frac{\partial}{\partial y}\left(\frac{\partial \rho}{\partial z}\right)$$

$$= -\frac{1}{f \rho_o} \frac{\partial}{\partial y}(+g\rho) = -\frac{g}{f \rho_o} \frac{\partial}{\partial y}(\sigma + \rho_o)$$

$$\frac{\partial u_g}{\partial z} = -\frac{1}{f \rho_o} \frac{\partial \sigma}{\partial y} \qquad \text{..... (16.5)}$$

and

$$\frac{\partial v_g}{\partial z} = \frac{1}{f \rho_o} \frac{\partial}{\partial z}\left(\frac{\partial \rho}{\partial x}\right) = \frac{1}{f \rho_o} \frac{\partial}{\partial x}\left(\frac{\partial \rho}{\partial z}\right)$$

$$= \frac{1}{f \rho_o} \frac{\partial}{\partial x}[-g(\sigma + \rho_0)]$$

(since ρ_o = constant)

$$\frac{\partial v_g}{\partial z} = -\frac{g}{f \rho_o} \frac{\partial \sigma}{\partial x} \qquad \text{..... (16.6)}$$

Eq. 16.5 and 16.6 together can be written as

$$\frac{\partial \vec{U}_g}{\partial z} = \frac{-g}{f \rho_o} kx \, \nabla \, \sigma \qquad \text{..... (16.7)}$$

Supposing density of water depends on temperature T linearly, L is the coefficient of thermal expsansion at T = Tδ.

T and ρ are related by equation of state

$$\rho = \rho_o [1 - \propto (T - T_o)]$$

or $$\sigma = \rho - \rho_o = \propto \rho_o (T - T_o) \qquad \text{.....(16.8)}$$

From (16.7) and (16.8) we have

$$\frac{\partial \overline{U}_g}{\partial z} = \frac{-g}{f\,\rho_o} k \times \nabla[-\propto \rho_o (T - T_o)]$$

$$\frac{\partial \overline{U}_g}{\partial z} = \frac{\propto g}{f} k \times \nabla T \quad (\because\ \nabla\, T_o = 0) \qquad .....(16.9)$$

In component from

$$\frac{\partial}{\partial z}(i\, u_g + j\, v_g) = \frac{\propto g}{f}\left[j\frac{\partial T}{\partial x} - i\frac{\partial T}{\partial y}\right]$$

This gives

$$\frac{\partial u_g}{\partial z} = -\frac{\propto g}{f}\frac{\partial T}{\partial y}$$

and

$$\frac{\partial v_g}{\partial z} = +\frac{\propto g}{f}\frac{\partial T}{\partial x} \qquad .....(16.10)$$

Eq. 16.10 is the thermal wind relation connecting the vertical shear of geostrophic current to horizontal temperature gradients.

16.1.3 Thermal Wind in Pressure Coordinates

Hydrostatic equation $\quad \delta p = -g\,\rho\,\delta z$

or

$$-\frac{\delta z}{\partial p} = -\frac{1}{gp} \qquad .....(16.11)$$

Geostrophic components are using above hydrostatic equation

$$u_g = -\frac{g}{f}\frac{\partial z}{\partial y}, \quad v_g = \frac{g}{f}\frac{\partial z}{\partial x} \qquad .....(16.12)$$

Differentiating u_g w.r.t p

$$\frac{\partial u_g}{\partial p} = -\frac{g}{f}\frac{\partial}{\partial p}\left(\frac{\partial z}{\partial y}\right) = -\frac{g}{f}\frac{\partial}{\partial y}\left(\frac{\partial z}{\partial p}\right)$$

$$= -\frac{g}{f}\frac{\partial}{\partial y}\left(-\frac{1}{g\rho}\right) \quad \left[\because p = \rho RT \text{ or } \frac{1}{\rho} = \frac{RT}{p}\right]$$

$$= -\frac{1}{f}\frac{\partial}{\partial y}\left(\frac{RT}{p}\right)_p \quad \text{(in pressure coordinates)}$$

$$\frac{\partial u_g}{\partial p} = \frac{RT}{f\,\rho}\left(\frac{\partial T}{\partial y}\right)_p \qquad .....(16.13)$$

Differentiating v_g w.r.t p

$$\frac{\partial v_g}{\partial p} = \frac{g}{f}\frac{\partial}{\partial p}\left(\frac{\partial z}{\partial x}\right) = \frac{g}{f}\frac{\partial}{\partial x}\left(\frac{\partial z}{\partial p}\right)$$

$$= \frac{g}{f}\frac{\partial}{\partial x}\left(-\frac{1}{g\rho}\right)$$

$$= -\frac{1}{f}\frac{\partial}{\partial x}\left(\frac{1}{p}\right)_p = -\frac{1}{f}\frac{\partial}{\partial x}\left(\frac{RT}{p}\right)_p$$

$$\frac{\partial v_g}{\partial p} = -\frac{R}{fp}\left(\frac{\partial T}{\partial x}\right)_p \qquad \text{.....(16.14)}$$

i.e.,

$$\frac{\partial u_g}{\partial p} = \frac{R}{fp}\left(\frac{\partial T}{\partial y}\right)_p \qquad \text{.....(16.15)}$$

$$\frac{\partial v_g}{\partial P} = \frac{R}{fp}\left(\frac{\partial T}{\partial x}\right)_p$$

Eq. 16.15 expresses the thermal wind relation in pressure coordinates.

16.2 Surface Geostrophic Currents

Consider a level surface slightly below sea surface, say about a couple of meters below the sea surface at $z = -h_l$ .....(16.16)

The pressure on the level surface is

$$p = \rho g(\zeta + h_l) \qquad \text{....(16.17)}$$

assuming ρ and g are constant in the upper two meters of ocean.

The geostrophic equations are

$$u_g = -\frac{1}{f\rho}\frac{\partial p}{\partial y}, \quad \upsilon_g = \frac{1}{f\rho}\frac{\partial p}{\partial x} \qquad \text{..... (16.18)}$$

$$u_{gs} = -\frac{1}{f\rho}g\rho\frac{\partial \zeta}{\partial y}, \quad v_{gs} = \frac{1}{f\rho}g\rho\frac{\partial \zeta}{\partial x}, \text{ since } h_l = \text{constant}$$

$$u_{gs} = -\frac{g}{f}\frac{\partial \zeta}{\partial y}, \quad v_{gs} = \frac{g}{f}\frac{\partial \zeta}{\partial x} \qquad \text{.....(16.19)}$$

Using eq.(16.17) write eq.(16.18) at the sea surface

where g = gravity, f = coriolis parameter, ζ = the height of sea surface above sea level.

16.2.1 Pressure Distribution – Geostrophic Stream Function

Let ψ be a stream function. Then $\psi = \psi_{(x,y)}$

$$\frac{\partial \psi}{\partial x} = -v, \quad \frac{\partial \psi}{\partial y} = u \text{ and } d\psi = \frac{\partial \psi}{\partial x}dx + \frac{\partial \psi}{\partial y}dy$$

$$d\psi = -vdx + udy \qquad \text{.....(16.20)}$$

Equation of streamline are given by $\frac{dx}{u} = \frac{dy}{v}$ .....(16.21)

For geostrophic flow $v_g = \frac{1}{\rho f}\frac{\partial p}{\partial x}; \; u_g = -\frac{1}{\rho f}\frac{\partial P}{\partial y}$

Integrating for stream function $v_g = -\frac{\partial \psi}{\partial x} = \frac{1}{\rho f}\frac{\partial \rho}{\partial x}$ .....(16.22)

Integrating 16.22 w.r.t x we have

$$\psi = \frac{1}{\rho_o f_o} p \qquad \text{.....(16.23)}$$

Eq. 16.23 shows that geostrophic stream function maps pressure distribution.

Stream function ψ varies from one streamline to another.

Equation of continuity in two dimensions (x,y) is given by

$$\frac{\partial u}{\partial x} + \frac{\partial v}{\partial y} = 0 \qquad \text{.....(16.24)}$$

For geostrophic flow, (by substituting (16.20 in 16.24)

$$\frac{\partial}{\partial x}\left(\frac{\partial \psi}{\partial y}\right) + \frac{\partial}{\partial y}\left(\frac{-\partial \psi}{\partial x}\right) = 0$$

i.e., $$\frac{\partial^2 \psi}{\partial x \partial y} - \frac{\partial^2 \psi}{\partial y \partial x} = 0$$

i.e., stream function ψ satisfies the equation of continuity.

16.2.2 The Main Features of Ocean Models

Ocean models are used to simulate oceanic flows with realistic and useful results.

In a frictionless flow, the horizontal motion that occurs when the pressure gradiant force is in balance with the coriolis force is called geostrophic fluid motion.

In the interior ocean that is excluding the top and bottom Ekman layers, the horizontal pressure gradient is practically balanced by coriolis force. This is geostrophic motion in ocean.

The equations of motion are:

$$\left.\begin{aligned} \alpha\frac{\partial p}{\partial x} &= +fv \\ \alpha\frac{\partial p}{\partial y} &= -fu \\ \alpha\frac{\partial p}{\partial z} &= -g \end{aligned}\right\} \quad \text{.....(16.25)}$$

where $\alpha = \frac{1}{\rho}$ = specific volume

$f = 2\Omega\sin\phi$, coriolis parameter

The above equation can be written as vector equation

$$\vec{V}_g = \frac{1}{\rho f} k \times \nabla p$$

where k = unit vector in z – direction.

i.e.,

$$iu_g + jv_g + k\omega_g = \frac{1}{\rho f} k \times \left(i\frac{\partial p}{\partial x} + j\frac{\partial p}{\partial y} + k\frac{\partial p}{\partial z} \right)$$

$$= \frac{1}{\rho f}\left(j\frac{\partial p}{\partial x} - i\frac{\partial p}{\partial y} \right)$$

Equating the like terms, we have

$$\left.\begin{aligned} u_g &= -\frac{\alpha}{f}\frac{\partial p}{\partial y} && \text{(a)} \\ v_g &= +\frac{\alpha}{f}\frac{\partial p}{\partial x} && \text{(b)} \\ \text{and} \quad -g\rho &= \frac{\partial p}{\partial z} \text{ or } \delta p = -g\rho\delta z && \text{(c) Hydrostatic equation} \end{aligned}\right\} \quad \text{.....(16.26)}$$

integrating (16.26c), z – component, with boundary conditions

z = –h when p = p_o and

z = ζ is the height p = p of the sea surface

$$\int_{p_o}^{p} \frac{\partial p}{\partial z} dz = \int_{-h}^{\zeta} -g\rho dz$$

or $$\int_{p_0}^{p} \partial p = \int_{-h}^{\zeta} g(\phi, z)\rho(z)dz$$

where $g = g(\phi, z)$ is a function of latitude ϕ and depth z

$\rho = \rho(z)$ is a function of depth.

$$P - P_0 = \int_{-h}^{\xi} g(\phi, z)\, \rho(z)dz$$

Sea surface ζ to be above or below the surface $z = 0$

Pressure gradient at the sea surface is balanced by surface current u_g [see Fig. 16.1].

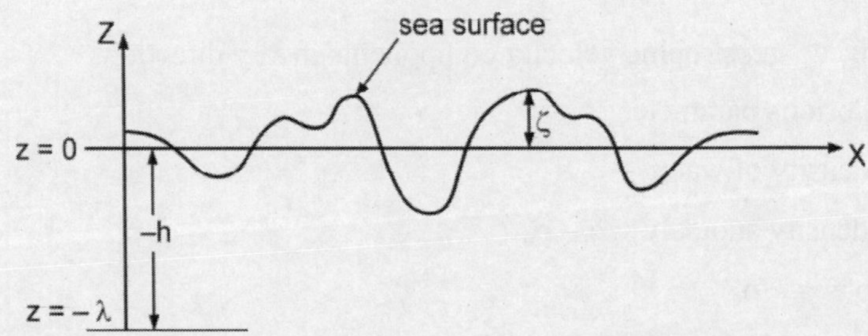

Fig. 16.1 Defining ζ and h

Under geostrophic balance it is assumed that the flow has no acceleration i.e.,

$$\frac{du}{dt} = 0, \frac{dv}{dt} = 0, \frac{dw}{dt} = 0$$

External force is gravity and friction is very small. Under these assumptions, the equations of motion given by

$(F_x = F_y = F_z = 0, e < < \text{small})$

$$\frac{du}{dt} = -\frac{1}{\rho}\frac{\partial p}{\partial x} + 2fv + F_x - ew$$

$$\frac{dv}{dt} = -\frac{1}{\rho}\frac{\partial p}{\partial y} - 2fu + F_y$$

$$\frac{dw}{dt} = -\frac{1}{\rho}\frac{\partial p}{\partial z} + ev_e - g + F_z$$

reduce to

$$\left.\begin{aligned} \frac{1}{\rho}\frac{\partial p}{\partial x} &= fv_g \\ \frac{1}{\rho}\frac{\partial p}{\partial y} &= -fu_g \\ \frac{1}{\rho}\frac{\partial p}{\partial z} &= -g \end{aligned}\right\} \quad \text{.....(16.27)}$$

These are the geostrophic motion equations

Geostrophic flow equations are

$$\frac{\partial p}{\partial x} = \rho f v_g, \; \frac{\partial p}{\partial y} = -\rho f u_g \text{ and } \frac{\partial p}{\partial z} = -\rho g \quad \text{.....(16.28)}$$

where u_g, v_g, geostrophic velocity components in x, y directions

f = Coriolis parameter

ρ = density of water

σ = density anomaly = $\rho - \rho_o$

or $\rho = \sigma + \rho_o$

where ρ_o = 1000 Kg/m^3, also called reference density.

The equation 16.28 can be put in vector equation as:

$$i\,u_g + jv_g + kw_g = \frac{1}{\rho f}K \times \left[i\frac{\partial p}{\partial x} + j\frac{\partial p}{\partial y} + k\frac{\partial p}{\partial z} \right]$$

or $$\vec{V}_g = \frac{1}{\rho f}(k \times \nabla p) \quad \text{.....(16.29)}$$

Another form of geostrophic equation

16.2.3 Geostrophic Equation x, y Direction Components under Normal Condition

$$u_g = -\frac{1}{\rho_o f}\frac{\partial p}{\partial y}$$

$$v_g = -\frac{1}{\rho_o f}\frac{\partial p}{\partial x}$$

Differentiating w.r.t z we get

$$\frac{\partial u_g}{\partial z} = -\frac{1}{f\rho_o}\frac{\partial}{\partial z}\left(\frac{\partial p}{\partial y}\right) = -\frac{1}{f\rho_o}\frac{\partial}{\partial y}\left(\frac{\partial p}{\partial z}\right)$$

$$= -\frac{1}{f\rho_o}\frac{\partial}{\partial y}(g\rho) \qquad \text{Since } \frac{\partial p}{\partial z} = g\rho$$

$$= \frac{g}{f\rho_o}\frac{\partial}{\partial y}(\sigma + \rho_o) \qquad \text{Since } \rho = \sigma + \rho_o$$

$$\frac{\partial u_g}{\partial z} = \frac{g}{f\rho_o}\frac{\partial \sigma}{\partial y} \qquad (\because \rho_o = \text{constant}) \qquad .....(16.30)$$

$$\frac{\partial v_g}{\partial z} = \frac{1}{f\rho_o}\frac{\partial}{\partial z}\left(\frac{\partial p}{\partial x}\right) = \frac{1}{f\rho_o}\frac{\partial}{\partial x}\left(\frac{\partial p}{\partial z}\right)$$

$$= \frac{1}{f\rho_o}\frac{\partial}{\partial x}(g\rho)$$

$$= \frac{g}{f\rho_o}\frac{\partial}{\partial x}(\sigma + \rho_o)$$

$$\frac{\partial v_g}{\partial z} = \frac{g}{f\rho_o}\frac{\partial \sigma}{\partial x} \qquad .....(16.31)$$

Eq. (16.30) and (16.31) together can be written as

$$\frac{\partial \vec{V}_g}{\partial z} = -\frac{g}{f\,\rho_o} k \times \nabla\sigma \qquad [\text{where } \overline{v}_g = iu_g + jv_g] \qquad .....(16.32)$$

16.2.4 Thermal Wind Relations

Let density of water depends on temperature (T) linearly,

$\propto$ is the coefficient of thermal expansion at $T = T_o$.

T and ρ are related by the equation of state

$$\rho = \rho_o\ [1 - \propto (T - T_o)]$$

or

$$\sigma = \rho - \rho_o = -\propto \rho_o (T - T_o) \qquad .....(16.33)$$

From (16.32) and (16.33) we get

$$\frac{\partial \vec{V}_g}{\partial z} = -\frac{-g}{f\,\rho_o} k \times \nabla[-\propto \rho_o (T - T_o)]$$

$$= \frac{g}{f\,\rho_o} k \times \nabla(\propto \rho_o T) \qquad \therefore\ \nabla T_o = 0;\ T_o = \text{constant}$$

$$\frac{\partial \vec{V}_g}{\partial z} = \frac{g \propto}{f} k \times \nabla T \qquad .....(16.34)$$

In component form eq. 16.34 is given by

$$\frac{\partial}{\partial z}(i\, u_g + j\, v_g) = \frac{g \propto}{f}\left[j\frac{\partial T}{\partial x} - i\frac{\partial T}{\partial y}\right], \text{ where } \nabla T = \left[\frac{i\partial T}{\partial x} + \frac{j\partial T}{\partial y}\right]$$

$$k \times \nabla T = \left[\frac{j\partial T}{\partial x} - \frac{i\partial T}{\partial y}\right]$$

This gives $\frac{\partial u_g}{\partial z} = -\frac{g \propto}{f}\frac{\partial T}{\partial y}$

$$\frac{\partial v_g}{\partial z} = \frac{g \propto}{f}\frac{\partial T}{\partial x} \qquad .....(16.35)$$

Eq. (16.35) is the thermal wind relation connecting the vertical shear of geostrophic current to horizontal temperature gradients.

$$\frac{\partial \vec{V}g}{\partial z} = \frac{\alpha g}{f} k \times \nabla T \qquad .....(16.36)$$

where α = thermal expansion coefficient

T = Temperature

V_g = geostrophic wind vector $(iu_g + jv_g)$

Eq. (16.36) is a simple form of thermal wind relation connecting vertical shear of the geostrophic current to the horizontal temperature gradient.

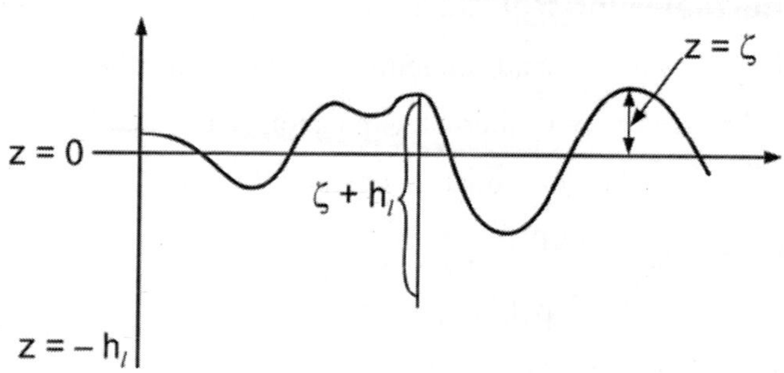

16.2.5 Thermal Wind in Pressure Coordinates, another Approach

$$u_g = -\frac{1}{f\rho}\frac{\partial p}{\partial y} \quad \text{.....(16.37)}$$

$$v_g = \frac{1}{f\rho}\frac{\partial p}{\partial x} \quad \text{.....(16.38)}$$

$$\frac{\partial z}{\partial p} = -\frac{1}{g\rho} \quad \text{.....(16.39)}$$

In pressure coordinates

$$\left(\frac{\partial p}{\partial x}\right)_z = g\rho\left(\frac{\partial z}{\partial x}\right)_{p,} ; \left(\frac{\partial p}{\partial y}\right)_z = g\rho\left(\frac{\partial z}{\partial y}\right)_p$$

$$\therefore \quad u_g = \frac{-g}{f}\frac{\partial z}{\partial y} \quad \text{.....(16.40)}$$

$$v_g = \frac{g}{f}\frac{\partial z}{\partial x} \quad \text{.....(16.41)}$$

Differentiating (16.40) & (16.41) w.r.t p

$$\frac{\partial u_g}{\partial p} = \frac{\partial}{\partial p}\left(\frac{-g}{f}\frac{\partial z}{\partial y}\right) = \frac{-g}{f}\frac{\partial^2 z}{\partial p \partial y} = -\frac{g}{f}\frac{\partial}{\partial y}\left(\frac{\partial z}{\partial p}\right)$$

$$= -\frac{g}{f}\frac{\partial}{\partial y}\left(-\frac{1}{g\rho}\right) \qquad \text{using (39)}$$

$$= \left(\frac{+1}{f}\frac{\partial}{\partial y}\frac{1}{\rho}\right) = \frac{1}{f}\frac{R}{p}\left(\frac{\partial T}{\partial y}\right)_p \qquad \frac{1}{\rho} = \frac{RT}{p} = \alpha$$

i.e.,

$$\frac{\partial u_g}{\partial p} = \frac{R}{fp}\left(\frac{\partial T}{\partial y}\right)_p \quad \text{.....(16.42)}$$

$$\frac{\partial v_g}{\partial p} = \frac{\partial}{\partial p}\left(\frac{g}{f}\frac{\partial z}{\partial x}\right) = \frac{g}{f}\frac{\partial^2 z}{\partial p \partial x} = \frac{g}{f}\frac{\partial}{\partial x}\left(\frac{\partial z}{\partial p}\right)$$

$$= \frac{g}{f}\frac{\partial}{\partial x}\left(-\frac{1}{g\rho}\right)_p \qquad \text{using (39)}$$

$$= -\frac{1}{f}\frac{\partial}{\partial x}\left(\frac{RT}{p}\right)_p$$

$$\frac{\partial v_g}{\partial p} = \frac{-R}{fp}\left(\frac{\partial T}{\partial x}\right)_p \qquad .....(16.43)$$

$$\therefore \quad i\frac{\partial u_g}{\partial P} + j\frac{\partial v_g}{\partial P} = \frac{R}{fP}\left[i\frac{\partial T}{\partial y} - j\frac{\partial T}{\partial x}\right]_p \qquad .....(16.44)$$

$$\text{or} \quad \frac{\partial \vec{V}_g}{\partial P} = \frac{R}{fP}\left[i\frac{\partial T}{\partial y} - j\frac{\partial T}{\partial x}\right]_p \qquad .....(16.45)$$

Eq. (16.44) represents thermal wind relationship in pressure coordinates.

16.3 Geostrophic Currents

Let the sea surface AB sloping from east (B) to west (A). Let θ be the angle AB makes with the horizontal line (surface) = θ. If we consider small columns of water at A & B, then at any horizontal surface under water at depth Z, the hydrostatic pressure P_A (below A) is the weight of water column above it, $P_A = \rho gz$ where ρ = density of water, g = gravity

In Fig. $AC = \delta x,\ CB = \delta z$

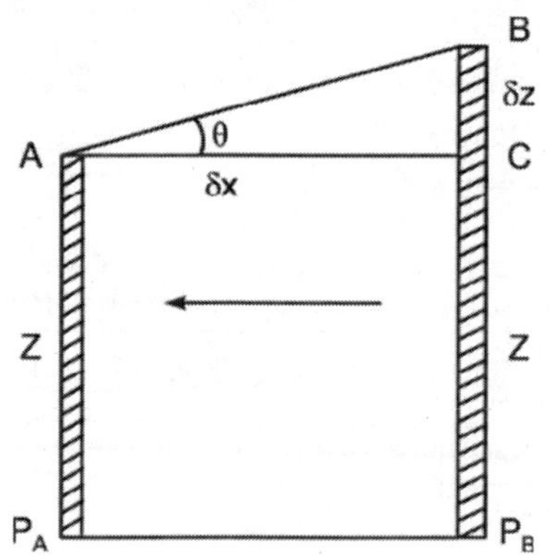

Fig. 16.2

Similarly the hydrostatic pressure P_B (below B) is the weight of water column above it.

$$P_B = \rho g\,(z + \delta z)$$

$P_B > P_A$. This implies water flows from B to A because of pressure gradient (High to Low). The force that gives rise to this motion is called horizontal pressure gradient force. When the pressure gradient force gives rise to the movement of surface waters, which becomes equilibrium motion on balance by wind stress (τ) and the coriolis force (f). This balanced movement of water (coriolis force = pressure gradient) is called geostrophic current.

Hydrostatic pressure: The hydrostatic pressure at any depth z in the ocean is the weight of the water acting on unit area (say 1 m^2).

Let P be the hydrostatic pressure at depth z.

$$p = \text{mass of overlying pressure sea water} \times g$$

$$= \rho z \times 1 \times g \qquad (\text{mass} \times g)$$

i.e., $p = +\rho g z$

This is hydrostatic equation [z is negative below water surface, g is also negative]

Generally we consider for small water column δz and

hence $\delta p = +\rho g\ \delta z$

Horizontal pressure gradient

From the Fig. 16.2, the hydrostatic pressure at a point A, depth z

$$P_A = \rho g z$$

and at B, at depth $(z + \delta z)$

$$P_B = \rho g\,(z + \delta z)$$

$P_B > P_A$, Let $\delta p = P_B - P_A$

Then $P = P_B - P_A = \rho g\,(z + \delta z) - \rho g z$

$$\delta p = \rho g \delta z \qquad \text{or} \qquad \frac{\delta p}{\delta x} = \rho g \frac{\delta z}{\delta x}$$

From the Fig.16.2, $\tan\theta = \dfrac{\delta z}{\delta x}$

$\therefore$ pressure gradient $\dfrac{\delta p}{\delta x} = \rho g \tan\theta$

In the limit $\dfrac{dp}{dx} = \rho g \tan\theta$ .....(16.46)

Eq. 16.46 denotes the rate of change of horizontal pressure or gradient in x-direction.

16.3.1 The Horizontal Pressure Gradient Force Acting on unit Volume of Sea Water

Force acting on unit mass of sea water is obtained by dividing by density ρ.

$\therefore$ Horizontal pressure gradient force

per unit mass $= \frac{1}{\rho}\frac{dp}{dx} = g\tan\theta$ ....(16.47)

using eq. 16.46

From Eq. 16.46, the horizontal pressure gradient force $\left(\frac{dp}{dx}\right)$ acting on a parcel of mass m is = mg tan θ. .....(16.48)

Let a parcel of mass m is moving with velocity u. Let f = coriolis force. Then the coriolis force acting on the parcel of water mass m and velocity 'u' is mfu .....(16.49)

In geostrophic equilibrium from (16.48) and (16.49) we have

$$mg\tan\theta = mfu$$

or $$\tan\theta = \frac{fu}{g}$$ .....(16.50)

This is gradient equation and in geostrophic flow it is true for any isobaric surface.

In Barotrophic conditions, the isopycnals are parallel to the isobars.

From the slope of the isopycnals θ can be determined (see fig below)

∴ the velocity u is determined by the above equation ..... (16.50)

i.e., $$u = \frac{g}{f}\tan\theta$$

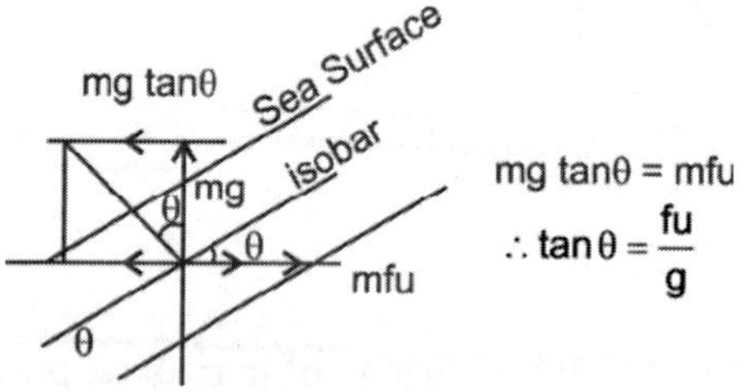

Geostrophic Flow Equations in Barotropic Condition

In barotrophic conditions (isobaric surfaces are parallel to isopymic surfaces), the variation of pressure over a horizontal surface (at any depth) is determined only by the slope of sea surface (since isobaric surfaces are parallel to the sea surface). Any variations (changes) in the density of sea water is also effect the weight overlying sea water.

In NH, a sea surface slope is up towards the east side, this results in a horizontal pressure gradient force toward west.

Geostrophic equation in barotropic condition

$$mg \tan\theta = mfu$$

or $$\tan\theta = \frac{fu}{g} \text{ at all depths.}$$

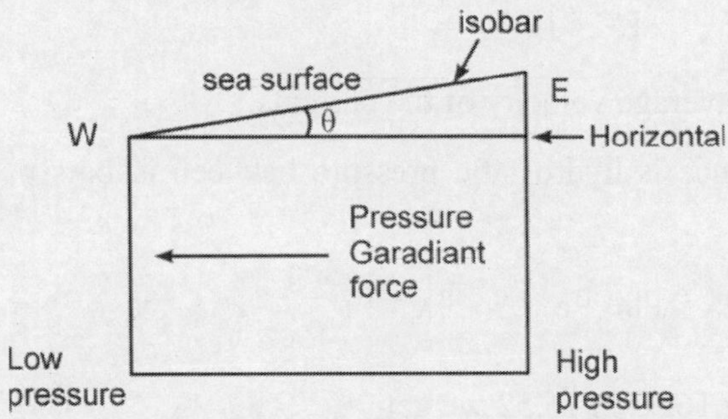

16.3.2 Geostrophic Equation in Baroclinic Conditions

In baroclinic conditions, isobaric surfaces intersect with isopycnic surfaces & pressure (isobars) changes with depth. The slope of the isobars change with depth.

Consider two stations A, B, on sea surface.

Let the reference surface be at depth z_o, which is horizontal with pressure P_o, with $A'B' = L$ where $A'B'$ are projections of A,B, respectively

Consider another surface at depth z_1, where the isobaric surface pressure P_1 makes an angle θ_1, with the horizontal.

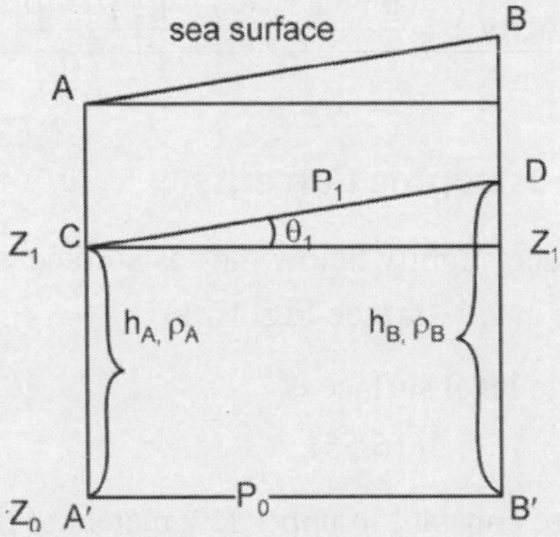

Let h_A, h_B be the depths CA' and DB' respectively. Let ρ_A, ρ_B be the average density of columns CA', DB' respectively.

The slope tan θ_1, is at CD with the horizontal is given by

$$\tan\theta_1 = \frac{h_B - h_A}{L} \quad \text{....(16.51)}$$

where $A'B' = L$

and $$u = \frac{g}{f}\left(\frac{h_B - h_A}{L}\right) \quad \text{....(16.52)}$$

where u is the average velocity of the currents.

The difference is hydrostatic pressure between isobars p_l and p_o is the same at C & D.

$$\therefore \quad \rho_A g h_A = \rho_B g\, h_B \quad \text{i.e., } h_A = h_B \frac{\rho_B}{\rho_A} \quad \text{.....(16.53)}$$

Substituting 16.53 in 16.52 we get

$$u = \frac{g}{f}\left(\frac{h_B - h_B \dfrac{\rho_B}{\rho_A}}{L}\right) = \frac{g}{f}\frac{h_B}{L}\left[1 - \frac{\rho_B}{\rho_A}\right]$$

$$\therefore \quad u = \frac{g}{f}\frac{h_g}{L}\left[1 - \frac{\rho_B}{\rho_A}\right] \quad \text{.....(16.54)}$$

Eq. 16.54 is the required geostrophic equation in Baroclinic condtion with average velocity current u.

Compare the above equation with two layer ocean result, which are very close viz.

$$\text{Surface velocity } v = \frac{g}{f}\frac{(h_A - h_B)}{\delta x} = \frac{g}{f}\left(\frac{\rho_2 - \rho_1}{\rho_1}\right)\left(\frac{H_A - H_B}{\delta x}\right)$$

16.3.3 Surface Geostrophic Currents

Consider a level surface slightly below the sea surface, say about two meters below the sea surface at $z = -h_l$ (see Fig. 16.1)

The pressure on the level surface is

$$p = \rho g(\zeta + h_1) \quad \text{.....(16.55)}$$

Assuming ρ, g are constant in upper few meters of the ocean, the geostrophic equations are

$$\left.\begin{aligned} u_g &= -\frac{1}{\rho f}\frac{\partial p}{\partial y} \\ v_g &= -\frac{1}{\rho f}\frac{\partial p}{\partial x} \end{aligned}\right\} \qquad .....(16.56)$$

Using (16.55) we write (16.56) at the sea surface

$$u_{g(s)} = -\frac{1}{\rho f}\rho g\frac{\partial \zeta}{\partial y}$$

$$v_{g(s)} = \frac{1}{\rho f}\rho g\frac{\partial \zeta}{\partial x} \qquad (\because\ h_1 \text{ is constant})$$

or

$$\left.\begin{aligned} u_{g(s)} &= \frac{-g}{f}\frac{\partial \zeta}{\partial y} \\ v_{g(s)} &= \frac{g}{f}\frac{\partial \zeta}{\partial x} \end{aligned}\right\} \qquad .....(16.57)$$

where g = gravity; f = coriolis parameter

ζ = the height of sea surface above level surface

e.g., (16.57) are the required surface geostrophic currents.

16.4 Ocean Surface Structure and Geostrophic Flow

In geostrophic balance:

geostrophic force = coriolis force acting on them

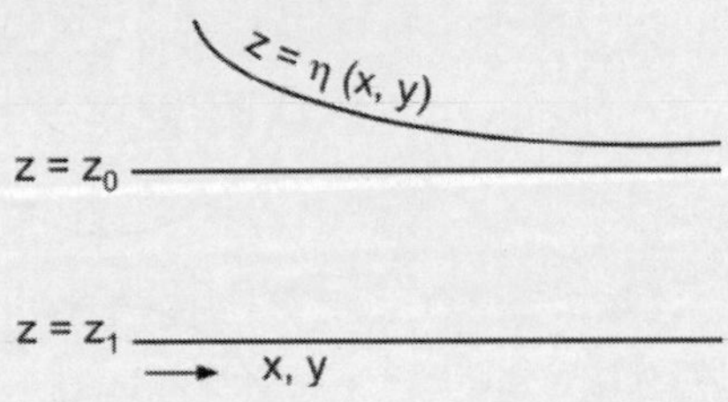

Consider the fig, the height of the free surface is $z = \eta(x, y)$

The depth of two reference surfaces one near the surface and another at a depth given by $z = z_o$ and $z = z_1$ respectively.

Integrating the hydrostatic relation $\int_{\rho(s)}^{\rho(z)} \delta p = \int_z^{\eta} g\rho\delta z$

$$\begin{aligned} P_{(z)} &= P_s + \int_z^{\eta} g\rho dz \\ &= P_s + g\bar{\rho}(\eta - z) \end{aligned} \qquad .....(16.58)$$

where $\bar{\rho} = \frac{1}{(\eta - z)} \int_z^{\eta} g\rho dz$ is the average density in the water column of depth $(\eta - z)$

We are interested in the near surface region $z = z_0$, fractional variations in column depth are much greater than those of density, so that we can disregard the later (density) by putting $\rho = \rho_{ref}$

The equation reduces to

$$P_{(z)} = P_s + g\rho_{ref}(\eta - z_0) \qquad \text{..... (16.59)}$$

Near surface, pressure gradient with gradients in surface elevation

$$i\frac{\partial p}{\partial x} + j\frac{\partial p}{\partial y} = g\rho_{ref}\left(i\frac{\partial \eta}{\partial x} + j\frac{\partial \eta}{\partial y}\right)$$

Thus the geostrophic flow just beneath the surface is

$$i\,u_{gs} + jv_{gs} = \frac{g}{f}\left(-i\frac{\partial \eta}{\partial y} + j\frac{\partial \eta}{\partial x}\right) \qquad \vec{U}_{gs} = iu_{gs} + j\,v_{gs}$$

= surface velocity

or $$\vec{U}_{gs} = \frac{g}{f} k \times \nabla\eta \qquad \text{.....(16.60)}$$

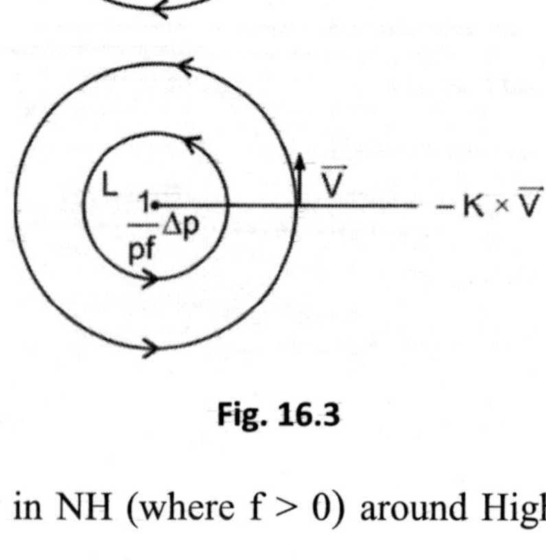

Fig. 16.3

Geostrophic flow in NH (where $f > 0$) around High pressure centre and Low pressure centre.

Note: $-k \times \vec{V} = \frac{1}{\rho f}\nabla p$

$$-k \times (iu_g + jv_g + kw_g) = \frac{1}{\rho f}\left(\frac{i\partial p}{\partial x} + j\frac{\partial p}{\partial y} + k\frac{\partial p}{\partial z}\right)$$

or $$-ju_g + iv_g + 0 = \frac{1}{\rho f}\left(i\frac{\partial p}{\partial x} + j\frac{\partial p}{\partial y} + k\frac{\partial p}{\partial z}\right)$$

equating like terms

$$\left.\begin{aligned} -u_g &= \frac{1}{\rho f}\frac{\partial p}{\partial y} \\ v_g &= \frac{1}{\rho f}\frac{\partial p}{\partial x} \\ w_g &= 0 \end{aligned}\right\} \qquad \text{..... (16.61)}$$

16.5 The Taylor-Proudman Theorem

Taylor-proudman theorem states that the geostrophic flow in two dimensions does not change in the direction of the rotation vector Ω.

The theorem is a case of fluid dynamics on the f – plane, i.e., the influence of vorticity due to earth's rotation. For geostrophic fluid flow with constant density (ρ_o) on a plane with constant rotation (coriolis parameter $f = f_o$)

Proof: The geostrophic fluid flow equations are

$$fv_g = \frac{1}{\rho_\circ}\frac{\partial p}{\partial x} \qquad \text{......(16.62)}$$

$$fu_g = -\frac{1}{\rho_\circ}\frac{\partial p}{\partial y} \qquad \text{..... (16.63)}$$

$$g = -\frac{1}{\rho_\circ}\frac{\partial p}{\partial z} \qquad \text{..... (16.64)}$$

And the equation of continuity

$$\frac{\partial u}{\partial x} + \frac{\partial v}{\partial y} + \frac{\partial w}{\partial z} = 0 \qquad \text{..... (16.65)}$$

Differentiating (partially) (16.62) w. r. t –z on f plane f = f_0 we have

$$\frac{\partial}{\partial z}\left(f_o v_g = \frac{1}{\rho_o}\frac{\partial p}{\partial x}\right)$$

$$v_g \frac{\partial f_o}{\partial z} + f_o \frac{\partial v_g}{\partial z} = \frac{1}{\rho_o}\frac{\partial}{\partial x}\left(\frac{\partial p}{\partial z}\right)$$

Note: [For continuos flow $\frac{\partial}{\partial z}\left(\frac{\partial p}{\partial x}\right) = \frac{\partial}{\partial x}\left(\frac{\partial p}{\partial z}\right)$]

$$f_o \frac{\partial v_g}{\partial z} = \frac{1}{\rho_o}\frac{\partial}{\partial z}(-\rho_o g)$$

[using equation (16.64) and $\frac{\partial f_o}{\partial z} = 0$]

$$f_o \frac{\partial V_g}{\partial z} = -\frac{\partial g}{\partial x}$$

or $$f_o \frac{\partial V_g}{\partial z} = 0$$

since $$\frac{\partial g}{\partial x} = 0 \text{ along } x - \text{axis}$$

i.e., $$\frac{\partial v_g}{\partial z} = 0 \qquad .....(16.66)$$

differentiating (16.63) w r t –z we have

$$\frac{\partial}{\partial z}(f_o u_g) = \frac{-\partial}{\partial z}\left(\frac{1}{\rho_o}\frac{\partial p}{\partial y}\right)$$

$$f_o \frac{\partial u_g}{\partial z} + u_g \frac{\partial f_o}{\partial z} = -\frac{1}{\rho_o}\frac{\partial}{\partial y}\left(\frac{\partial p}{\partial z}\right)$$

or $$f_o \frac{\partial u_g}{\partial y} = -\frac{1}{\rho_o}\frac{\partial}{\partial y}(-\rho_o g)$$

[using (16.64) and putting $\frac{\partial f_o}{\partial z} = 0$

$f_o \frac{\partial u_g}{\partial z} = \frac{\partial g}{\partial y} = 0$] (treating g constant locally $\frac{\partial g}{\partial y} = 0$)

i.e., $$\frac{\partial u_g}{\partial z} = 0 \qquad .....(16.67)$$

from (16.66) & (16.67)

$$\frac{\partial v_g}{\partial z} = 0 \text{ and } \frac{\partial u_g}{\partial z} = 0$$

that is geostrophic motion does not change in the direction $f\hat{k}$.

This implies that the vertical derivative of the horizontal velocity of constant density fluid is zero. The T-P theorem is applicable to slowly varying flows of homogeneous, rotating inviscid fluid. Thus the theorem imposes (places) strong constraint on the flow.

16.6 Consequences of Taylor-Proudman Theorem

The equations of geostrophic fluid motion are:

$$f\nu_g = \frac{1}{\rho_o}\frac{\partial p}{\partial x} \qquad .....(16.68)$$

or $$\nu_g = \frac{1}{f\rho_o}\frac{\partial p}{\partial x}$$

$$fu_g = -\frac{1}{\rho_o}\frac{\partial p}{\partial y} \qquad .....(16.69)$$

or $$u_g = -\frac{1}{f\rho_o}\frac{\partial p}{\partial y}$$

$$g = -\frac{1}{\rho_o}\frac{\partial p}{\partial z} \qquad .....(16.70)$$

or $$g\rho_o = -\frac{\partial p}{\partial z}$$

Differentiate (16.68) w. r . t y and (16.69) w . r. t x, we have

$$\frac{\partial v_g}{\partial y} = \frac{\partial}{\partial y}\left(\frac{1}{f\rho_o}\frac{\partial p}{\partial x}\right)$$

or $$\frac{\partial v_g}{\partial y} = \frac{1}{f\rho_o}\frac{\partial^2 p}{\partial x \partial y}$$

$$\frac{\partial u_g}{\partial x} = -\frac{\partial}{\partial x}\left(\frac{1}{f\rho_o}\frac{\partial p}{\partial y}\right) \text{ or } \frac{\partial u_g}{\partial x} = -\frac{1}{f\rho_o}\frac{\partial^2 p}{\partial x \partial y}$$

Adding we eliminate p, we get

$$\frac{\partial u_g}{\partial x}+\frac{\partial v_g}{\partial y}=0 \qquad [\text{or } \nabla_H \cdot \vec{V}_g = 0] \qquad .....(16.71)$$

For incompressible fluid, the equation of continuity is

$$\frac{\partial u_g}{\partial x}+\frac{\partial v_g}{\partial y}+\frac{\partial w_g}{\partial z}=0$$

Using (16.71) we get

$$\frac{\partial w_g}{\partial z}=0 \qquad .....(16.72)$$

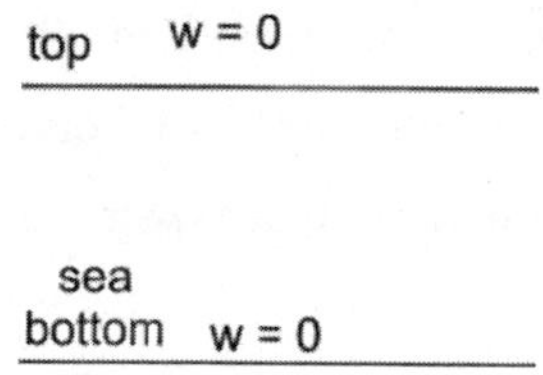

Since w = 0 at the surface of sea and at the bottom of the sea (floor), if the bottom is level, there can be no vertical velocity on f – plane. Further $\frac{\partial w}{\partial z}=0$ is not dependent of (ρ) density that is constant. The constraints are: slow steady motion, in incompressible fluid in two dimension and does not change in direction of the rotation vector(Ω).

Note: if we consider momentum equations in rotating coordinate axes, viz.

$$\frac{d\vec{V}}{dt}=\frac{1}{p}\nabla p-\nabla\phi-2\vec{\Omega}\times\vec{V}+\vec{F}$$

where $\phi = gz - \frac{\Omega^2 r^2}{2}$ (modified to gravitational potential)

If the flow is slow and steady, frictionless barotropic flow, the velocity $\vec{V}$ = iu + jv + kw, then u, v, w (horizontal and vertical components) cannot change in the direction of the rotation vector $\vec{\Omega}$. This amounts to the flow is two dimensional with vertical fluid column remaining vertical (that is fluid acts as stiff in the direction of Ω) as shown in Fig. 16.4.

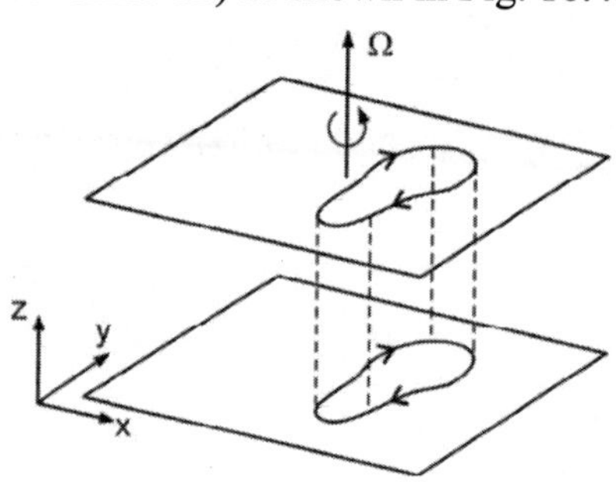

Fig. 16.4

Taylor-Proudman theorem states that, slow, steady frictionless flow of a barotropic, incompressible fluid is two dimensional, which does not change in the direction of rotation vector $\vec{\Omega}$.

The columns are called Taylor Columns.

Fluid column rigidity is provided by rotation.

16.7 Limitations of the Geostrophic Equations Ocean Currents

1. Geostrophic balance applies to flows that exceed a few tens of kilometers and with periods greater than a few day.
2. The geostrophic balance cannot be perfect.
3. Geostrophic currents cannot evolve or change with time, because the balance ignores acceleration of the flow.
4. If the horizontal flow is less than 50 km and time less than a few days, then acceleration dominates. However over longer times and distances, acceleration is negligible but not zero.
5. Near the equator (within about 20°lat) the geostrophic balance is not applicable, because the coriolis parameter $f = 2\Omega \sin\phi \rightarrow 0$ as $\phi \rightarrow 0$.
6. The geostrophic balance does not take into account the influence of friction.

16.7.1 Two-Layer Ocean

Suppose the ocean is composed of two layers in the vertical. Let ρ, be the density of water in the upper layer (above Pycnocline) and ρ_2 in the lower layer ($\rho_1 < \rho_2$) below pycnocline. Let 'h' be the height of the layer above ideal sea level surface and 'H' be the height of the lower level below the ideal layer and h << H.

Consider two stations A & B on the ocean bottom of sea surface and MN an ideal level surface.

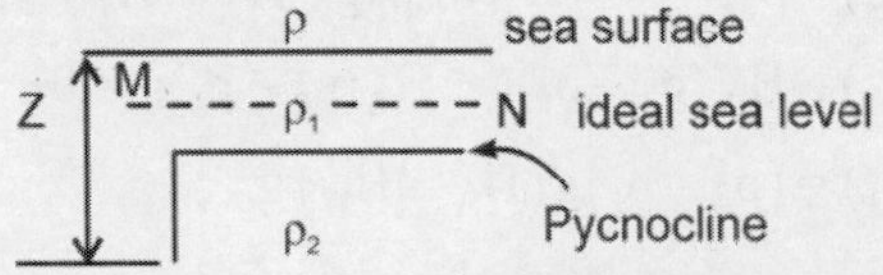

ρ_1= density in the upper layer above pycnocline

ρ_2= density in the lower layer below pycnocline

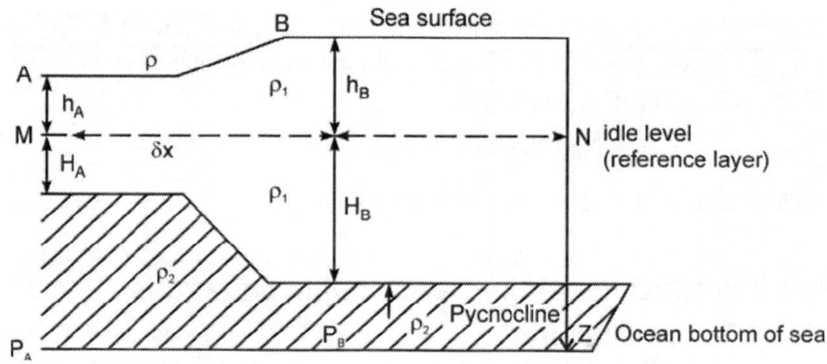

Let h_A, H_A and h_B, H_B be the height & depth of top layer, above and below ideal surface MN at the stations A and B as shown in the figs.

Let common depth be z, which is the level of no motion in the ocean.

Let p_A = pressure at depth z at station A and

p_B = pressure at depth z at station B

Hydrostatic equation is $\delta p = -g\rho\delta z$

Using hydrostatic equation with above notations we have

$$P_A = \rho_1 g(h_A + H_A) + \rho_2 g(z - H_A)$$

$$P_B = \rho_1 g(h_B + H_B) + \rho_2 g(z - H_B) \quad \text{.....(16.73)}$$

If z is level of no motion, then $P_A = P_B$.

Surface slope $\frac{\rho_1(h_A - h_B)}{\delta x} = \frac{(\rho_2 - \rho_1)(H_A - H_B)}{\delta x}$

or $$\frac{h_A - h_B}{\delta x} = \left(\frac{\rho_2 - \rho_1}{\rho_1}\right)\left(\frac{H_A - H_B}{\delta x}\right) \quad \text{.....(16.74)}$$

Since $P_A = P_B \Rightarrow P_A - P_B = 0$

From Eq. 16.73 we have

$$\rho_1 g(h_A - h_B) + \rho_1 g(H_A - H_B) + \rho_2 g(-H_A + H_B) = 0$$

or $$\rho_1 g(h_A - h_B) - H_A(\rho_2 - \rho_1)g + H_B(\rho_2 - \rho_1)g = 0$$

or $$\rho_1 g(h_A - h_B) = (\rho_2 - \rho_1)g(H_A - H_B)g$$

or $$g\frac{(h_A - h_B)}{\delta x} = \left(\frac{\rho_2 - \rho_1}{\rho_1}\right)\left(\frac{H_A - H_B}{\delta x}\right)g$$

If v be the surface velocity, then

$$fv = g\frac{(h_A - h_B)}{\delta x} = \left(\frac{\rho_2 - \rho_1}{\rho_1}\right)g\frac{(H_A - H_B)}{\delta x} \quad \text{.....(16.75)}$$

Eq. 16.75 implies that the slope of the sea surface ($h_A - h_B$) will be known from the knowledge of subsurface slope,

that is slope of density interface ($H_A - H_B$).

This allows to estimate the surface flow velocity from the shape of the pycnocline.

16.7.2 Cyclonic and Anticyclonic Eddy in Geostrophic Circulation

The two layer ocean analysis discussed above helps to depict cyclonic eddy in geostrophic circulation and anticyclonic eddy. Two layer assumption results in mapping of geostrophic currents only and not the entire field current.

In the Fig. 16.5(a) s_1, s_2 dipicted ideal sea surface, L_1 L_2 level surface and P_1P_2 sub surface thermocline. In this Fig. 16.5 subsurface pycnocline slope is greater than the surface change of cyclone feature.

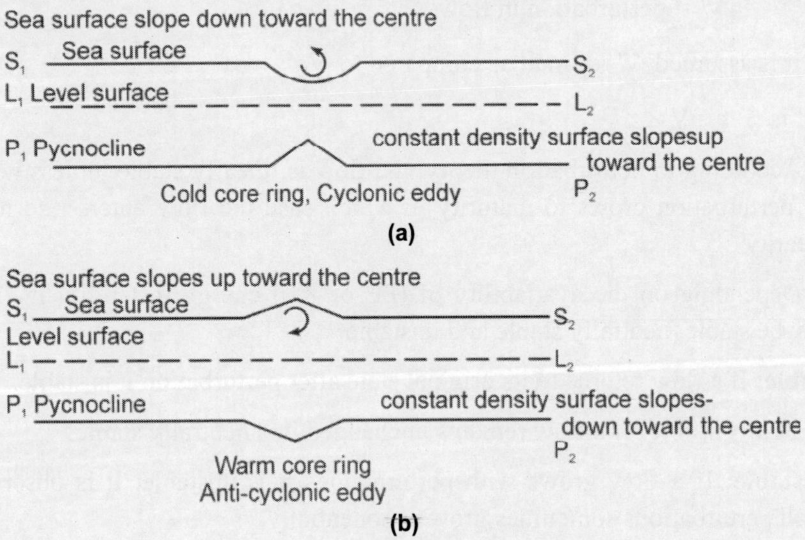

Fig. 16.5

These features show cyclonic eddy or cold features due to upwelling of the central isopycals in the centre.

Similarly warm feature looks like in Fig. 16.5(b). A warm feature rotates anticyclonic eddy due to down welling of central isopycnocline in the centre.

Eddies in Gulf Stream of Atlantic Ocean

The ocean observations indicate that ocean is not steady as laminar but it is highly turbulent. Ocean eddies are similar to baroclinic eddies in the atmosphere. The ocean eddies are about 100 km in lateral scale with life period ranging months to a year.

Satellite observations of Atlantic ocean indicate there is a strong SST gradient across the Gulf stream. (Gulf, Stream accelerates meanders and undulates which break off and form closed eddies). These eddies are "Anticyclonic warm core rings" to the north of the stream and "Cyclonic cold core rings to the south of the Stream. The Gulf Stream region has baroclinic instability.

Instability of Geostrophic Ocean Currents

In general all oceanic flows are unsteady.

According to perturbation theory.

$$V = \overline{V} + V'$$

where V = Speed of flow

$\overline{V}$ = Average speed of flow

V' = perturbation in flow

It is assumed V' is small as compared to $\overline{V}$

that is $V' \angle\angle \overline{V}$

According to perturbation theory, the flow is linearly stable, unless when the perturbation grows to maturity in which case the flow enters into non-linearity.

Depending on the availability of (PE or KE) energy instability in flow may be stable, neutrally stable and unstable.

Stable: If a flow returns to its original state after perturbation it is stable.

Neutrally stable: If a flow remains unchanged it is neutrally stable.

Unstable: If a flow grows with perturbations it is unstable. It is observed small perturbations sometimes grow exponentially.

In oceans if isopycnals are flat (in strati fluid) there is available potential energy and hence it is stable.

If isopycnals are little or available potential energy exists, then there would be instability.

Barotropic instability: If the isopycnals are parallel to isobars, the flow is barotropic. If the flow is not barotropic (isopycnals are intersecting the isobars) the flow is called baroclinic.

If there exists shear in horizontal flow (then there exists kinetic energy) then there would be barotrophic instability. In the Gulf Stream or jet like currents, there exists horizontal shear, as a consequence there exists barotropic instability.

Baroclinic instability: Baroclinic instability develops in the presence of available potential energy of the flow. Sub-Meso sclae eddies develop instabilities in the oceans mixed layer when density fronts develop. The density fronts are strongly tilted isopycnals, which are subjected to potential energy release.

Baroclinic instability exists in ocean geostrophic flow because of earth's rotation. Earth's rotation creates mean geostrophic flow tilted to isopycnals. It means barotropic instability is similar to instabilities of all sheared flows.

16.7.3 Margules Theory

Cross-section of currents at times show sharp fall in density surfaces, which are called lines of discontinuity, that is large change in density on either side of the current. Baroclinic current in the sector can be estimated by Margules theory. This technique can be used in oceanography to estimate the speed and direction of currents perpendicular to the frontal section (like in the atmospheric different air masses front).

Let OF be the frontal section (that is discontinuity). OX horizontal axis and OZ perpendicular axis.

Let Υ be the slope of the front (as shown in fig.)

Let ρ_1, ρ_2 be the densities on each side of the front (where $\rho_1 > \rho_2$).

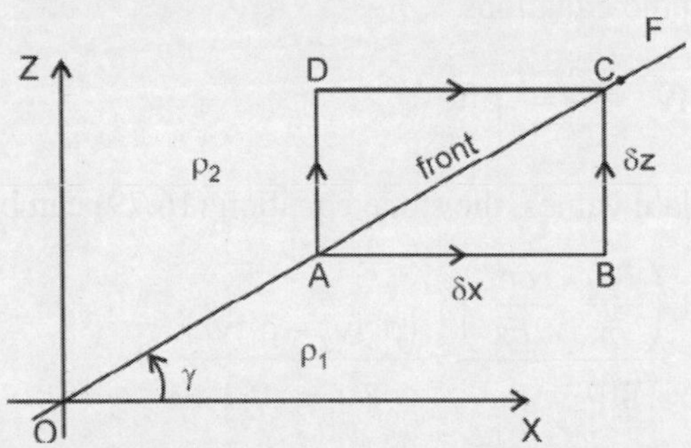

Consider the paths ABC and ADC on either side of the front AC segment. Along the frontal surface (AC) pressure must be same on both sides.

Since the points A and C are common on frontal surface, pressure change along the path ADC (on front)

= pressure change along the path ABC (on front)

that is,

$$\left(\frac{\partial \rho}{\partial z}\delta z+\frac{\partial p}{\partial x}\delta x\right)_{ADC}=\left(\frac{\partial \rho}{\partial x}\delta x+\frac{\partial p}{\partial z}\delta z\right)_{ABC} \qquad .....(16.76)$$

where δx, δz are small

Hydrostatic balance expressed $\frac{\partial p}{\partial z}=-\rho g$ .....(16.77)

(both sides of the front)

$$\tan\gamma=\frac{\delta z}{\delta x} \qquad .....(16.78)$$

From 16.76 [using 16.77] we have

$$-g\rho_2\delta z+\frac{\partial p_2}{\partial x}\delta x=\frac{\partial p_1}{\partial x}\delta x-g\rho_1\delta z$$

or $$g(\rho_2-\rho_1)\delta z=-\left(\frac{\partial p_1}{\partial x}-\frac{\partial p_2}{\partial x}\right)\delta x$$

or $$\frac{\delta z}{\delta x}=\frac{\left(\frac{\partial p_2}{\partial x}-\frac{\partial p_1}{\partial x}\right)}{g(\rho_2-\rho_1)}$$

∴ $$\tan r=\frac{\delta z}{\delta x}=\left(\frac{\frac{\partial p_2}{\partial x}-\frac{\partial p_1}{\partial x}}{g(\rho_2-\rho_1)}\right) \qquad .....(16.79)$$

From geostrophic equations

$$\frac{\partial p}{\partial x}=\rho f v\,;\quad \frac{\partial p}{\partial y}=-\rho f u$$

With appropriate values, the slope equation (16.79) can be written as

$$\tan\gamma=\frac{\left(\frac{\partial p_2}{\partial x}-\frac{\partial p_1}{\partial x}\right)}{g(\rho_2-\rho_1)}=\frac{(\rho_2 f v_2-\rho_1 f v_1)}{g(\rho_2-\rho_1)}$$

Approximating $\rho_1 \simeq \rho_2$ in the numerator in the above relation,

we have $$\tan\gamma=\frac{f}{g}\frac{(v_2-v_1)\rho_1}{(\rho_2-\rho_1)} \qquad .....(16.80)$$

which is Margules equation

16.7.4 Application of Margules theory to the Gulf Stream

Constant density surfaces in the Gulf Stream, slope downward to the east, while the sea surface topography slopes upward to the east. Constant pressure and constant density surfaces have opposite slopes.

If the sharp interface (line surface of discontinuity) between two water masses reaches the surface (oceanic front) which has properties similar to atmospheric air mass fronts.

Eddies in the proximity of Gulf Stream can have warm or cold cores.

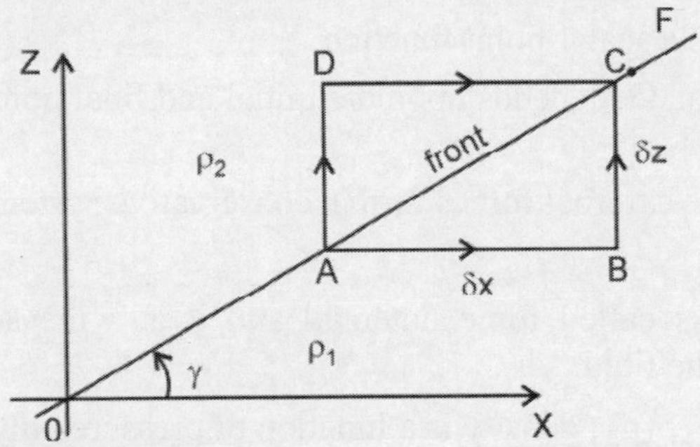

Application of Margules method to such meso-scale eddies gives the direction of the flow. Anticyclonic eddies have warm cores (ρ_1 is deeper in the centre of the eddy than elsewhere) and the constant pressure surfaces slope upwards. The sea surface is higher at the centre of the ring. Cyclonic eddies are the reverse is of cold core.

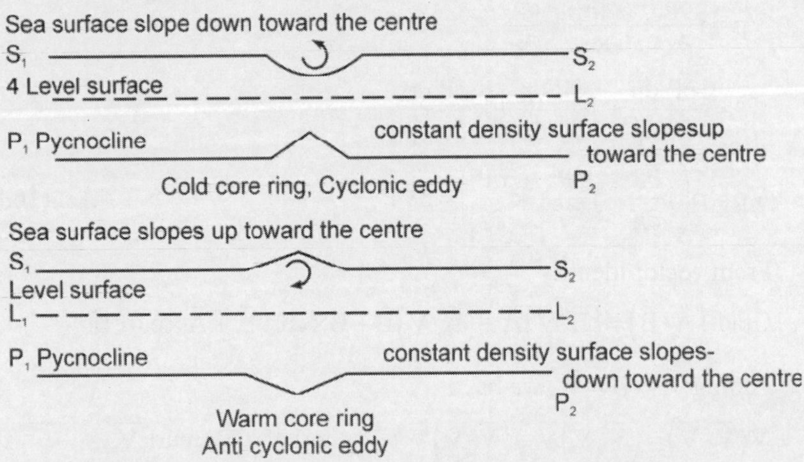

Note: In ocean water discontinuous transition from one type of water to another is never found, that is lines of discontinuity does not exist in ocean. However in several regions a transition takes place in short distance, which for practical purposes is the layer of transition or discontinuity.

16.8 Transformation of Equation of Motion

1. **Conservative Field of Force:** In a conservative field of force, the work done by the force $\vec{F}$ of the field in taking a unit mass (or transferring a unit mass) from some point A to another point B is independent of the path

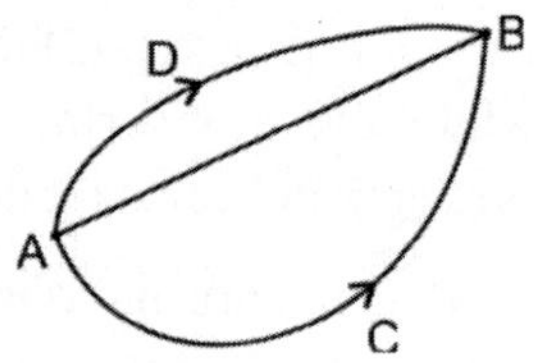

that is, $\int_{ACB} \vec{F}\cdot d\vec{r} = \int_{ADB} \vec{F}\cdot d\vec{r} = -\Omega$ (say)

where Ω is a scalar point function.

The value of Ω depends upon the initial and final points (positions) A and B.

Suppose the external forces form a conservative system say

$$\vec{F} = -\nabla\Omega \quad \text{..... (16.81)}$$

where Ω is called force potential and it is a measure of potential energy of the field.

2. Suppose $\rho = \rho(p)$, density is a function of pressure only

Let $P = \int_c \frac{dp}{\rho}$, where c is some constant.

$$\text{Then } \frac{dP}{dp} = \frac{1}{\rho} \quad \text{..... (16.82)}$$

$$\nabla p = i\frac{\partial p}{\partial x} + j\frac{\partial p}{\partial y} + k\frac{\partial p}{\partial z}$$

$$= i\frac{\partial P}{\partial x}\frac{dp}{dP} + j\frac{\partial P}{\partial y}\frac{dp}{dP} + k\frac{\partial P}{\partial z}\frac{dp}{dP}$$

$$\nabla p = \rho\left(i\frac{\partial P}{\partial x} + j\frac{\partial P}{\partial y} + k\frac{\partial P}{\partial z}\right) = \rho\nabla P \quad \text{..... (16.83)}$$

From vector identity

$$\text{Grad}\left(\vec{A}\cdot\vec{B}\right) = \left(\vec{B}\cdot\nabla\right)\vec{A} + \left(\vec{A}\cdot\nabla\right)\vec{B} + \vec{B}\times\text{curl }\vec{A} + \vec{A}\times\text{curl }\vec{B}$$

Putting $\vec{A} = \vec{B} = \vec{V}$, we have

$$\nabla\left(\vec{V}\cdot\vec{V}\right) = \left(\vec{V}\cdot\nabla\right)\vec{V} + \left(\vec{V}\cdot\nabla\right)\vec{V} + \vec{V}\times\text{curl }\vec{V} + \vec{V}\times\text{curl }\vec{V}$$

$$\nabla\left(\vec{V}\cdot\vec{V}\right) = 2\left[\left(\vec{V}\cdot\nabla\right)\vec{V} + \vec{V}\times\text{curl }\vec{V}\right]$$

$$\text{or } \frac{1}{2}\nabla\vec{V}^2 = \left(\vec{V}\cdot\nabla\right)\vec{V} + \vec{V}\times\left(\nabla\times\vec{V}\right) \qquad \left[\text{ curl } \vec{V}=\nabla\times\vec{V}\right]$$

$$\therefore \quad \left(\vec{V}\cdot\nabla\right)\vec{V} = \frac{1}{2}\nabla\vec{V}^2 - \vec{V}\times\left(\nabla\times\vec{V}\right) \qquad \text{..... (16.84)}$$

We know the equation of motion is given by

$$\frac{\partial\vec{V}}{\partial t} + \left(\vec{V}\cdot\nabla\right)\vec{V} = -\frac{1}{\rho}\nabla p + \vec{F} \qquad \text{(in rotation coordinate).}$$

If we assume $\vec{F} = -\nabla\Omega$, and $\overrightarrow{W} = \frac{1}{2}\nabla\times\vec{V}$

The above form of equation of motion reduces to

$$\frac{\partial\vec{V}}{\partial t} + \frac{1}{2}\nabla\vec{V}^2 - \vec{V}\times\left(\nabla\times\vec{V}\right) = -\nabla(\Omega + P)$$

$$\text{or } \frac{\partial\vec{V}}{\partial t} - \vec{V}\times\left(\nabla\times\vec{V}\right) = -\nabla\left(\Omega + P + \frac{1}{2}\vec{V}^2\right)$$

$$\text{or } \frac{\partial\vec{V}}{\partial t} - 2\vec{V}\times\overrightarrow{W} = -\nabla\left(\Omega + P + \frac{1}{2}\vec{V}^2\right)$$

$$\text{or } \frac{\partial\vec{V}}{\partial t} - 2\vec{V}\times\overrightarrow{W} + \nabla\left(\Omega + P + \frac{1}{2}\vec{V}^2\right) = 0 \qquad \text{..... (16. 85)}$$

where $\overrightarrow{W}$ = vorticity vector.

(i) If fluid motion is irrotational, $\overrightarrow{W} = 0$ and ϕ is velocity potential such that $\vec{V} = -\nabla\phi$

Then the above equation(16.85) reduces to

$$-\frac{\partial}{\partial t}(\nabla\phi) + \nabla\left(\Omega + P - \frac{1}{2}\vec{V}^2\right) = 0$$

$$\text{or } \nabla\left(\Omega + P + \frac{1}{2}\vec{V}^2 - \frac{\partial\phi}{\partial t}\right) = 0 \qquad \text{.....(16. 86)}$$

$\therefore \Omega + P + \frac{1}{2}\vec{V}^2 - \frac{\partial\phi}{\partial t} =$ constant, eq. (16.86) which may be taken as a function of time.

(ii) If the fluid motion is both irrotational and steady, then $\frac{\partial\vec{V}}{\partial t} = 0$ & $\overrightarrow{W} = 0$

The above equation (16.85) reduces to

$$\nabla\left(\Omega + P + \frac{1}{2}\vec{V}^2\right) = 0$$

$$\therefore \Omega + p + \frac{1}{2}\vec{V}^2 = \text{constant (absolute)} \qquad \text{..... (16. 87)}$$

The above equation (16.87) is called Bernoulli's Theorem.

16.9 Alternate Proof of Bernoulli's Theorem

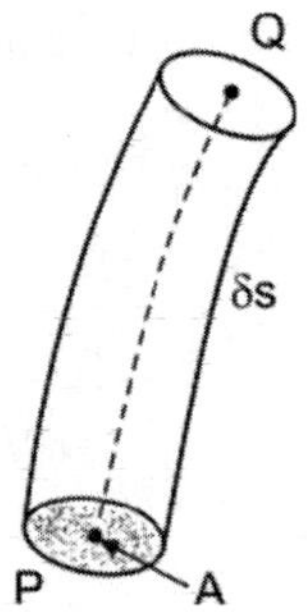

Consider the fluid motion in a small cylinder with crossectional area A along a stream line of length δs.

Let $\vec{F}$ = component of body force per unit mass acting in the direction of the stream line and let $\vec{V}$ be the velocity of the fluid.

The equation of motion (by Newton's second law)

We have

$$\rho A, \delta s \frac{d\vec{V}}{dt} = \rho A.\delta s.\vec{F} - A\frac{\partial p}{\partial s} ds$$

or $$\frac{d\vec{V}}{dt} = \vec{F} - \frac{1}{\rho}\frac{\partial p}{\partial s}$$

or $$\frac{\partial \vec{V}}{\partial t} + \vec{V}.\frac{\partial \vec{V}}{\partial s} = \vec{F} - \frac{1}{\rho}\frac{\partial p}{\partial s}$$

for steady motion $\frac{\partial \vec{V}}{\partial t} = 0$

$$\therefore \ \vec{V}.\frac{\partial \vec{V}}{\partial s} = \vec{F} - \frac{1}{\rho}\frac{\partial p}{\partial s} \qquad \text{.....(16.88)}$$

Assuming that the external body forces (are derivable from some potential function Ω), then

$$\vec{F} = -\frac{\partial \Omega}{\partial s} \qquad \text{.....(16.89)}$$

From (16.88) & (16.89) we get

$$\vec{V}.\frac{\partial \vec{V}}{\partial s} = -\frac{\partial \Omega}{\partial s} - \frac{1}{\rho}\frac{\partial p}{\partial s} \qquad \text{.....(16.90)}$$

Integrating (16.90) along a streamline

$$\int\left(\frac{1}{2}\frac{\partial\vec{V}^2}{\partial s}+\frac{\partial\Omega}{\partial s}+\frac{1}{\rho}\frac{\partial p}{\partial s}\right)ds=0$$

$\frac{1}{2}\vec{V}^2+\Omega+\int\frac{dp}{\rho}=$ constant, which depends on particular streamline

i.e., $\int\frac{dp}{\rho}+\frac{1}{2}\vec{V}^2+\Omega=$ constant, is another form of Bernoulli's equation.

The above Bernoulli equation (also called Bernoulli integral) characterizes steady flows along trajectories of its fluid particles.

Note: Eurler's equation describes stream lines, while lagranges describes pathlines or trajectories of fluid particles.

Bernoulli equation holds for an arbitrary steady flow of an incompressible stratified fluid.

16.10 The Fluid Dynamics for the Fly of a Plane

The dynamics of a plane fly is the application of Bernoulli integral or the KUTTA-JOUKOWSKI theorem.

Theorem: When aerofoil (or cylinder of any shape) is placed in a uniform stream (or wind) of speed $\vec{V}$; then the resultant thurst on the cylinder is a lift of magnitude $K\rho V$ per unit of length and is at right angles to the stream, where K is circulation around the cylinder (or aerofoil).

Proof: let the fluid density $\rho=\rho_o$ constant and approach velocity (constant) $\vec{V}=\vec{V}_o$

Let the initial moment velocity circulation

$K=\int_C\vec{V}.\vec{dr}$ over a closed fluid contour 'C' .....(16.91)

Covering the wing to its boundary (which is not zero).

The fluid contour C keeps the properties for a long period of time. The value of K will be equal to its initial value.

The Bernoulli equation $\Omega+P+\frac{1}{2}\vec{V}^2=$ constant, can be written as

$$\frac{P}{\rho}+\frac{\vec{V}^2}{2}=\text{constant (initially)}, \qquad .....(16.92)$$

since Ω scalar point function called force potential which is a measure of potential energy is constant for a level flow.

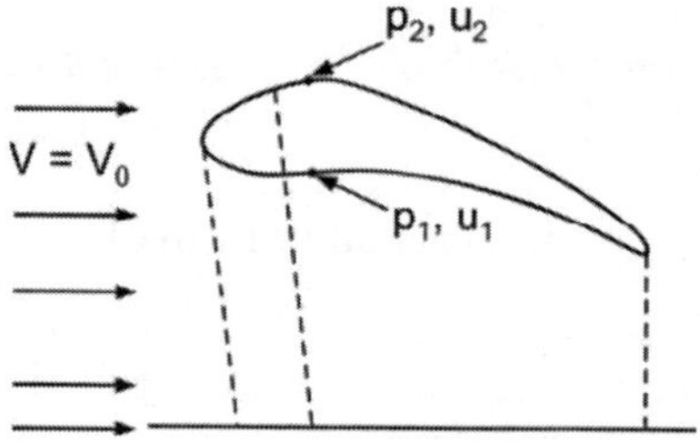

A constant flow of incompressible fluid past an aerofoil or cylinder of any shape

By applying (16.92) we have

$$p_1=-\rho_o\frac{u_1^2}{2}+\text{const}$$

$$p_2=-\rho_o\frac{u_2^2}{2}+\text{const}$$

$$p_1-p_2=\frac{1}{2}\rho_o\left(u_2^2-u_1^2\right)$$

$$p_1-p_2=\frac{1}{2}\rho_o(u_2+u_1)(u_2-u_1) \qquad ......(16.93)$$

where p_1, p_2, u_1,u_2 are pressure and velocities at two levels bottom and top edges of aerofoil as shown in fig.

Since the wing is narrow, the ratio $=\frac{\text{width}}{\text{length}}<<1$

By conservation of mass, let $\frac{u_1+u_2}{2}=u_o$, then

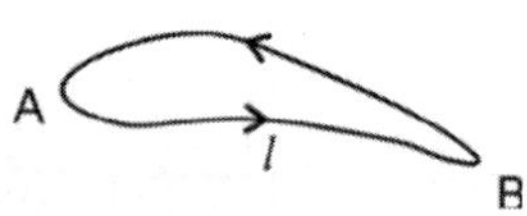

(16.93) becomes $p_1-p_2=\rho_o\ u_o(u_2-u_1)$

The vertical component of the resulting force is

$$\vec{F}=\rho_o u_o\int_0^l(u_2-u_1)\ dx \qquad .....(16.94)$$

where l = length of the wing.

The circulation K in respect of aerofoil wing becomes $K = \oint_C \vec{V}.d\vec{r} = \int_o^l u_1 dx - \int_l^o u_2 dx$

$$K = -\int_o^l (u_2 - u_1)\, dx \qquad \text{......(16.95)}$$

From (16.94) and (16.95) we have

$$\vec{F} = -\rho_o\, u_o K \qquad \text{(–ve sign is left)} \qquad \text{......(16.96)}$$

Eq.(16.96) is the Kutta-Joukowski theorem on the lifting force on the wing.

CHAPTER 17

THE VORTICITY IN OCEAN

Definition vorticity: If $\vec{V}$ be the velocity vector of a fluid particle, then the quantity

$$\vec{W} = \frac{1}{2}\nabla \times \vec{V} = \frac{1}{2}\text{curl}\vec{V}$$

is called vorticity.

17.1 The Vorticity in Ocean

The velocity rotation (angular velocity) of an infinitesimal (volume) element is the vorticity. In simple terms, vorticity is the rotation of the fluid. Vorticity is of two types. (i) Planetary vorticity and (ii) Relative vorticity.

Planetary vorticity: Everything on the earth rotates with the earth. This includes water bodies (oceans/seas), atmosphere and other fluid objects. This is called planetary vorticity (f).

Its value is twice the rotation of the earth

$$f = 2\Omega \sin\phi \text{ radians}$$

or $\quad f = 2\sin\phi$ cycles/day.

where $\quad \Omega$ = angular speed of the earth

ϕ = latitude

$$\Omega = \frac{2\pi}{86164} = 7.293\times10^{-5}\ /\ \text{sec}$$

f = coriolis parameter.

f has maximum value (2Ω) at the poles (where $\phi = 90^\circ$) and minimum value zero at the equator (where $\phi = 0^\circ$).

Relative Vorticity: The rotation of oceans and atmosphere are not the same as the rate of rotation of the earth. Oceans and atmosphere have same rotation relative to the earth.

17.2 Ocean Currents and Atmospheric Wind

Relative vorticity (ζ) is the vorticity due to currents in the ocean and winds in the atmosphere.

Ocean vorticity is given by

$$\zeta = \nabla_z \times \vec{V} = \text{curl}_z \vec{V}$$

curl_z = local vertical component of the relative vorticity $= \dfrac{\partial v}{\partial x} - \dfrac{\partial u}{\partial y}$

where $\vec{V} = iu + jv$ = horizontal velocity vector

For a rigid body rotating at Ω,

$$\text{curl}\vec{V} = 2\Omega$$

Fluid flow need not rotate as rigid body to gain relative vorticity. It may result from shear. In case of a north/south western boundary in the ocean

$$u = 0,\ v = v_{(x)} \text{ and } \zeta = \frac{\partial v_{(x)}}{\partial x},\ \ \&\ \zeta << f$$

17.3 Absolute Vorticity

The sum of the planetary and relative vorticity is called absolute vorticity.

Absolute vorticity $= f + \zeta$

17.4 Vorticity Equation

The equations of motion for frictionless flow are given by

$$\frac{du}{dt} = fv - \frac{1}{\rho}\frac{\partial p}{\partial x}$$

$$\frac{dv}{dt} = -fu - \frac{1}{\rho}\frac{\partial p}{\partial y}$$

or
$$\frac{\partial u}{\partial t} + u\frac{\partial u}{\partial x} + v\frac{\partial u}{\partial y} + w\frac{\partial u}{\partial z} = fv - \frac{1}{\rho}\frac{\partial p}{\partial x} \qquad \text{..... (17.1)}$$

$$\frac{\partial v}{\partial t}+u\frac{\partial v}{\partial x}+v\frac{\partial v}{\partial y}+w\frac{\partial v}{\partial z}=-fu-\frac{1}{\rho}\frac{\partial p}{\partial y} \qquad .....(17.2)$$

Differentiating (17.1) w.r.t y and (17.2) w.r.t x

we have,

$$\frac{\partial}{\partial x}\left(\frac{\partial v}{\partial t}+u\frac{\partial v}{\partial x}+v\frac{\partial v}{\partial y}\right)=-\frac{\partial}{\partial x}\left(\alpha\frac{\partial p}{\partial y}+fu\right) \qquad .....(17.3)$$

$$\frac{\partial}{\partial y}\left(\frac{\partial u}{\partial t}+u\frac{\partial u}{\partial x}+v\frac{\partial u}{\partial y}\right)=-\frac{\partial}{\partial y}\left(\alpha\frac{\partial p}{\partial x}-fv\right) \qquad ..... (17.4)$$

Subtracting (17.3-17.4), we have (p will be eliminated)

$$\frac{\partial}{\partial t}\left(\frac{\partial v}{\partial x}-\frac{\partial u}{\partial y}\right)+u\left(\frac{\partial^2 v}{\partial x^2}-\frac{\partial^2 u}{\partial x\partial y}\right)+\left(\frac{\partial v}{\partial x}\frac{\partial u}{\partial x}-\frac{\partial u}{\partial x}\frac{\partial u}{\partial y}\right)$$

$$+v\left(\frac{\partial^2 v}{\partial x\partial y}-\frac{\partial^2 u}{\partial y^2}\right)+\left(\frac{\partial v}{\partial y}\frac{\partial v}{\partial x}-\frac{\partial u}{\partial y}\frac{\partial v}{\partial y}\right)$$

$$=-\frac{\alpha\partial^2 p}{\partial x\partial y}+\frac{\alpha\partial^2 p}{\partial x\partial y}-f\frac{\partial u}{\partial x}-f\frac{\partial v}{\partial y}-u\frac{\partial f}{\partial x}-v\frac{\partial f}{\partial y}$$

Since $\alpha=\text{constant}$ & $\frac{\partial f}{\partial x}=0$

or $$\frac{\partial}{\partial t}(\zeta)+u\frac{\partial}{\partial x}\left(\frac{\partial v}{\partial x}-\frac{\partial u}{\partial y}\right)+\frac{\partial v}{\partial y}\left(\frac{\partial v}{\partial x}-\frac{\partial u}{\partial y}\right)+\nu\frac{\partial}{\partial y}\left(\frac{\partial v}{\partial y}-\frac{\partial u}{\partial y}\right)+\frac{\partial u}{\partial x}\left(\frac{\partial v}{\partial x}-\frac{\partial u}{\partial y}\right)$$

$$=-f\left(\frac{\partial u}{\partial x}+\frac{\partial v}{\partial y}\right)-v\beta$$

or $$\frac{\partial\zeta}{\partial t}+u\frac{\partial\zeta}{\partial x}+v\frac{\partial\zeta}{\partial y}+\frac{\partial u}{\partial x}\zeta+\frac{\partial v}{\partial y}\zeta=-f\left(\frac{\partial u}{\partial x}+\frac{\partial v}{\partial y}\right)-v\beta$$

$$\frac{\partial\zeta}{\partial t}+u\frac{\partial\zeta}{\partial x}+v\frac{\partial\zeta}{\partial y}+\zeta\left(\frac{\partial u}{\partial x}+\frac{\partial v}{\partial y}\right)+f\left(\frac{\partial u}{\partial x}+\frac{\partial v}{\partial y}\right)=-v\beta$$

$$\frac{d\zeta}{dt}+(f+\zeta)\left(\frac{\partial u}{\partial x}+\frac{\partial v}{\partial y}\right)+v\beta=0$$

or $$\frac{d\zeta}{dt}+(f+\zeta)\left(\frac{\partial u}{\partial x}+\frac{\partial v}{\partial y}\right)+v\frac{\partial f}{\partial y}+u\frac{\partial f}{\partial x}=0$$

$$\because \beta = \frac{\partial f}{\partial y} \;\&\; \frac{\partial f}{\partial x} = 0,\; \frac{\partial f}{\partial t} = 0$$

$$\frac{d\zeta}{dt} + (f+\zeta)\left(\frac{\partial u}{\partial x} + \frac{\partial v}{\partial y}\right) + \frac{\partial f}{\partial t} + u\frac{\partial f}{\partial x} + v\frac{\partial f}{\partial y} = 0$$

$$\frac{d\zeta}{dt} + (f+\zeta)\left(\frac{\partial u}{\partial x} + \frac{\partial v}{\partial y}\right) + \frac{df}{dt} = 0$$

or $$\frac{d}{dt}(f+\zeta) + (f+\zeta)\left(\frac{\partial u}{\partial x} + \frac{\partial v}{\partial y}\right) = 0 \qquad .....(17.5)$$

Eq. 17.5 is the vorticity equation for two dimensional flow.

Eq. (17.5) is Rate of change of absolute vorticity $\left[\frac{d}{dt}(f+\zeta)\right]$ + generation of vorticity on horizontal divergence $\left[(f+\xi)\left(\frac{\partial u}{\partial x} + \frac{\partial v}{\partial y}\right)\right] = 0$

From the equation of continuity for incompressible fluid, we have

$$\frac{\partial u}{\partial x} + \frac{\partial v}{\partial y} + \frac{\partial w}{\partial z} = 0 = \nabla.\vec{V} \qquad .....(17.6)$$

Integrating (17.6) in z – direction with conditions $z = 0$ to $z = h$

$$\int_0^h \left(\frac{\partial u}{\partial x} + \frac{\partial v}{\partial y}\right) dz + \int_0^h \left(\frac{\partial w}{\partial z}\right) dz = 0$$

or $$\left(\frac{\partial u}{\partial x} + \frac{\partial v}{\partial y}\right)\int_0^h dz + \int_0^h \partial w = 0$$

$$\left(\frac{\partial u}{\partial x} + \frac{\partial v}{\partial y}\right)h + \int_0^h \partial w = 0 \qquad \because\; w = \frac{dz}{dt} = \frac{dh}{dt}$$

$$\left(\frac{\partial u}{\partial x} + \frac{\partial v}{\partial y}\right)h + w\int_0^h = 0$$

$$\left(\frac{\partial u}{\partial x} + \frac{\partial v}{\partial y}\right)h = \frac{-dh}{dt}$$

or $$\frac{\partial u}{\partial x} + \frac{\partial v}{\partial y} = \frac{-1}{h}\frac{dh}{dt} \qquad .....(17.7)$$

substituting (17.7) in (17.5) we get

$$\frac{d(f+\zeta)}{dt}+(f+\zeta)\times\frac{-1}{h}\frac{dh}{dt}=0$$

or $$\frac{d(f+\zeta)}{dt}-\frac{(f+\zeta)}{h}\frac{dh}{dt}=0$$

or $$\frac{d}{dt}\left(\frac{f+\zeta}{h}\right)=0 \qquad .....(17.8)$$

or $$\frac{f+\zeta}{h}=\text{constant}=\Pi \text{ (say)}$$

The quantity $\frac{f+\zeta}{h}$ or Π is called potential vorticity

Potential vorticity is conserved along a fluid trajectory.

For Baroclimic flow in a stratified fluid, the potential vorticity can be written as

$$\Pi=\frac{f+\zeta}{\rho}\nabla\lambda$$

where λ = any conserved quantity for each fluid element

If $\lambda=\rho$, then $\Pi=\frac{f+\zeta}{\rho}\frac{\partial\rho}{\partial z}$ .....(17.9)

Assuming the horizontal gradient of density are small compared to vertical density gradients. In general in interior of the ocean $f >> \zeta$ and hence eq (17.9) can be written as

$$\Pi=\frac{f}{\rho}\frac{\partial\rho}{\partial z} \qquad .....(17.10)$$

Eq. (17.10) allows the potential vorticity of various (stratified) layers of the ocean to be determined directly from hydrographic data.

Potential vorticity is a dynamically important quantity related to ζ (relative vorticity) and f (planetary vorticity). Conservation of potential vorticity is an important concept in fluid dynamics like conservation of angular momentum in solid body mechanic.

Potential vorticity takes into account the height 'h' of water column together with local spin (vorticity).

$$\frac{f+\zeta}{h}=\text{constant}$$

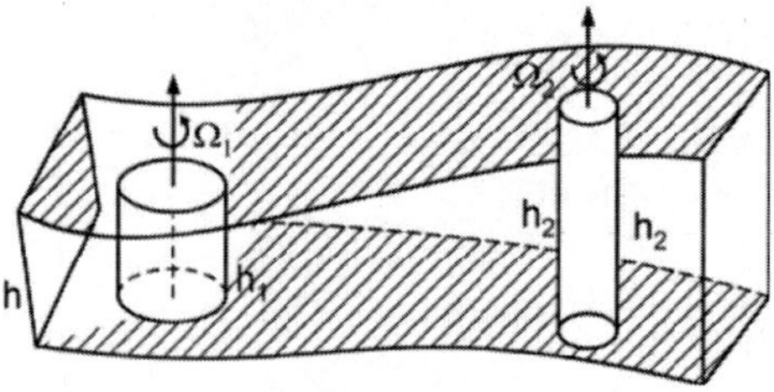

Fig. 17.1

Case (i): f = constant, h, ζ increase or decrease together. If 'h' decreases and flattened, its spin decreases. If 'h' increases, column is stretched (thinned) results in spin increase (because of conservation of m), when it is on the same latitude.

Case (ii): h = constant, if f increase, ζ decreases and vice versa. Changes in latitude needs change in ζ. As water column moves to equator ward, f decreases hence ζ must increase. If water column moves toward pole f increases, this requires ζ decreases.

In the ocean $f >> \zeta$ and $\frac{f}{h} =$ constant. This requires that flow in the ocean at constant depth to be zonal. ζ is positive for cyclonic rotation and negative for anticyclonic rotation.

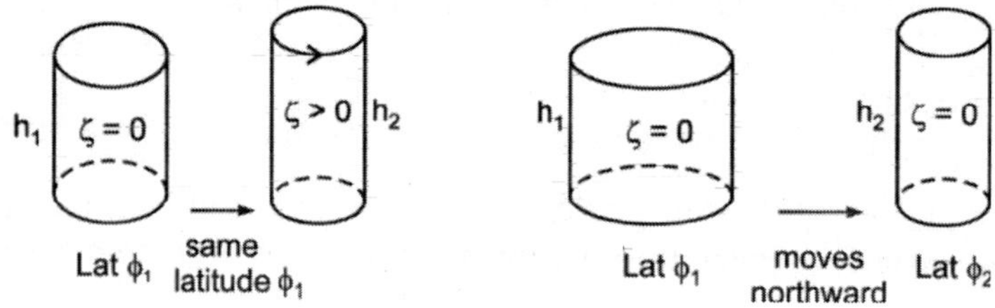

Fig. 17.2

Case (iii): ζ = constant if f increases (parcel moves towards pole), h increases and vice versa.

17.5 Helmholtz's Vorticity Equation

If the body forces are conservative and density (ρ) is a function of pressure only

$$\rho = \rho(p), \text{ then } \frac{d}{dt}\left(\frac{\vec{W}}{\rho}\right) = \left(\frac{\vec{W}}{\rho} \cdot \nabla\right)\vec{V}$$

where $\vec{W} = \frac{1}{2}\nabla \times \vec{V}$ is the vorticity vector.

Proof: The equation of motion can be written as (vector),

$$\frac{\partial \vec{V}}{\partial t}+\left(\vec{V}\cdot\nabla\right)\vec{V}=-\nabla P+\vec{F}$$

$$=-\nabla P-\nabla\Omega \qquad \text{.....(a)}$$

where $\alpha\nabla p=\nabla p$

$\vec{F}=\nabla\Omega$ scalar point function

$$\frac{d\rho}{dp}=\frac{1}{\rho}$$

Since $\nabla(\vec{V}\cdot\vec{V})=2\left(\vec{V}\cdot\nabla\right)\vec{V}+2\vec{V}\times\left(\nabla\times\vec{V}\right)$

or $$\left(\vec{V}\cdot\nabla\right)\vec{V}=\frac{1}{2}\nabla\vec{V}^2-\vec{V}\times\left(\nabla\times\vec{V}\right) \qquad \text{.....(b)}$$

(Derivation see in Bernoulli equation)

$$\frac{\partial \vec{V}}{\partial t}-\vec{V}\times\left(\nabla\times\vec{V}\right)=-\nabla\left(\Omega+P+V^2\right) \qquad \text{from(a)\&(b)} \qquad \text{.....(17.11)}$$

The equation of continuity is

$\therefore$ $$\frac{d\rho}{dt}+\rho\left(\nabla\cdot\vec{V}\right)=0 \qquad \text{.....(17.12)}$$

Taking curl of Eq. (17.11) (both sides), we get

$$\nabla\times\left[\frac{\partial \vec{V}}{\partial t}-\vec{V}\times\left(\nabla\times\vec{V}\right)\right]=0, \qquad \text{since } \nabla\times\nabla\phi=0$$

$\therefore$ $$\frac{\partial}{\partial t}\left(\nabla\times\vec{V}\right)-\nabla\times\left[\vec{V}\times\left(\nabla\times\vec{V}\right)\right]=0$$

or $$\frac{\partial \vec{w}}{\partial t}-\nabla\times\left(\vec{V}\times\overrightarrow{W}\right)=0,\ \because \vec{W}=\nabla\times\vec{V} \qquad \text{.....(17.13)}$$

But $$\nabla\times\left(\vec{V}\times\vec{W}\right)=\left(\nabla\cdot\vec{W}\right)\vec{V}-\left(\nabla\cdot\vec{V}\right)\vec{W}+\left(\vec{W}\cdot\nabla\right)\vec{V}-\left(\vec{V}\cdot\nabla\right)\vec{W}$$

or $$\nabla\times\left(\vec{V}\times\vec{W}\right)=-\left(\nabla\cdot\vec{V}\right)\vec{W}+\left(\vec{W}\cdot\nabla\right)\vec{V}-\left(\vec{V}\cdot\nabla\right)\vec{W} \qquad \text{.....(17.14)}$$

since $$\nabla\cdot\vec{W}=\frac{1}{2}\nabla\cdot\nabla\times\vec{V}=0$$

Also $$\frac{\partial \vec{W}}{\partial t}=\frac{d\vec{w}}{dt}-\left(\vec{V}\cdot\nabla\right)\vec{W} \qquad \text{.....(17.15)}$$

From eq. (17.13), (17.14) & (17.15) we get

$$\frac{d\vec{W}}{dt}+\left(\nabla\cdot\vec{V}\right)\vec{W}-\left(\vec{W}\cdot\nabla\right)\vec{V}=0$$

or $$\frac{d\vec{W}}{dt}-\frac{\vec{W}}{\rho}\frac{d\rho}{dt}-\left(\vec{W}\cdot\nabla\right)\vec{V}=0 \quad \text{using eq. (17.12)}$$

or $$\frac{1}{\rho}\frac{d\vec{W}}{dt}-\frac{\vec{W}}{\rho^2}\frac{d\rho}{dt}-\left(\frac{\vec{W}}{\rho}\cdot\nabla\right)\vec{V}=0$$

or $$\frac{d}{dt}\left(\frac{\vec{W}}{\rho}\right)=\left(\frac{\vec{W}}{\rho}\cdot\nabla\right)\vec{V}$$

which is the required result.

17.6 The Helmholtz Theorem

Theorem: Any vector field $\vec{V}$ can be represented in terms of (i) a divergent or irrotational component and (ii) a solenoidal or rotational component.

$$\vec{V}=\nabla\chi+\nabla\times\vec{\psi} \qquad \text{(A)}$$

where χ is a scalar potential (an analogous to the potential function) and $\vec{\psi}$ is a vector potential.

Divergent component $\vec{V}_d=\nabla\chi$ .....(17.16)

Posses zero verticity $\nabla\times\vec{V}_d=0$ .....(17.17)

The solenoidal component $\vec{V}_s=\nabla\times\vec{\psi}$ .....(17.18)

Possess zero divergence $\nabla\cdot\vec{V}_s=0$ .....(17.19)

For two dimensional field of horizontal

Velocity (A) reduces to

$$\vec{V}_h=\nabla\chi+k\times\nabla\vec{\psi} \qquad .....(17.20)$$

where χ is the velocity potential and ψ is stream function.

The horizontal velocity field has divergence

$$\nabla\cdot\vec{V}_h=\nabla^2\chi \qquad .....(17.21)$$

and vorticity $\nabla\times\vec{V}_h=\nabla^2\psi\, k$ .....(17.22)

The divergent component of motion $\vec{V}_d$ is orthogonal to contours of χ

The solenoidal component $\vec{V}_s$ is tangential to contours of ψ

17.7 Kelvin Waves

Ocean waves which are trapped by coast-lines (or equator) are called Kelvin waves.

Kelvin waves include gravity waves and effect of coriolis (force). The amplitude of Kelvin wave is highest at the coastline (or equator) and the wave amplitude decays exponentially with off-shore (pole ward) distance.

Kelvin waves propagate with the coast to the right in the northern hemisphere and to the left in the southern hemisphere. At the equator (which acts as a boundary) Kelvin waves propagate only eastward.

Kelvin waves behave like surface gravity waves alongshore direction of propagation (but only in one direction). Kelvin waves propagate in only me direction. Wavelengths are of the order 10 – 1000 of kilometers. Wave propagation speed is high, but however it can take days to weeks for the transformation of wave-crest to wave-trough at a given point of observation.

In the cross – shore direction Kelvin waves differ from surface gravity waves. The amplitude is largest at the crest but the off - shore decay follows Rossby wave deformation radius scale. Kelvin wave water velocities in the direction of propagation to the coast is zero, that is, water velocities are parallel to the coast. Alongshore velocities are geostrophic.

Note: surface waves or gravity waves are the waves that ultimately break on the beach. The restoring force is due to the large density difference between air and water at the sea surface.

17.8 Rossby Waves

The waves which are markedly influenced by the meridional gradient of the coriolis parameter ($f = 2\Omega \sin\phi$) is called Rossby waves.

The differential equation of the Rossby waves may be expressed as

$$\frac{\partial \nabla^2 H}{\partial t} + u\frac{\partial \nabla^2 H}{\partial x} + \beta\frac{\partial H}{\partial x} = 0$$

where H is the geopotential of an isobaric surface.

u = zonal flow in sub-tropics

x – axis denotes East

y – axis denotes North

Rossby parameter $\beta = \frac{\partial f}{\partial y} = \frac{2\Omega\cos\phi}{R}$

where ϕ = Latitude

R = radius of the earth

∇^2 = Laplace operator

$$= \nabla \cdot \nabla = \frac{\partial^2}{\partial x^2} + \frac{\partial^2}{\partial y^2}$$

$$\nabla^2 \equiv \frac{\partial^2}{\partial x^2} + \frac{\partial^2}{\partial y^2}$$

The phase velocity of simple Rossby waves is given by

$$C = u - \frac{\beta L^2}{4\Pi^2},$$

L = wave length.

17.9 Rossby Wave Properties

Pure Rossby waves or Kelvin waves never found except in simplified modes. However ocean's variability can be understood by Rossby wave properties.

Rossby waves, wave length ranges 10-1000 km, while the ocean depth is only 5-10 km and is stratified. Particle motion in Rossby waves is nearly transverse (horizontal, parallel to the earth's surface).

Restoring force for Rossby waves is the variation in coriolis parameter (f) with latitude (ϕ).

Rossby wave crests and troughs move only west wards (both in NH & SH), that is phase velocity is westwards. However group velocity of Rossby waves can be westwards or eastwards.

Rossby waves group velocity is westwards for long wavelengths (more than 50 km) and eastwards for short wavelengths. Rossby wave velocities are nearby geostrophic.

17.10 Barotropic Flow

Where levels of constant pressure in the ocean are always parallel to the constant density surfaces (isopycnals), the flow is termed Barotropic.

Example: If the density of ocean water changes (varies) with depth (vertically) but does not change horizontally, then constant pressure surfaces are always parallel to the levels of constant density (isopycnals). The relative velocity would be zero.

If the ocean is homogeneous with constant density, in this case constant pressure surfaces would always be parallel to the sea surface and the geostrophic velocity would be independent of depth (that is constant). In this case relative velocity will be zero.

The above two cases represent barotropic flow.

17.11 Baroclinic Flow

The ocean flow that is not barotrophic is called baroclinic.

When levels of constant pressure are inclined to the surfaces of constant density (of sea water) the flow is called Baroclinic flow.

In baroclinic flow, the sea water density changes (varies) both vertically and horizontally. Baroclinic flow varies with depth. If the fluid is at rest, the constant density surfaces would be parallel to the constant pressure.

In general, the variation of (ocean) flow in vertical would have two components. One Barotropic component (which is independent of depth) and another Baroclinic component (which varies with depth).

CHAPTER 18
GEOPHYSICAL HYDRODYNAMICS

18.1 The Helicity Invariant

Helicity is defined as

$$\chi = \vec{V}\cdot\vec{\Omega}$$

$$\chi = \vec{V}\cdot\left(\nabla\times\vec{V}\right) \quad \text{.....(18.1)}$$

where $\vec{\Omega} = \text{rot}\vec{V}$

$$= \nabla\times\vec{V}$$

Helicity is important for the description of Tornades, Tropical cyclones (Typhoons, Hurricanes).

According to Kelvin theorem, the vortex lines are frozen into the fluid. For a non-stationary processes, the mutual locations of vortex and stream lines (that is, structure and topology of the flow) change in time.

The value of χ surves as a measure of this local structure change. According to Kelvin theorem, the vortex lines neither they born nor disappear. If vortex lines are knotted (or linked) the total number of such linkages should not change (remain same) during the evolution, at least for an unbounded volume of fluid. Helicity equation resolves the question on the existence of an integral topological invariant.

The Euler's equation of motion in the Bernoulli form is

$$\frac{\partial\vec{V}}{\partial t} - \vec{V}\times\text{rot}\vec{V} = -\alpha\nabla p - \frac{1}{2}\nabla\left(\vec{V}^2\right)$$

or

$$\frac{\partial\vec{V}}{\partial t} = \vec{V}\times\vec{\Omega} - \alpha\nabla p - \nabla\left(\frac{\vec{V}^2}{2}\right) \quad \text{.....(18.2)}$$

(where $\vec{\Omega} = \nabla \times \vec{V} = \text{curl } \vec{V} \text{o r rot } \vec{V}$)

The vorticity equation can be put in the following form,

$$\frac{\partial \vec{\Omega}}{\partial t} = \text{rot}\left(\vec{V} \times \vec{\Omega}\right) - \text{rot}\left(\alpha \nabla p\right) \qquad .....(18.3)$$

Multiplying eq. (18.2) by $\vec{\Omega}$ and eq. (18.3) by $\vec{V}$ and adding them we obtain

$$\frac{\partial \vec{V}}{\partial t} \cdot \vec{\Omega} = \left(\vec{V} \times \vec{\Omega}\right) \cdot \vec{\Omega} - \left(\alpha \nabla p\right) \cdot \vec{\Omega} + \left[\nabla\left(\frac{\vec{V}^2}{2}\right)\right] \cdot \vec{V}$$

$$\frac{\partial \vec{\Omega}}{\partial t} \cdot \vec{V} = \left[\text{rot}\left(\vec{V} \times \vec{\Omega}\right)\right] \cdot \vec{V} - \left[\text{rot}\left(\alpha \nabla p\right)\right] \cdot \vec{V}$$

On adding

LHS : $\dfrac{\partial\left(\vec{V} \cdot \vec{\Omega}\right)}{\partial t} = \dfrac{\partial \chi}{\partial t}$

where $\chi = \vec{V} \cdot \vec{\Omega}$

$= \vec{V} \cdot \left(\nabla \times \vec{V}\right)$

R.H.S: since the term $\left(\vec{V} \times \vec{\Omega}\right) \cdot \vec{\Omega} = 0$,

then the expression $\vec{V} \cdot \text{rot}\left(\vec{V} \times \vec{\Omega}\right)$ can be written as

$$\text{div}\left[\vec{\Omega}\left(\nabla \cdot \vec{V}\right) - \vec{V}\left(\vec{V} \cdot \vec{\Omega}\right)\right]$$

i.e., $\vec{V} \cdot \text{rot}\left(\vec{V} \times \vec{\Omega}\right) = \text{div}\left[\vec{\Omega}\left(\vec{V} \cdot \vec{V}\right) - \vec{V}\left(\vec{V} \cdot \vec{\Omega}\right)\right]$

This expression arrived using the formulas

$$\text{div}\left(\vec{A} \times \vec{B}\right) = \vec{B} \cdot \text{rot } \vec{A} - \vec{A} \cdot \text{rot } \vec{B} \qquad .....(18.4)$$

$$\left(\vec{A} \times \vec{B}\right) \times \vec{C} = \vec{B}\left(\vec{A} \cdot \vec{C}\right) - \vec{C}\left(\vec{A} \cdot \vec{B}\right) \qquad .....(18.5)$$

where $\vec{A}, \vec{B}, \vec{C}$ are arbitrary sufficiently smooth vector fields.

Thus we obtain the expression

$$\vec{V} \cdot \text{rot}\left(\vec{V} \times \vec{\Omega}\right) = \text{div}\left[\vec{\Omega}\left(\vec{V} \cdot \vec{V}\right) - \vec{V}\left(\vec{V} \cdot \vec{\Omega}\right)\right] \text{v}$$

Adding RHS of the equation for the helicity evolution assume the form

$$\operatorname{div}\left[\vec{\Omega}\left(\vec{V}\cdot\vec{V}\right)-\vec{V}\cdot\chi\right]-\operatorname{div}\left[\alpha p+\left(\frac{\vec{V}\cdot\vec{V}}{2}\right)\cdot\vec{\Omega}\right]$$

$$-\vec{\Omega}\cdot\alpha\nabla p+\nabla(\alpha p)\cdot\vec{\Omega}-\vec{V}\cdot\operatorname{rot}(\alpha\nabla p) \qquad \text{.....(18.6)}$$

Further using the vector identity

$$\operatorname{rot}\left(\alpha\vec{A}\right)=\nabla\alpha\times\vec{A}+\alpha\operatorname{rot}\vec{A} \qquad \text{.....(18.7)}$$

where α is any scalar field to transform the term $\vec{V}\cdot\operatorname{rot}(\propto\nabla p)$

The RHS simplifies to:

$$-\operatorname{div}\left[\vec{V}\cdot\chi+\vec{\Omega}\cdot\left[\propto p-\left(\frac{\vec{V}\cdot\vec{V}}{2}\right)\right]\right]-\alpha^2 p\cdot\left(\vec{\Omega}\cdot\nabla\rho\right)$$

$$+\vec{V}\cdot\left[\alpha^2\nabla\rho\times\nabla p\right] \qquad \text{.....(18.8)}$$

We observe that RHS contains potential vorticity $\Pi=\vec{\Omega}\cdot\nabla\rho$ of an incompressible stratified fluid field. We now obtain the evolution equation

For helicity χ **:**

$$\frac{\partial\chi}{\partial t}=-\operatorname{div}\left[\vec{V}\cdot\chi+\vec{\Omega}\cdot\left\{\alpha p-\left(\frac{\vec{V}\cdot\vec{V}}{2}\right)\right\}\right]-\alpha^2 p\cdot\Pi+\vec{V}\cdot\left[\alpha^2\nabla\rho\times\nabla p\right]$$

Using formula $\vec{A}\cdot\left(\vec{B}\times\vec{C}\right)=\vec{C}\cdot\left(\vec{A}\times\vec{B}\right)$

We have

$$\nabla p\cdot\left(\nabla\rho\times\vec{V}\right)=\vec{V}\cdot(\nabla p\times\nabla\rho)$$

$$=-\vec{V}\cdot(\nabla\rho\times\nabla p)$$

Which eliminates the last two terms from the RHS of the expression for the helicity evolution and changes to the divergent form

$$\frac{\partial\chi}{\partial t}=-\operatorname{div}\left[\vec{V}\cdot\chi+\vec{\Omega}\cdot\left(\alpha p-\frac{\vec{V}\cdot\vec{V}}{2}\right)+\alpha^2 p\cdot\left(\nabla\rho\times\vec{V}\right)\right]-2^2\propto^2 p\Pi$$

$$\text{.....(18.9)}$$

This is the required expression that the source of helicity is the potential vorticity of the fluid. If the later is absent (that is, in the case of homogeneous medium), the RHS of (18.9) has the divergent form. In turn, this means that the total helicity

$$H=\int_V \chi\, dv$$

$$\text{Satisfies}\ \frac{dH}{dt} = -\int_V \text{div}\left[\vec{V}\cdot\chi + \vec{\Omega}\left(\alpha\cdot p - \frac{1}{2}\vec{V}^2\right)\right]d\mu$$

$$= -\oint_{\partial v}\left[\vec{V}\cdot\chi + \vec{\Omega}(\alpha_o p - \frac{1}{2}\vec{V}^2\right]d\sigma$$

where $\quad d\sigma = \frac{\delta\mu}{\delta h}\vec{n}$

is preserved if the stream lines and the vortex lines are tangents to the surface ∂V which bounds the flow domain V.

The value of H is called the helicity invariant, which characterizes the degree of knottedness or linking of the vortex lines.

18.2 Equations of an Ideal Compressible Fluid

Some definitions

An **adiabatic process** is a thermodynamic process in which no heat exchange takes place with external bodies.

A quantity is called **state variable** if the integral of that quantity around a closed path is zero.

If $\quad \oint d\phi = 0$, then ϕ is a state variable.

Entropy: The physical quantity which describes the ability of a system to do work is termed entropy of the system.

If $\quad \oint \frac{dQ}{T} = 0$ or $\oint d\phi = 0$ where $\phi = \frac{dQ}{T}$, then ϕ is called entropy.

$Q =$ quantity of heat

$T =$ temperature

$$\phi = C_p \ln\theta + \text{constant}$$

where $\quad \theta =$ potential temperature, $\theta = T\left(\frac{p}{p_o}\right)^k$,

$p_o = 1000\ hP_a$, $C_p - C_v = R$

$$k = \frac{r-1}{r},\ r = \frac{C_p}{C_v}$$

Isentropic Process: A thermodynamics process in which entropy of the system remains constant is called isentropic process.

A thermo – isolated thermodynamical system, fluid parcel is described by two independent parameters (say) pressure (p), density (ρ)) in the absence of chemical reactions.

Equations of motion can use second law of thermal dynamics, namely, the entropy of a thermo – insulated system remains constant.

Entropy of an individual ideal fluid parcel (element or particle) is a Lagrangian invariant.

An ideal incompressible fluid in a steady state is described by, the only density (physical quantity) which measures the degree of inertia of the medium. Because of the equations of motion of an incompressible fluid can be developed within the mechanical frame work. (where the pressure in the medium is taken as the constraint reaction, and vanishes when there is no relative motion).

The laws of equilibrium thermodynamics are applicable locally and to close up the equations of motion one can use second law of thermodynamics. According to which the entropy of a thermo-isolated system remains constant. This implies that the entropy of an individual parcel (particle) of an ideal compressible fluid is a lagrangian invariant.

The equations of motion can be written in the form

$$\frac{d\vec{V}}{dt} = \frac{\partial \vec{V}}{\partial t} + \left(\vec{V}\nabla\right)\vec{V} = -\alpha\nabla p \qquad .....(18.10)$$

$$\frac{d\rho}{dt} = \frac{\partial \rho}{\partial t} + \left(\vec{V}\nabla\right)\rho = -\rho \text{ div } \vec{V} = -\rho\left(\nabla \cdot \vec{V}\right) \qquad .....(18.11)$$

$$\frac{dS}{dt} = \frac{\partial S}{\partial t} + \left(\vec{V}\nabla\right)S = 0 \qquad .....(18.12)$$

$$S = S(\rho, p) \qquad .....(18.13)$$

where S is a specific entropy and is a function of density & pressure.

S = specific entropy, entropy of an individual fluid parcel of unit mass. ρ and p are functions of state of thermodynamic equilibrium system.

The equations (18.10) to (18.13) are called equations of an ideal gas dynamics.

Equation (18.12) can be written as local mass conservation law

$$\frac{\partial \rho s}{\partial t} + \text{div}\left(\rho s\vec{V}\right) \equiv \frac{d(\rho s)}{dt} + \rho s \text{ div } \vec{V} = 0 \qquad .....(18.14)$$

The Eulers equation $\dfrac{\partial \vec{V}}{\partial t} + \left(\vec{V}\nabla\right)\vec{V} = -\alpha\nabla p$ .....(18.15) be written as, [using vector identity

$$\text{grad}\left(\vec{A}\cdot\vec{B}\right)=\left(\vec{B}\cdot\nabla\right)\vec{A}+\left(\vec{A}\cdot\nabla\right)\vec{B}+\vec{B}\times\text{rot }\vec{A}+\vec{A}\times\text{rot }\vec{B}$$

put $\vec{A}=\vec{B}=\vec{V}$, then]

$$\text{grad}\left(\vec{V}\cdot\vec{V}\right)=\left(\vec{V}\cdot\nabla\right)\vec{V}+\left(\vec{V}\cdot\nabla\right)\vec{V}+\vec{V}\times\text{rot }\vec{V}+\vec{V}\times\text{rot }\vec{V}$$

i.e.,
$$\frac{1}{2}\text{grad}\left(\vec{V}\cdot\vec{V}\right)=\left(\vec{V}\cdot\nabla\right)\vec{V}+\vec{V}\times\text{rot }\vec{V}$$

Substituting this in eq (18.15) we have

$$\frac{\partial\vec{V}}{\partial t}+\frac{1}{2}\nabla\left(\vec{V}^2\right)-\vec{V}\times\text{rot}\vec{V}=-\alpha\nabla p$$

or
$$\frac{\partial\vec{V}}{\partial t}-\vec{V}\times\text{rot}\vec{V}=-\alpha\nabla p-\frac{1}{2}\nabla\left(\vec{V}^2\right) \qquad .....(18.16)$$

18.3 Isentropic Motion of a Compressible Fluid

In an isentropic motion, the entropy of the system has the same value at all points of the fluid.

The condition of entropy's constancy is:

$$S(\rho,p)=\text{constant} \qquad .....(18.17)$$

Eq. (18.17) determines density and any other thermodynamical quantity as a function of pressure only.

In a **barotropic motion** density is a function of pressure only, consequently isentropic motion, is barotropic. Conversely in an isentropic motion, the density is a function of pressure and hence barotropic.

$\rho=\rho$ (p) implies the dependence $S=S(p)$

The equation of motion,

$$\frac{dS}{dt}=\frac{\partial S}{\partial t}+\left(\vec{V}\cdot\nabla\right)S=0,\ S=S(\rho,p) \qquad .....(18.18)$$

$$\frac{\partial S}{\partial t}+\left(\vec{V}\cdot\nabla\right)S=0$$

$$\frac{dS}{dt}=\frac{dS}{dp}\frac{dp}{dt}=0$$

This implies $\frac{dS}{dp}=0$, hence $S=S(p)=\text{constant}$

Since P is not a passive scalar (physically) hence $\frac{dp}{dt} \neq 0$.

Consider the function $f(p) = \frac{1}{\rho(p)}$ .....(18.19)

its primitive $W(p) = \int f(p)dp = \int \frac{dp}{\rho}$ .....(18.20)

Differentiating (18.20) w.r.t 'p' & using (18.19)

$$\frac{dw}{dp} = \frac{1}{\rho} \quad .....(18.21)$$

$$\left(\frac{dw}{dp} = f(p) = \frac{1}{\rho} \right)$$

The Euler's equation viz.

$$\frac{d\vec{V}}{dt} = \frac{\partial \vec{V}}{\partial t} + \left(\vec{V}\nabla\right)\vec{V} = -\frac{1}{\rho}\nabla p$$

$$= -\frac{dw}{dp}\nabla p, \qquad \text{using (18.21)}$$

$$= -\frac{1}{\rho}\nabla p = -\nabla w \qquad \left(\because \nabla p \simeq dp\right)$$

$\therefore$ For an isentropic or barotropic motion of a compressible fluid, the equations of gas dynamics assume a simpler form

$$\left.\begin{aligned} &\frac{\partial \vec{V}}{\partial t} + \left(\vec{V}\nabla\right)\vec{V} = -\nabla w \\ &\frac{\partial \rho}{\partial t} + \operatorname{div}\left(\rho\vec{V}\right) = 0 \end{aligned}\right\} \quad .....(18.22)$$

and $W = w(p)$

$\therefore$ The equation of motion for a barotropic fluid is eq. (18.22), where w is a known function of pressure p.

W is called specific enthalapy or the heat function of the fluid unit mass. From definition, Enthalopy of a thermoclynamical system of unit mass is

$$dw = T ds + \alpha dp,$$

which for an isentropic process coincides with definition (of the primitive)

$$W(p) = \int f(p) dp$$

$$= \int \frac{1}{\rho}\, dp$$

$$W(p) = \frac{1}{\rho} \qquad \text{[using eq. (18.19)]}$$

Here T = temperature of the medium in Kelvins

i.e.,

$$\left.\begin{aligned} &\frac{\partial \vec{V}}{\partial t} + (\vec{V}\nabla)\vec{V} = -\nabla w \\ &\frac{\partial \rho}{\partial t} + \operatorname{div}(\rho\vec{V}) = 0 \\ &W = W(p) \end{aligned}\right\} \quad \text{.....(18.23)}$$

Eq. (18.23) are called equations of motion for barotropic fluid.

Barotropic fluid is analogus to homogeneous incompressible fluid.

Note:

1. In a barotropic motion of an ideal compressible fluid, the velocity circulation over a closed contour on an isentropic surface is constant (preserved).
2. In case of stationary flow of an ideal compressible fluid, the quantity $B = \frac{1}{2}v^2 + W(p, s_o)$ is constant (preserved) along the trajectories of fluid particles,

   where $W(p, s_o) = \int \frac{dp}{\rho(p, s_o)}$, depends on $S(\rho, p) = s_o$, where s_o, is a parameter defining the surface to which this trajectory belongs.

18.4 Ertel Invariant

Definition: Potential vorticity of the equations of gas dynamics or the Ertel invariant Π_E is defined as

$$\Pi_E = \frac{\text{rot } \vec{V} \cdot \nabla S}{\rho}$$

where the isentropic surface the value of $\delta s = \nabla s \cdot \vec{n}\, h$,

$\vec{n}$ is normal to the surface, h is cylindrical height.

Definition: Potential vorticity of an incompressible fluid Π is given by

$$\Pi = \vec{\Omega} \cdot \nabla \rho \qquad \text{where } \vec{\Omega} = \nabla \times \vec{V}.$$

18.5 The Equations of Shallow-Water Theory

Theory: Consider a two dimensional motion of a thin (vertical liquid column) layer of ideal incompressible fluid of constant density ρ, with a free surface in a gravitational field, as shown in Figure 18.1.

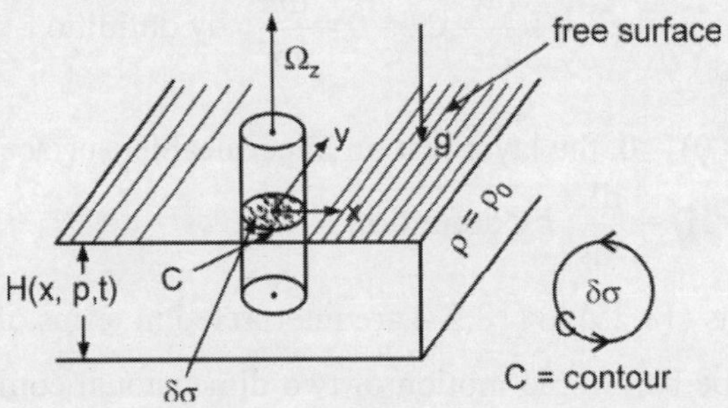

Fig. 18.1 Two dimensional motion of vertical liquid column in the field of gravity with free surface, density ρ_o = constant

Condition: Thin layer, thickness H, horizontal scale L of the flow is much larger (L > > H)

Horizontal flow in x-y plane, vertical axis z. vertical acceleration $\left(w = \frac{dz}{dt} = 0\right)$ is negligible compared to horizontal components u, v.

Pressure gradient satisfied by hydrostatic relation $\frac{\partial p}{\partial z} + g\rho_o = 0$

Integrating $p = \rho_o\, g(H - z)$ (limits z = z to Z = H) .....(18.24)

where H = H (x, y, t)

The equations of motion are

$$\frac{\partial u}{\partial t} + u\frac{\partial u}{\partial x} + v\frac{\partial u}{\partial y} = -g\frac{\partial H}{\partial x} \qquad .....(18.25)$$

$$\frac{\partial v}{\partial t} + u\frac{\partial v}{\partial x} + v\frac{\partial v}{\partial y} = -g\frac{\partial H}{\partial y} \qquad .....(18.26)$$

$$\frac{dH}{dt}+H\left(\frac{\partial u}{\partial x}+\frac{\partial v}{\partial y}\right)\equiv\frac{\partial H}{\partial t}+\frac{\partial(Hu)}{\partial x}+\frac{\partial(Hv)}{\partial y}=0 \qquad .....(18.27)$$

Here u = u (x, y, t), v = v (x, y, t) (are independent of z) are velocity components in x, y directions.

Eq. (18.27) obtained by integrating (continuity equation) in z the divergence free condition

$$\int_{w(z=0)=0}^{w(z=H)}\left(\frac{\partial u}{\partial x}+\frac{\partial v}{\partial y}+\frac{\partial w}{\partial z}\right)dz=0=\frac{dH}{dt} \text{ by definition}$$

w (z = 0) = 0, the layer rest on impermeable surface and

$w\,(z = H) = \frac{dH}{dt}$ by definition.

The equations (18.25) to (18.27) are interpreted in terms of gas dynamics, viz. they describe barotropic motion of two dimensional compressible fluid, whose density and pressure are subjected to (constrained by) the polynomial relation (polytropic gas), $p=\frac{1}{2}\alpha\, g\, \rho^2$ .....(18.28)

where α a dimensional constant.

Substituting (18.28) and $\alpha\, p = H$ in equations of motion

$$\text{Viz } \frac{d\vec{V}}{dt}=\frac{\partial\vec{V}}{\partial t}+\left(\vec{V}\nabla\right)\vec{V}=-\frac{1}{\rho}\nabla p$$

and

$$\frac{d\rho}{dt}=\frac{\partial\rho}{\partial t}+\left(\vec{V}\nabla\right)\rho=-\rho \text{ div } \vec{V}$$

we get the shallow – water equations

18.6 Alternate Method

Hydrostatic equation in equilibrium is

$$g\,\bar{\rho}(h-z)=p \qquad .....(18.29)$$

where h = height of free surface

$$h = H + h' \;\&\; p=\bar{p}+p',\; g\,\bar{\rho}\,h=p'$$

$$\text{Leads to } g\,\frac{\partial h}{\partial x}=-\frac{1}{\rho}\frac{\partial p}{\partial x} \qquad .....(18.30)$$

Using (18.30), the equations of motion (in two dimensions) may be written as

$$\left.\begin{aligned} \frac{\partial u}{\partial t} + u\frac{\partial u}{\partial x} + v\frac{\partial u}{\partial y} &= fv - g\frac{dh}{dx} \\ \frac{\partial v}{\partial t} + u\frac{\partial v}{\partial x} + v\frac{\partial v}{\partial y} + w\frac{\partial v}{\partial z} &= -fu - g\frac{dh}{dy} \end{aligned}\right\} \qquad .....(18.31)$$

Under hydrostatic assumption and $\rho = \rho_o$, constant, horizontal pressure gradient is independent of height.

Equation of continuity of incompressible fluid is

$$\frac{\partial u}{\partial x} + \frac{\partial v}{\partial y} + \frac{\partial w}{\partial z} = 0 \qquad .....(18.32)$$

Integrating (18.32) w. r. t. z, we have

$$\int_o^h \left(\frac{\partial u}{\partial x} + \frac{\partial v}{\partial y}\right) dz + \int_o^h \frac{\partial w}{\partial z}\, dz = 0$$

boundary conditions $w_o = 0$, $w_h = \dfrac{dh}{dt}$

$$\left(\frac{\partial u}{\partial x} + \frac{\partial v}{\partial y}\right) h + w_h - w_o = 0$$

$$\left(\frac{\partial u}{\partial x} + \frac{\partial v}{\partial y}\right) h = -\frac{dh}{dt} \qquad \text{[using boundary conditions]}$$

or
$$\left(\frac{\partial u}{\partial x} + \frac{\partial v}{\partial y}\right) h = -\left(\frac{\partial h}{\partial t} + u\frac{\partial h}{\partial x} + v\frac{\partial h}{\partial y}\right) \qquad .....(18.33)$$

Equations (18.31), (18.32) and (18.33) are called the shallow-water equations. These three equations have three unknows u, v and h.

18.7 The Rossby – Obukhov – Potential Vortex in Shallow-Water Theory

According to Kelvins theorem for barotropic two dimensional gas

$$\overline{k} = \zeta d\sigma \qquad .....(18.34)$$

is a Lagrangian invariant

where $\zeta = \dfrac{\partial v}{\partial x} - \dfrac{\partial u}{\partial y}$ (vertical component of vorticity $\vec{W}$)

18.7.1 Lagrangian Invariant

For scalar field ϕ, the equation

$$\frac{d\phi}{dt}=\frac{\partial\phi}{\partial t}+\left(\vec{V}\cdot\nabla\right)\phi=0$$

is called Lagrangian invariant.

In case of incompressible fluid $\rho=\rho(x,t)$ [density is a function of location x and time (t)] is constant. With these conditions, we have

$$\frac{d\rho}{dt}=\frac{\partial\rho}{\partial t}+\left(\vec{V}\cdot\nabla\right)\rho=0 \qquad \text{.....(18.35)}$$

where $\vec{V}=\vec{V}(x,t)=\frac{dx}{dt}$

[The quantities of the state of fluid namely, weight and velocity, at any point of the space occupied by the fluid as field satisfy the lagrangian invariant].

Let $d\sigma$ = an infinitesimal area of any horizontal cross – section of a liquid column on a closed contour C, which is on x – y plane.

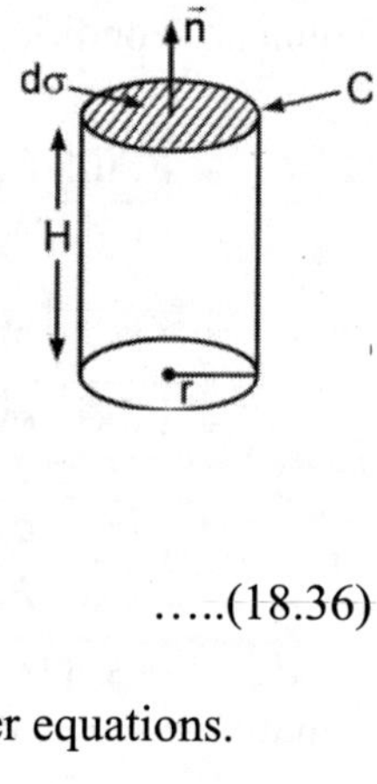

m = mass of the column

Then $d\sigma H\rho_o = m$

The lagrangian invariance of $\overline{k}$ implies the lagrangian variance of the quantity.

$$\Pi_{Ro}=\frac{\zeta}{H} \qquad \text{.....(18.36)}$$

where Π_{Ro} is called the potential vorticity for shallow-water equations.

[**Note:** The concept of potential vorticity was introduced by Rossby (in 1939) geophysical hydrodynamics for oceans, and for atmosphere by Obukhov (in 1949) independently]

$$\zeta=2\omega \qquad \text{.....(18.37)}$$

where ω = angular velocity of local rotation of the field.

Π_{Ro} in the shallow-water theory is equivalent to the conservation of angular momentum of the liquid cylinder, whose base area $d\sigma$, height = H

The momentum of inertia of the cylinder is

$$I = \frac{1}{2}mr^2 = \frac{1}{2\Pi} m\, d\sigma$$

$$= \frac{1}{2\Pi} m \frac{v}{H} = \frac{1}{2\Pi} \frac{m^2}{\rho_o} \frac{1}{H} \qquad \text{.....(18.38)}$$

where r = radius of the cylinder

V = volume = $d\sigma H$

Its angular momentum

$$M = Iw = \frac{1}{4\Pi} \frac{m^2}{\rho_o} \frac{\zeta}{H}$$

The theorem of conservation of potential vorticity provides some insight. The relation (18.38) states that areas of higher concentration above the surface (i.e., area of smaller value of H) have lower vorticity (smaller ζ) or lower concentration of the surfaces (areas of larger values of H) have higher vorticity (ζ)

$\therefore$ The turbulent spots observed both in atmosphere and in ocean and chaotically arranged at different heights or depths are formed due to irregular stratification of compressible or incompressible fluid.

18.8 The Obukhov – Charney Basis

For global geophysical flows the Rossby-Kibel number $\varepsilon = \dfrac{U}{2\Omega_o L}$ is small.

where Ω_o = angular velocity

$\vec{U}$ = velocity of fluid motion

L = geometrical scale of flow.

The principal feature of the large scale dynamics of an ideal rotating fluid is that this dynamics is almost completely determined by four fundamental properties which were applicable to synoptic scale atmospheric motions.

A.M Obukhov and J.G. Charney studied these motions and observed the following properties.

Properties

1. Geophysical flows are quasi-hydrostatic. The hydrostatic relations for them holds up to $0(\varepsilon)$:

$$\frac{\partial p}{\partial z} + g\rho = 0(\varepsilon) \qquad .....(18.39)$$

Here z is the vertical coordinate (measured in the direction opposite to the vector of the gravity acceleration)

2. The currents are in quasi - geostrophic equilibrium i.e., Coriolis force with accuracy of order ε is balanced by the horizontal 2D – gradiant of pressure (vertical component of the coriolis force is not taken into account)

$$2\vec{\Omega}_0 \times \vec{V} = -\frac{1}{\rho}\nabla p + O(\varepsilon) \qquad .....(18.40)$$

where $\nabla = i\frac{\partial}{\partial x} + j\frac{\partial}{\partial y}, \vec{V} = iu + jv$ horizontal wind vector.

3. The potential vorticity of individual fluid particles is preserved.

In the rotating reference frame it can be written as

$$\rho\Pi = \left(\vec{\Omega} + 2\vec{\Omega}_0\right).\ \text{grad}\ \theta \qquad .....(18.41)$$

where $\vec{\Omega} = \text{rot}\vec{U} = \nabla \times \vec{U}$

$$\text{grad} = \nabla + k\frac{\partial}{\partial z} = i\frac{\partial}{\partial x} + j\frac{\partial}{\partial y} + k\frac{\partial}{\partial z}$$

$$\frac{d\Pi}{dt} = 0, \quad \vec{u} = iu + jv + kw = \vec{V} + kw$$

θ = potential temperature, which plays important role of specific entropy

$$\frac{d\theta}{dt} = 0$$

$$s = C_p l_n \theta$$

4. $$\theta = T\left(\frac{p_o}{p}\right)^k \qquad .....(18.42)$$

where T= temperature (^{o}k)

P = pressure

p_o= 1000 hP_a

$$k = \frac{R}{C_p}$$

R = gas constant

C_p = specific heat capacity at constant pressure

For various reasons in geophysical dynamics, the θ of an individual parcel is used for specific entropy s. $s = C_p \ln\theta$.

The value of θ coincides with temperature T, which the fluid parcel attains under adiabatic compression to the value of surface pressure P_o. (hence the term potential temperature)

The equations (18.43) & (18.44)

$$\frac{\partial p}{\partial z} + g\rho = 0(\varepsilon) \quad .....(18.43)$$

$$2\vec{\Omega}\times\vec{V} = -\frac{1}{\rho}\nabla p + 0(\varepsilon) \quad .....(18.44)$$

Can be written as single equation

$$\vec{G} \simeq 2\vec{\Omega}_o \times \vec{U} + \frac{\text{grad } p}{\rho} - \vec{g} = 0(\varepsilon) \quad .....(18.45)$$

$\vec{G}$ can be called as adiabatic invariance, an approximate "persistence" of the zero value of the vector $\vec{G}$.

Properties (1) to (4) form a 'four dimensional basis" (see fig), which span the entire space of geophysical flows, and which can be used as basis (or a launching pad) for studying their structure and reductions for the original three dimensional equations

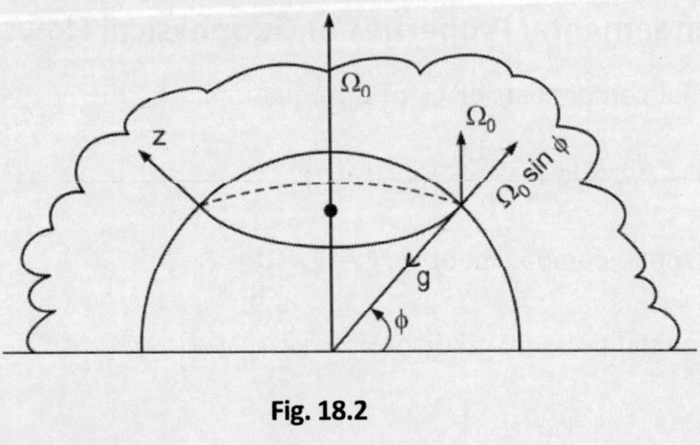

Fig. 18.2

Schematic representation of the earth's meridian section in NH, direction x-is East, y–North, Z-vertically upward and ϕ-Latitude.

18.8.1 The Concept of Geophysical Flow for a Barotropic Atmosphere

A relative motion of an inviscid rotating barotropic fluid in a gravitational field is called Geophysical, provided the following conditions are satisfied.

$$\frac{\partial}{\partial t} \leq \vec{U}\nabla, \qquad M_a \leq \frac{\vec{U}}{c} << 1 \qquad (M_a = \text{Mach number})$$

$$\epsilon \simeq \frac{\vec{U}}{2\Omega_0 L} = 0\left(\frac{\omega}{2\Omega_0}\right) << 1$$

$$\delta \simeq \frac{\Omega^2{}_0 L}{g} << 1$$

where C = speed of sound

$\vec{U}$ = speed of flow

L = geometric scale of flow

g = gravitational acceleration

δ = centrifugal forces entering the equations of motion in the same as gravitational forces having virtually no effect on the behaviour of geophysical flows (w. r. t this parameter the motions under consideration are similar).

Inviscid geophysical flows assumed to be large scale, slowly changing in time and particularly subsonic motions of rotating fluids that have small values of the Rossby – Kibal number $\varepsilon = \dfrac{U}{2\Omega_0 L}$

18.8.2 Fundamental Properties of Geophysical Flows

The horizontal components of $\vec{G}$ of the equation,

$$\vec{G} = 2\vec{\Omega}_0 \times \vec{U} + \frac{\text{gradp}}{\rho} - \vec{g} = O(\varepsilon) \qquad \text{.....(18.46)}$$

The horizontal component of $\vec{g}$, $g_x = g_y = 0$

The horizontal pressure gradient $\nabla p = i\dfrac{\partial p}{\partial x} + j\dfrac{\partial p}{\partial y}$,

$$\vec{\Omega}_0 \times \vec{U} = -\frac{\nabla p}{\rho} + o(\varepsilon)$$

where $\vec{U} = iu + jv + kw$

∴ The components of eq (18.46) are

$$-2\Omega_o v + \frac{1}{\rho}\frac{\partial p}{\partial x} = 0$$

$$+2\Omega_o u + \frac{1}{\rho}\frac{\partial p}{\partial y} = 0$$

where $\Omega_o = \Omega_{oz}$ or ζ is the vertical component of the vector Ω_o

$$\left.\begin{aligned} \therefore u &= -\frac{1}{2\Omega_o\rho}\frac{\partial p}{\partial y} + 0(\varepsilon) \\ \therefore u &= -\frac{1}{2\Omega_o\rho}\frac{\partial p}{\partial y} + 0(\varepsilon) \end{aligned}\right\} \quad \text{.....(18.47)}$$

Eq (18.47) is equivalent to

$$\vec{V}_g = \frac{1}{2\Omega_o\rho} k \times \nabla p \quad \text{.....(18.48)}$$

$\vec{V}_g$ is called geostrophic wind (which is balance between pressure gradient and coriolis force). Geostrophic wind blows parallel to the isobars, looking downstream of wind, low will be on left in NH (and to the right in SH).

18.8.3 The Criterian of Incompressibility of Rotating Fluid

We have proved $\delta\rho = \frac{1}{\left(\frac{\partial p}{\partial \rho}\right)_s}\delta p$

and $\left(\frac{\partial p}{\partial \rho}\right)_{s=so} = C^2$

$$\therefore \delta\rho = \frac{\delta p}{C^2} \quad \text{.....(18.49)}$$

For weak compressibility $\frac{\delta\rho}{\rho} << 1$

For slowly (relatively) evolving flows $\frac{\partial}{\partial t} \le \left(\vec{U}\nabla\right)$ are divergence free up to the $\left(M_a\right)^2$; where M_a = Mach number. This situation is different in a fluid rotating as a whole.

According to $2\vec{\Omega}_o \times \vec{V} = -\frac{1}{\rho}\nabla p + O(\varepsilon)$

or $$\vec{G} = 2\vec{\Omega}_o \times \vec{U} + \frac{\text{grad } p}{\rho} - \vec{g} = 0(\varepsilon)$$

The variation in pressure caused by the relative fluid motion

$$\delta p \simeq 2\Omega_o UL\rho$$

where U = wind flow velocity

L = horizontal scale of change

Neglecting pressure variations due to vertical components and since hydrostatic relation

$$\frac{\partial p}{\partial z} + g\rho = O(\varepsilon) \qquad \text{(satisfies)}$$

we have $$\frac{\delta p}{\rho} \sim \frac{2\Omega_o UL}{c^2} = \frac{2\Omega_o}{c}\frac{LU}{c} = \frac{L}{c/2\Omega_o} M_a$$

$$= \frac{L}{L_o} M_a$$

where $L_o = \dfrac{c}{2\Omega_o}$ is Rossby – Obukhov scale.

It follows that for geophysical flows, the three dimensional divergence satisfies

$$\text{Div}\vec{U} = \frac{\partial u}{\partial x} + \frac{\partial v}{\partial y} + \frac{\partial w}{\partial z} = \frac{U}{L} O\left(\frac{L}{L_o} M_a\right) \qquad \text{.....(18.50)}$$

According to eq (18.50) the scale of geophysical flows is of the O (L_o) is linear, hence a rotating fluid is less compressible to the next order of magnitude when compared to a non - rotating fluid.

18.8.4 Geophysical Flows ar1e a Quasi – Two Dimensional

If we assume the atmosphere barotropic, then

$$u = -\frac{1}{2\Omega_o\rho}\frac{\partial p}{\partial y} + o(\varepsilon)$$

$$v = +\frac{1}{2\Omega_o\rho}\frac{\partial p}{\partial x} + o(\varepsilon)$$

These equations can be written in a (single) form

$$\vec{V}=\frac{1}{2\Omega_o}k\times\nabla W+o(\varepsilon),\quad W(p)=\int\frac{dp}{\rho(p)} \qquad \text{.....(18.51)}$$

where $W(p)$ is a function of $f(p)=\dfrac{1}{\rho(p)}$, $\qquad \rho=\rho(p)$

[Using the formula,

$\operatorname{div}(\vec{A}\times\vec{B})=\vec{B}\text{ rot }\vec{A}-\vec{A}\text{ rot }\vec{B}$]

Referring to eq (18.47)

$$\frac{\partial u}{\partial x}+\frac{\partial v}{\partial y}=\frac{1}{2\Omega_0\rho}\left(\frac{-\partial^2 p}{\partial y\partial x}+\frac{\partial^2 p}{\partial x\partial y}\right)+o(\varepsilon)$$

$$\text{Div }\vec{V}=O(\varepsilon) \qquad \text{.....(18.52)}$$

Comparing eq (52) with eq (18.50) we find that

$$\frac{\partial\omega}{\partial z}=O(\varepsilon)$$

[since $\omega(z=0)=0$ on the bottom boundary of the solid atmosphere]

$$w(x,\ y,\ z)=O(\varepsilon) \qquad \text{.....(18.53)}$$

We know $\vec{G}=2\vec{\Omega}_0\times\vec{U}+\dfrac{\text{grad }p}{\rho}-\vec{g}=O(\varepsilon)$

$$-2\Omega_0\frac{\partial\vec{U}}{\partial z}+2\Omega_z\text{ Div U}=O(\varepsilon)\ \ [\text{Since }\vec{\Omega}_0=i\times o+j\times o+k\Omega_0]$$

From eq. (18.50) Div $\vec{U}=\dfrac{\partial u}{\partial x}+\dfrac{\partial v}{\partial y}+\dfrac{\partial w}{\partial z}=\dfrac{U}{L}\,O\left(\dfrac{L}{L_o}M_a\right)$

We have (from eq. (18.53)

$$\frac{\partial\vec{u}}{\partial z}=O(\varepsilon)\Rightarrow\frac{\partial\vec{v}}{\partial y}=O(\varepsilon),\frac{\partial w}{\partial z}=O(\varepsilon) \qquad \text{.....(18.54)}$$

Because $M_a<\varepsilon$ [for atmosphere $M_a\simeq\dfrac{1}{30}$, while $\varepsilon\simeq 0.1$]

The eq. (18.53) and eq (18.54) statements together called Proudman – Taylor theorem. Proudman – Taylor theorem is that the general rotation suppresses the vertical velocity of baratropic geophysical flows and the dependence of horizontal velocity components on the vertical coordinate.

Further it is noted that w(x, y, z) = O (ε) holds for baroclinic geophysical flows under certain conditions. This means that the geophysical flows are quasi – two dimensional.

18.9 Thermal Wind

For baroclinic geophysical flows consider rot (eq 18.45) $= \nabla \times \vec{G}$

Viz $$\nabla \times [2\vec{\Omega}_o \times \vec{U} + \frac{\nabla p}{\rho} - \vec{g} = O(\varepsilon)]$$

$$= -2\Omega_o \frac{\partial \vec{U}}{\partial z} + 2\Omega_o \text{Div } \vec{U} + \nabla \times \left(\frac{\nabla p}{\rho} - \vec{g} \right) = O(\varepsilon)$$

Given $$\text{Div } \vec{U} = \nabla \cdot \vec{U} = \frac{\partial u}{\partial x} + \frac{\partial v}{\partial y} + \frac{\partial w}{\partial z} = \frac{U}{L} o \left(\frac{L}{L_o} M_a \right)$$

and $$\text{rot} \left(\phi \, \vec{A} \right) = \text{ grad } \phi \times \vec{A} + \phi \text{ rot}\vec{A}$$

we have $$2\Omega_o \frac{\partial \vec{U}}{\partial z} - \frac{1}{\rho^2} (\text{grad } \rho \times \text{grad } p) = O(\varepsilon)$$

Since the vertical scale of geophysical flow is much smaller than the horizontal flow (quasi-two-dimensionality), the formulae

$$\left. \begin{aligned} &\frac{\partial p}{\partial z} + g\,\rho = O(\varepsilon), \\ &(\nabla p =) \text{grad } p = \vec{g}\rho + O(\varepsilon) \end{aligned} \right\}$$

can be written as

$$\left. \begin{aligned} \frac{\partial \vec{U}}{\partial t} &= -\frac{1}{2\Omega_o \rho} (\text{grad } \rho \times \vec{g}) + O(\varepsilon) \\ &= -\frac{1}{2\Omega_o} (\text{grad } \text{ln}\rho \times \vec{g}) + O(\varepsilon) \end{aligned} \right\} \quad \text{.....(18.55)}$$

is equivalent Cartesian coordinates

$$\frac{\partial u}{\partial z} = \frac{g}{2\Omega_o \rho} \frac{\partial \rho}{\partial y} + o(\varepsilon) = \frac{g}{2\Omega_o} \frac{\partial \text{ lu } \rho}{\partial y} + o(\varepsilon) \quad \text{.....(18.56a)}$$

$$\frac{\partial v}{\partial z} = \frac{-g}{2\Omega_o \, \rho} \frac{\partial \rho}{\partial x} + o(\varepsilon) = \frac{-g}{2\Omega_o} \frac{\partial \text{ lu } \rho}{\partial x} + o(\varepsilon) \quad \text{.....(18.56b)}$$

In modeling experiments of baroclinic geophysical flows, the distribution of density and temperature are given by $\rho = \overline{\rho} + \rho'$, $T = \overline{T} + T'$

where $\overline{\rho}, \overline{T}$ are average values independent of time and coordinates, while ρ', T' are functions of (x, y, z, t).

The deviations are related by

$$\frac{\rho'}{\bar{\rho}} = -\frac{T'}{\bar{T}}, \begin{pmatrix} \rho = \bar{\rho} + \rho' \\ T = \bar{T} + T' \end{pmatrix} \quad .....(18.57)$$

Using eq. (18.57), the equation 18.56(a), 18.56(b), component forms can be written as

$$\left.\begin{aligned} \frac{\partial u}{\partial z} &= \frac{-g}{2\Omega_0 \bar{T}} \frac{\partial T}{\partial y} + o(\varepsilon) \\ \frac{\partial v}{\partial z} &= \frac{+g}{2\Omega_0 \bar{T}} \frac{\partial T}{\partial x} + o(\varepsilon) \end{aligned}\right\} \quad ..... (18.58)$$

In the above equations the prime is omitted, since $\bar{T}$ is independent of coordinates.

These formulae (eq. 18.58) applied to the oceans. The baroclinic atmosphere formula (18.58) are replaced by the following using

θ = potential temperature for T and

$\theta_s = \bar{T}$ The standard atmospheric potential temperature.

$$\left.\begin{aligned} \frac{\partial u}{\partial z} &= \frac{-g}{2\Omega_0 \theta_s} \frac{\partial \theta}{\partial y} + o(\varepsilon) \\ \frac{\partial v}{\partial z} &= \frac{+g}{2\Omega_0 \theta_s} \frac{\partial \theta}{\partial x} + o(\varepsilon) \end{aligned}\right\} \quad ..... (18.59)$$

Eq. (18.58) and (18.59) are called thermal wind equations. These formulas are the main mechanism of external energy drive, that feeds on the general circulation of atmosphere and ocean.

Eq. (18.58) and (18.59) show the vertical wind shear that is induced by horizontal temperature gradient (generated by pole - equator temperature difference created by uneven heating of the atmosphere and earth's surface by insolation).

Notation: Two dimensional operator $\frac{d}{dt} = \frac{\partial}{\partial t} + u\frac{\partial}{\partial x} + v\frac{\partial}{\partial y}$

Three dimensional operator $\frac{D}{Dt} = \frac{\partial}{\partial t} + u\frac{\partial}{\partial x} + v\frac{\partial}{\partial y} + w\frac{\partial}{\partial z}$

$$\frac{D}{Dt} = \frac{d}{dt} + w\frac{\partial}{\partial z}$$

$$\nabla\phi = \text{grad}\phi = \left(i\frac{\partial}{\partial x} + j\frac{\partial}{\partial y} + k\frac{\partial}{\partial z}\right)\phi = i\frac{\partial\phi}{\partial x} + j\frac{\partial\phi}{\partial y} + k\frac{\partial\phi}{\partial z}$$

ϕ is scalar.

$$\nabla \cdot \vec{A} = \text{Div}\vec{A} = \frac{\partial A_x}{\partial x} + \frac{\partial A_y}{\partial y} + \frac{\partial A_z}{\partial z}, \text{ where } \vec{A} = iA_x + jA_y + kA_z$$

18.9.1 Shallow-Water Theory for Barotropic Atmosphere (Rotating Ideal Fluid of Constant Density)

If we assume Rossby – Kibel number $\varepsilon = \frac{U}{2\Omega_0 L} \to 0$, this corresponds $\vec{G} = 0$ and geostrophic relation and implies hydrostatic equilibrium.

The equation $\vec{G} = 2\vec{\Omega}_0 \times \vec{U} + \frac{\text{grad p}}{\rho} - \vec{g} = o(\varepsilon)$,

under the assumption $\varepsilon \to o$, $\vec{G} = o$ gives

$$2\left(\vec{\Omega}_0 \times \vec{U}\right) = -\frac{\text{grad P}}{\rho} + \vec{g} \qquad \text{.....(18.60)}$$

Eq. (18.60) equating is [zero approximation along with divergence property of 2D velocity field eq. (18.52)

$$\text{div } \vec{V} = o(\varepsilon) \text{ and } \frac{\partial w}{\partial z} = o(\varepsilon) \text{or } w = o$$

This describes stationary climate state of barotropic atmosphere.

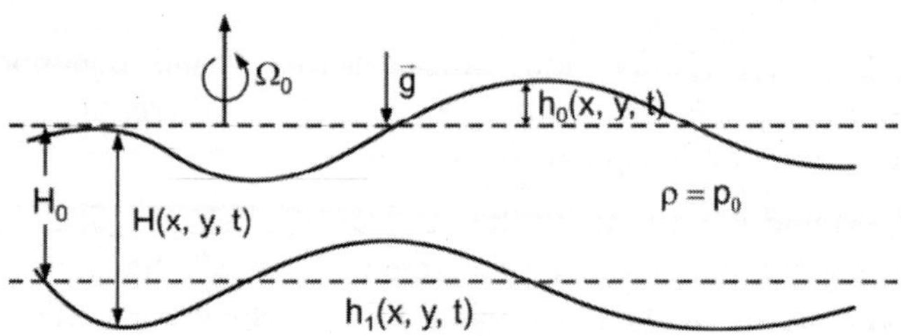

Fig. 18.3

Vector $\vec{G}$ is an adiabatic invariant that correspond to various solutions of hydrodynamical equations including climate states of the atmosphere.

The original formulation of the problem is based on three dimensional hydrodynamical equations. It was subsequently simplified by using the Obukhov – Charney basis.

For example a layer of an ideal fluid of constant density $\rho = \rho_0$ of variable depth with a free surface that rotates about vertical z-axis with constant angular velocity Ω_0 [see fig. 18.3]

A layer of fluid with constant density $\rho = \rho_o$ with free surface and uneven bottom, rotating about z-axis in a gravity field with constant angular velocity (Ω_0).

Here the bottom relief is defined by a smooth function

$z = h_l(x, y) << H_o$ of horizontal coordinates.

First: Based on $w(x, y, z) = O(\varepsilon)$ (Eq. 18.53)

and $\frac{\partial \vec{U}}{\partial z} = o(\varepsilon)$(implies) $\frac{\partial \vec{V}}{\partial z} = o(\varepsilon), \frac{\partial w}{\partial z} = o(\varepsilon)$ in the Euler equation.

$$\frac{\partial \vec{V}}{\partial t} + u\frac{\partial \vec{V}}{\partial x} + v\frac{\partial \vec{V}}{\partial y} + w\frac{\partial \vec{V}}{\partial z} + 2\vec{\Omega}_0 \times \vec{V} = -\frac{\nabla p}{\rho} \quad .....(18.61)$$

Here for horizontal velocity $\vec{V}$, we can neglect $w\frac{\partial \vec{V}}{\partial z}$ as a quantity of second order in ε, while keeping only terms of order o (1) & $o(\varepsilon)$.

Secondly: The relation $\frac{\partial p}{\partial z} + g\,\rho = o(\varepsilon)$ .....(18.61)

implies precise validity of the quasi – static condition as compared to quasi – geostrophic equilibrium condition

$$2\vec{\Omega}_0 \times \vec{V} = -\frac{\nabla p}{\rho} + o(\varepsilon) \quad .....(18.61)$$

(here $\nabla = i\frac{\partial}{\partial x} + j\frac{\partial}{\partial y}$; two dimensional)

The difference between the above two equations is the vertical velocity component.

~~In the present condition we take~~

$$p = \rho_o g\left(H(x, y, t) - z\right) \quad .....(18.62)$$

According to eq (18.62), the hydrodynamic pressure component is given by the deviation.

$h(x, y, z, t) = H(x, y, z, t) - H_o + h_l(x, y)$ of the height of the free surface from its equilibrium value as shown in the Fig. 18.3.

After substituting eq (18.62) in eq (18.61), the equation for the horizontal velocity $\vec{V} = iu + jv$ of the flow takes the following form

$$\frac{d\vec{v}}{dt} = \frac{\partial \vec{v}}{\partial t} + (\vec{v}\nabla\vec{v}) + 2\vec{\Omega}_0 \times \vec{V} = -g\nabla H(x, y, t) \quad .....(18.63)$$

Keeping the equation of continuity in three dimensional incompressibility of the medium $\frac{\partial u}{\partial x} + \frac{\partial v}{\partial y} + \frac{\partial w}{\partial z} = 0$,

assuming small terms of order $o(\varepsilon)$ as $\frac{\partial w}{\partial z}$ are not excluded.

The RHS of eq (18.63) is independent of z and Proudman – Taylor theorem viz general rotation suppresses the vertical velocity of barotropic geophysical flows, as well as the dependence of their horizontal velocity component on the vertical coordinate.

Assuming u = u (x, y, t), and v = v (x, y, t) and integrating the last equation along the layer height with w ($z = h_1$) = 0 and w (z = H) $\simeq \frac{dH}{dt}$, we can rewrite the equation of mass conservation as

$$\frac{dH}{dt} + H \text{ div } \vec{V} \equiv \frac{\partial H}{\partial t} + \text{div}\left(H\vec{V}\right) = 0 \quad .....(18.64)$$

Breaking eq. (18.63), $\frac{\partial \vec{V}}{\partial t} + \left(\vec{V}\nabla\vec{V}\right) + 2\vec{\Omega}_0 \times \vec{V} = -g\nabla H$

into two (u, v component) equations, differentiating the first w. r. t x and the second w. r. t y and taking the difference we get an equation for vorticity $\Omega_z(\zeta)$.

Then using eq. (18.64), finally we get

$$\frac{d}{dt}\left(\frac{\Omega_z + 2\Omega_o}{H}\right) = 0 \quad .....(18.65)$$

$$[\text{where } \zeta = \Omega_z = \frac{\partial v}{\partial x} - \frac{\partial u}{\partial y}]$$

Thus, the fundamental properties:

(i) Geophysical flows are quasi – hydrostatic, and the hydrostatic relation for them holdz up to $o(\varepsilon)$

$$\frac{\partial p}{\partial z} + g\,\rho = o(\varepsilon)$$

(ii) And the currents in the quasi – geostrophic equation [(i.e., the coriolis force with the accuracy of $o(\varepsilon)$ is balanced by the horizontal two dimensional pressure gradient $(\nabla_H p)$]

$$2\vec{\Omega}_o \times V = -\frac{\nabla p}{\rho} + o(\varepsilon)$$

where $\nabla = i\frac{\partial}{\partial x} + j\frac{\partial}{\partial y}$ and $\vec{V} = iu + jv$ is the horizontal wind

Above two properties of geophysical flows allows one to reduce the problem of describing three – dimensional motions of an incompressible fluid to the study of two dimensional motions of a barotropic gas with the help of equations

Eq. (18.63), $\frac{d\vec{V}}{dt} = \frac{\partial \vec{V}}{\partial t} + \left(\vec{V}\nabla\vec{V}\right) + 2\vec{\Omega}_o \times \vec{V} = -g\nabla H$

And eq. (18.64), $\frac{dH}{dt} + H \text{ div } \vec{V} \equiv \frac{\partial H}{\partial t} + \text{div}\left(H\vec{V}\right) = 0$

These equations (18.63) and (18.64) are called the equations of a rotating shallow-water.

In this regard it must be noted that, in contrast to the classical shallow-water theory, application of a shallow-water approximation of a rotating fluid is not limited by the condition $\frac{H}{L} << 1$ (where L is a horizontal scale of flow). The reason is that in this case the two dimensionality of the motion is a consequence of rotation rather than layer thinness.

18.10 The Advection Equation

A simple linear differential equation with constant coefficients is called advection equation.

Let $\frac{\partial F}{\partial t} + c\frac{\partial F}{\partial x} = 0$ .....(18.66)

where F = F (x, t), c > o

Analytical solutions to the differential equations can be compared with the corresponding difference equations, whereby numerical accuracy can be compared.

A simple wave equation is of the form

$$F(x, o) = A\ e^{i\mu x} \quad .....(18.67)$$

where $\mu = \frac{2\Pi}{L}$, initial condition at t = o

Let F (x, t) = G (t) H(x) .....(18.68)

be the separation of variables

$$\text{then} \quad \frac{\partial F}{\partial t} = G'(t)H(x)$$

$$\frac{\partial F}{\partial x} = G(t)H'(x)$$

Substituting these values in eq (18.66). We have

$$G'(t)H(x) + c\left[G(t)H'(x)\right] = 0$$

$$\text{or} \quad \frac{G'(t)}{G(t)} = -C\,\frac{H'(x)}{H(x)} = -\lambda(\text{say}) \text{ (here variables are seperable)}$$

Integrating $G = A_1\ e^{-\lambda t}$

$$H = A_2\ \rho^{\frac{\lambda x}{e}}$$

Applying initial condition $\left[F(x,o) = A\ e^{i\mu x}\right]$

We get $\lambda = i\mu c$

$$\therefore F(x,t) = A\ e^{i\mu(x-ct)} \qquad .....(18.69)$$

From the arbitrary initial condition we have F(x, o) = G (x), the general solution is G (ξ)

where $(\xi) = x - ct$

$$F(x, t) = G(x - ct) \qquad .....(18.70)$$

It follows from this, the solution of F will have same value at all coordinates (x, t)

$$x - ct = a \text{ (constant)} \qquad .....(18.71)$$

for which $t = \dfrac{x}{c} - a$, $\quad$ slope $= \dfrac{1}{c}$

and $F = G(a)$ at those points (x, t)

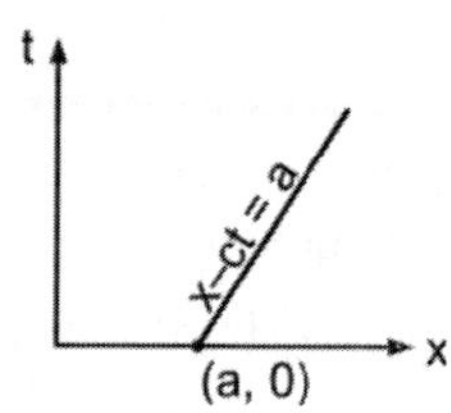

The lines eq (18.71) x – ct = a in the (x, t) plane are characteristics of the advection equation as shown in Fig.

The line x – ct = a, where t = 0, x = a is the characteristic of advection equation.

18.11 Numerical Solution

To find the numerical solution to the (advection equation) initial problem,

$$\text{Let} \quad \left.\begin{array}{ll} x = m\ \Delta x, & m = 0, \pm 1, \pm 2..... \\ t = n\ \ \Delta t, & n = 0,\ 1,\ 2..... \end{array}\right\} \qquad .....(18.72)$$

which replaces the continuous (x, t).

Space by a mesh or grid of discrete points:

Using second order finite difference approximation

$$\text{of} \quad f'(x) = \frac{f(x+\Delta x) - f(x-\Delta x)}{2\Delta x} + o\left(\Delta^2 x\right) \qquad .....(18.73)$$

for the derivation in $\dfrac{\partial F}{\partial t} + C\dfrac{\partial F}{\partial x} = o$ gives

$$\frac{F_{(m,n+1)} - F_{(m,n-1)}}{2\Delta t} + C\frac{\left[F_{(m+1,n)} - F_{(m-1,n)}\right]}{2\Delta x} = 0$$

$$\text{or} \quad F_{m,n+1} = F_{m,n-1} - C\ \frac{\Delta t}{\Delta x}\left[F_{m+1,n} - F_{m-1,n}\right] \qquad .....(18.74)$$

The above equation (18.74) is referred to a three – level scheme, because three time levels are involved. This equation (18.74) represents a simple marching procedure called the Leap-frog scheme, whereby the value of F at some point $m\ \Delta\ x$ and time $(n+1)\ \Delta\ t$ is derived from the values at previous times n & m – 1.

CHAPTER 19

THE DEEP OCEAN CIRCULATION

Incoming solar energy is transferred to the ocean through

(i) Buoyancy fluxes (i.e., heat fluxes and water vapour fluxes) and

(ii) Through the wind.

Tides create internal waves, that break and create turbulence & mixing in ocean water. Earth's rotation effect – Rotating fluids behave differently from non-rotating fluids. In non-rotating fluids pressure difference between two points in the fluid drives (moves) the fluid towards low pressure. In rotating fluids the fluid flow, there will be a balance between coriolis force and geostrophic force.

Ocean circulation has two components (parts)

(i) Wind driven,

(ii) Thermohaline (or buoyancy dominated).

Wind blowing on the ocean surface in the beginning cause small capilary waves and then a number of waves and swell in the ocean.

Impulsive wind changes cause short period (term) inertial currents and Langmuir cells. Steady slow winds create ocean frictional Ekman layer. In persistent wind momentum transfer, geostrophic circulation development takes place.

Thermohaline circulation is associated with heating and cooling, and evaporation, precipitation, run – off and sea ice formation, (all these change salinity). Therm means temperature and haline means salinity.

Thermohaline circulation is generally weak and slow as compared to wind-driven circulation. Thermohaline forcing is from local to broad scale. Broad scale buoyancy forcing results in vertical diffusion and this effects temperature and salinity structure. Weak forcing maintains ocean vertical stratification. Thermohaline effects Meridional overturning circulation (MOC).

The energy source of thermohaline circulation include the wind effect and tidal effect, that create turbulence. [Turbulence is responsible for diffusivity/upwelling across the isopycnals that stop (closes) the thermohaline overturning]. Wind driven and thermohaline circulations are in geostrophic balance.

The ocean circulation that is induced by convection (deep reaching) partly drives by the surface buoyancy loss (that is densing or increasing density) in polar latitudes is called the thermohaline circulation (see fig.19.1). Fig.19.1 shows the deep ocean is ventilated by localized convection at polar latitudes, induced by loss of buoyancy (because of intense cooling and/or addition of salt) causing surface waters to sink to great depth. Compensating upwelling occurs on large scale which is indicated by arrows at mid-depth.

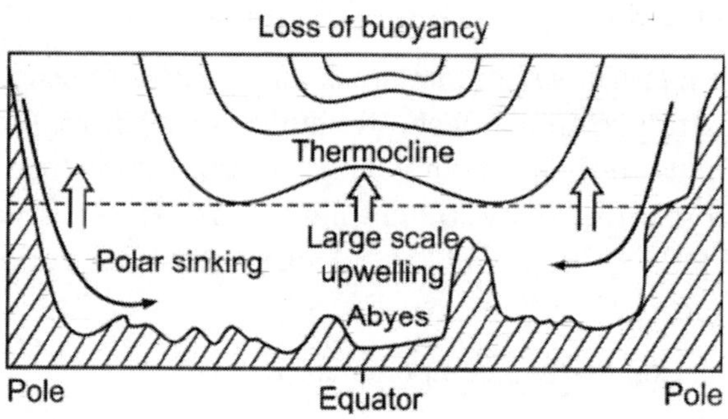

Fig. 19.1

Deep convection in the ocean is mainly localized in space and occurs only in a few key locations (northern North Atlantic Ocean and around Antarctica). This localized forcing causes on a global scale. Wind driven circulation is faster and its effect is confined to top one kilometer depth.

Thermohaline circulation plays a major role in setting properties of the Abyssal Ocean.

Both wind-driven and buoyancy driven circulations play an important role in meridional ocean heat transport, Air-sea heat and fresh water fluxes drive thermohaline circulation and influences the Abyssal flow pattern and gauged (rated) from observations of interior distribution of salinity and oxygen.

Taylor-Proudman theorem on sphere explains wind-driven circulation. This model explains the existence of deep western boundary currents carrying fluid away from their source regions. Wind-driven and thermohaline circulations have role in the meridional flow of heat and fresh water.

19.1 Heat, Freshwater and Buoyancy Fluxes

Atmospheric convection is triggered by warming at the surface. Vertical mass transport is confined to a few regions of strong updrafts [which are associated with deep convection over the warmest oceans and land masses in tropics and vast area of subsidence in between]. In contrast to the atmosphere, the ocean is forced from above by air-sea fluxes.

Buoyancy (b) has been discussed earlier and given by the formula

$$b = \frac{-g(\sigma - \sigma_o)}{\rho}$$

where

g = acceleration due to gravity

σ = density anomaly = $\rho - \rho_r$; $\rho_r = 1000$ kg/m^3

ρ = density of parcel

ρ_r = reference density

$\sigma - \sigma_o$ = difference between the density of the parcel and its surroundings

19.2 Thermohaline Circulation as the Circulation of Temperature and Salt

Three dimensional distributions and surface boundary conditions of temperature and salt are different which must separate the thermal calculation from the salt or fresh water circulation.

The meridional overturning circulation (MOC) is the zonal average of the flow plotted as a function of depth and latitude. Plots of the circulation show where vertical flow is important, but they do not show any information about how circulation in the gyres influences the flow.

Broadly, the deep circulation is the circulation of mass. Note that the mass circulation also carries heat, salt, oxygen and other properties. However the circulation of the other properties is not the same as the mass transport.

The deep circulation is mostly wind-driven. However tidal mixing is also important. The wind effects in several ways. It cools the surface water and evaporates water. This determines where deep convection takes place and produces turbulence (in the deep ocean) which mixes cold water upward.

19.3 Importance of the Deep Circulation

Deep circulation carries:

(i) Heat,
(ii) Salinity,
(iii) Oxygen,
(iv) Carbondioxide and
(v) Other properties from high latitudes (polar regions) to lower latitudes (equatorial region) in water throughout the world (global ocean water).

19.4 Important Consequences of Deep Ocean Circulation

1. The difference between the cold deep water and comparatively warm surface water determines the stratification of the ocean. Stratification effects ocean dynamics. It modulates climate.
2. The volume of deep water (V_d) is far larger than the volume of surface water (V_s). $V_d >> V_s$

   Currents in deep ocean are relatively weak but they have transports (MT_d) comparable to the surface transport (MT_s).

   $$M_d \ T_d \simeq M_s \ T_s$$

   M_d = Deep water mass

   M_s = Surface water mass

   T_a = Transport speed of deep water

   T_s = Transport speed of surface water
3. Fluxes of heat and other variables carried by the deep circulation effects earth's heat budget and climate.

   The fluxes vary from decades to centuries to millennia and this variability is considered to moderate climate over such periods of time intervals. The ocean may be the primary cause of variability over times ranging from years to decades and it might have helped to moderate ice-age climate.

   Note: If density of ocean water changes with depth (vertically) but does not change horizontally, then constant pressure surfaces will be parallel to the ocean surface and also parallel to the levels of constant density or isopycnal surfaces.
4. The deep circulation is caused by vertical mixing. Vertical mixing is largest above mid-ocean ridges (near sea mouths) and also in strong boundary currents.

Isopycnals & Diapycnals: Horizontal mixing of water (in sea/oceans) along the same density surface is called isopycnal, while vertical mixing of sea water across (or perpendicular) to two or more isopycnals (different density surfaces) is called diapycnal.

A_H (horizontal eddy viscosity) is along the isopycnals and A_v (vertical viscosity) is normal to the A_H is along diapycnals.

19.5 Diapycnal Downwelling or Buoyancy Loss Process

Sea water becomes denser due to net cooling, net evaporation and brine rejection during sea – ice formation. Convection is created by net buoyancy loss (densing) in the open ocean as a result of surface water cooling and advects mixes downward. Convection creates mixed layer, whose depth can be hundreds of meters by end of winter, whereas a wind-stirred mixed layer could be about 150 m of depth (by wind driven turbulence).

Surface water cooling and excessive evaporation are the causes of ocean convection. Deep convection stands for surface mixed layer extending to more than 1000 m. Deep convection results due to reduction in stratification (called pre-conditioning), and violent mixing (strong convection). Finally denser water sinking and spreading takes place.

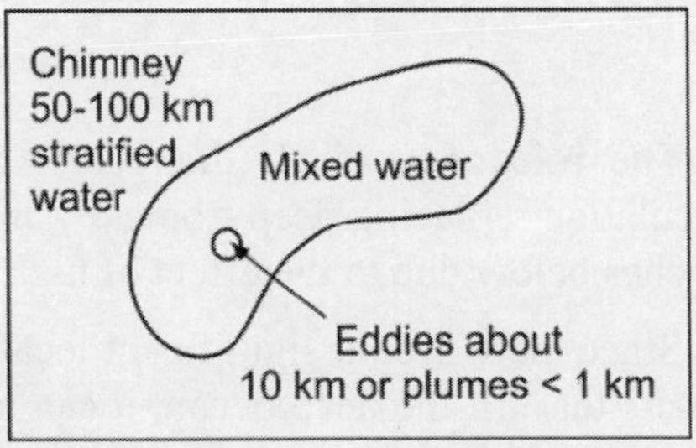

Fig. 19.2 Process model

During strong convection large heat loss takes place because of steady cold dry winds blowing from land. Re-stratification or adjustment followed by spreading leads to convection loss or diapycnal downwelling.

A model of deep convection region is shown in Fig. 19.2. The structure of convective region has a chimney. It is a patch of 10 to more than 100 km across within this column (chimney) stratification weakens, results in deep convection. It contains convective plumes which are real sites of convection (dimension less than 1 km across). The vertical structure within the convective plumes is uncertain.

Only few locations around the world are favorable for deep convection, like Greenland sea, Labrador sea, Mediterranean sea, Weddle sea, Ross sea and East sea (Japan sea). These locations barring Japan sea ventilate most of the deep global ocean waters. In southern hemisphere brine rejection process leads to denser bottom waters.

19.6 Diapycnal Upwelling or Buoyancy Gain Process

The structure of basin and global scale overturning circulations depend on the amount of density rise in the convective source regions, and the existence of heat source (a buoyancy source) in the deep ocean at low latitudes. The fig.19.3 represents the process.

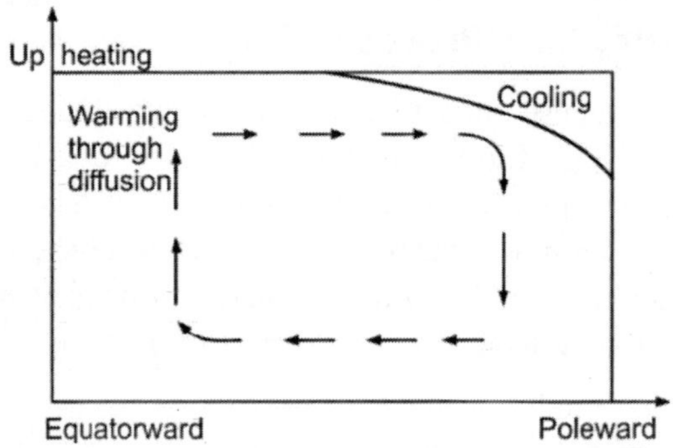

Fig. 19.3

The role of vertical (diapycnal) diffusion in meridional over turning circulation replacing deep tropical warm source with diapycnal diffusion that reaches below due to the effect of high latitude cooling.

Since there are no siginificant local deep heat sources in the world ocean waters that fill the deep ocean, it can only return to the surface as a result of diapycnal eddy diffusion of buoyance (heat and fresh water) downward from sea surface.

According to Munk, most of the ocean is dominated by the balance.

Vertical advection ≃ vertical diffusion

$$w\frac{\partial T}{\partial z} = \frac{\partial}{\partial z}\left(KV\frac{\partial T}{\partial z}\right)$$

19.7 Density Thermohaline Circulation–Discussion

Density of ocean water (ρ) is a function of temperature (t), salinity (s), evaporation (e), condensation (c), precipitation (P) & freshwater (r_{ω}) and denoted as $\rho = \rho(t, s, e, c, p, r_{\omega})$ .....(19.1)

In open ocean, changes in density is determined by

$$\rho > P \text{ or } P > \rho \qquad \text{.....(19.2)}$$

changes in density direction as that of

heating → cooling ⇒ $\rho_h < \rho_c$

or cooling to heating ⇒ $\rho_c > \rho_h$

ρ_h density on heating, ρ_c density on cooling. Circulation that result because of external factors influencing the density of surface waters temperature and salinity must be accounted and also thermohaline circulation.

19.8 Bjerknes Theorem

It states that, if a thermohaline circulation to produce energy, the expansion must take place at a greater depth than the contraction.

This theorem can be applied to determine whether (within any) given circulation energy is gained or lost because of thermal changes.

If thermal (t) and haline (s) circulations are separated, it is found in some cases they work together and in other cases they counteract each other.

(i) In equatorial region greater heating takes place, where $P > e$ and density decreases by reduction of salinity.

(ii) In subtropical anticyclones heating is less (or cooling is more), density of sea water increases because $e > P$.

In between (i) & (ii) (between equatorial region and sub-tropical anticyclones), conditions are favourable for development of a strong themohaline circulation.

North and South of these latitudes, the haline circulation will counteract the thermal (because decrease in 'e' is directed by excess precipitation, simultaneously density increased by cooling). A weak thermohaline circulation might be expected.

In the absence of wind system (forcing), we expect at surface a slow thermohaline circulation directed from equator to poles (in opposite direction). This circulation would be modified by the rotation of the earth (coriolis force) and by the shape (form of boundaries) of the ocean basins.

In other words, the wind system tends to bring about a distribution of density, that is inconsistent with the effect of heating and cooling and the actual distribution tend to balance between the two forces. These two factors, namely the wind and the process of heating & cooling are variable both in time and space. Because of this a stationary distribution of density with accompanying stationary currents does not exist. Only average conditions over a long time and over a large area considered, they can be regarded as stationary.

19.8.1 Vertical Convection Currents

In horizontal flow, the thermohaline circulation is of little significance. However the thermohaline circulation is of great importance for the vertical convection.

19.9 Convection Currents

When surface water density increases due to cooling and evaporation and when this density becomes greater (heavier) than the density of the underlying water strata, the surface water sinks and the sinking water is replaced by the water from some sub-surface depth. The vertical currents that develop in this way are called convection currents. The depth to which vertical convection currents descend (penetrate) depends on the stratification of the water.

$$\rho = \text{cold surface water density}$$
$$\rho' = \text{strata density}$$
$$\rho > \rho'$$

The homogeneity of upper layer develops when vertical convection currents are active for some time. The thickness of homogeneous upper layer depends upon

(i) The original stratification of water

(ii) The intensity of convection currents and

(iii) The time of the process lasted.

Homogeneous layer is formed in two ways.

(i) By mechanical stirring due to wind or

(ii) By the effect of the thermohaline vertical convection.

The vertical convection currents are of great importance in higher latitudes, where excessive evaporation is found. Heating of the surface layer is so great that the decrease of density (by heating) more than balances the increase by evaporation. As a result surface salinity will be greater than the salinity at a short depth below the surface.

Density (ρ) is a function of temperature (t) and salinity (s) and is denoted $\rho = \rho(t,s)$. Homogeneous water layer surface density (of water) decreases by

(i) Heating,

(ii) Precipitation

(iii) Water melt of ice

(iv) Runoff water from land.

Surface water density increases by:

(i) Cooling

(ii) Evaporation

(iii) Ice formation.

If the density of surface water is increased beyond that of the underlying water strata vertical convection currents develop, which lead to the formation homogeneous water layer.

19.10 The Bjerknes Hypothesis

As noted in ELNino – Southern Oscillation (ENSO), studies, Bjerknes first hypothised (in 1969) that interaction between the ocean and atmosphere was essential to ENSO.

During 1980 and 1990s many investigators contributed to making this conjecture a full fledged theory. Modestly on the part of these scientists, "The Bjerknes' hypothesis" is used as the name for this theory. In essence ENSO developed as a coupled cycle in which anomalies of SST in the Pacific cause the trade winds to strengthen or weaken, and this in turn drives the ocean circulation changes that produce anomalous SST.

19.11 Equatorial Ocean Process

The chief feature of Tropical Ocean is a thin permanent lens of shallow layer of warm water lying over deep colder water.

NE – trade winds in the NH and SE – trade winds in the SH meet near equator (ITCZ). The SE – trade winds blow along equator and are strongest in the east.

North of the equator, Ekman transport is northward and south of the equator, Ekman transport is southward. The divergence of Ekman flow creates upwelling on the equator. In the west, the upwelled water is warm and in the east, the upwelled water is cold (because of shallow thermocline).

North-South mixing with warm waters observed on both sides of the equator and heat fluxes through the sea surface along the equator (ITCZ). The east-west temperature gradient on the equator drives the zonal atmospheric circulation (also called Walker circulation).

Thunderstorm activity over warm pool of sea water lifts air upward (in west) and associated sinking (subsidence) of air in the east. This feedback leads to ELNino – Southern Oscillation.

In summary, heat released by rain in the equatorial region drives the atmospheric circulation, equatorial currents help redistribute the heat equator to poleward. This modulates the oceanic forcing of the atmosphere. During ELNino, trade winds weaken in the western Pacific and makes shallow thermocline. This causes Kelvin waves move eastward along the equator and deepens the thermocline in the eastern Pacific.

The following figures a, b, c represent the pacific ocean walker circulation, ELNino circulation and LaNina circulation

D = Divergence

H = High

C = Convergence

L = Low

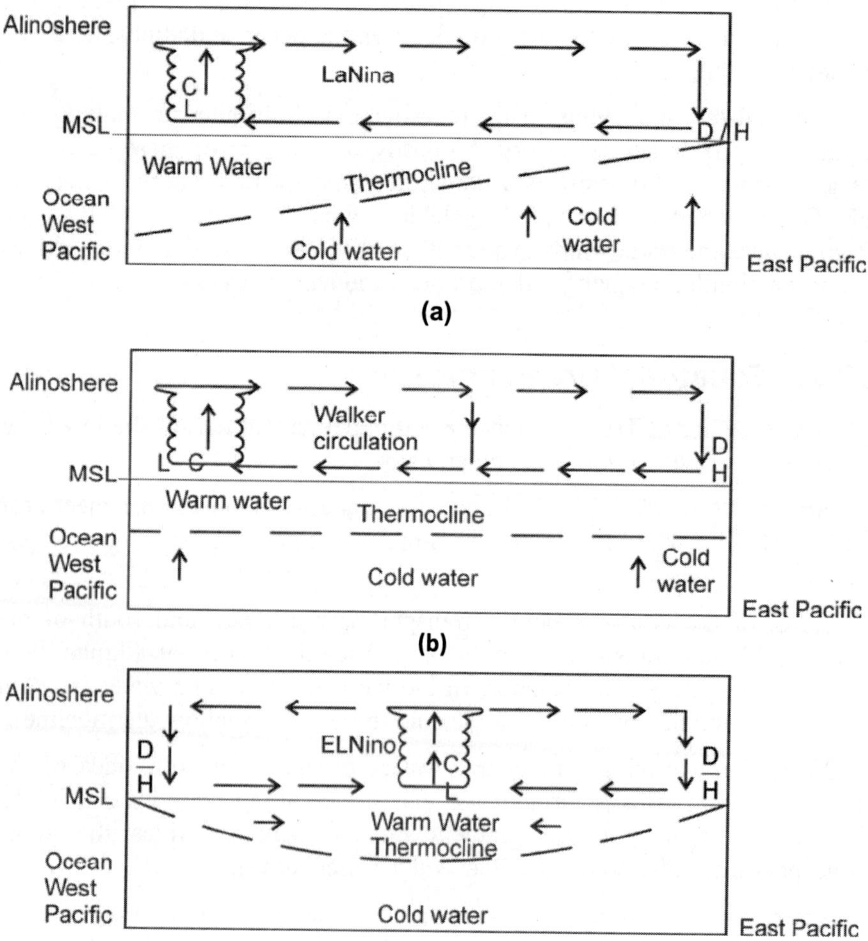

(a)

(b)

(c)

Fig. 19.4

CHAPTER 20

THE ROLE OF OCEAN IN CLIMATE FLUCTUATION

According to IPCC, the heat content of the entire global system has increased since 1950s. The oceans have absorbed 90% of the heat increase. The heat content of upper ocean (0 – 700 m slab) increased from 1961 to 2003 was about 16×10^{22} J. This heat caused average increase in sea temperature by about 0.1 °C in the upper 700 m slab. The SST changes (1961-2003) by about 0.4 °C at the surface and 0.1 °C rise in temperature in the upper ocean depth 700 m. This heat amount would be equivalent to an almost 100 °C change (rise) in atmospheric temperature. The oceans have much greater capacity to store heat than the atmosphere. The ocean is one thousand times more dense than air and the specific heat of ocean water is about four times that of air. The thermal adjustment of the atmosphere alone is about a month. The thermal adjustment time scales in the ocean are very much longer. The oceans have a much greater capacity to store heat than the atmosphere. This explains the buffering of atmospheric temperature by the oceans enormous heat capacity. (Buffer meaning protection from damaging impact).

The time scale for adjustment of the deep ocean is 1000 years, because the slow circulation of the abyssal ocean limits (obstructs) rate at which its heat can be brought to the surface. These long time scales buffer atmospheric temperature changes, consequently ocean would play a very important role in climate. Oceans reduce the amplitude of seasonal extremes of temperature and buffering atmospheric climate changes.

The world's oceans occupy about 71% of the surface area while the land occupies only about 29% area. The following properties about atmosphere and oceans must be noted.

The mass of the atmosphere $\simeq 5.26 \times 10^{18}$ kg

Global mean surface temperature $\simeq$ 288 °k

Global mean atmospheric pressure $\simeq 1.013\times10^5 P_a = 1013$ hpₐ

Global mean atmospheric density $\simeq$ 1.225 kg/m^3

Global ocean surface area 3.61 $\times10^{14}$m^2

Global ocean mean depth 3.7 km

Global ocean volume 3.2 $\times10^{17}$m^3

Global ocean mean density 1.035$\times10^3$ kg/m^3

Global ocean mass 1.3 $\times$ 10^{21} kg

Specific heat of water 4.18 $\times$ 10^3 J/kg/°K

Latent heat of fusion (of water) 3.33 $\times10^5$J/kg

Latent heat of evaporation (of water) 2.25 $\times10^6$ J/kg

Density of fresh water 0.999 $\times10^3$ kg/m^3

Viscosity of sea water 10^{-3} kg/m/s

Specific heat of ocean is about 4 times that of (atmospheric) air.

Process	**Time Scales**
Weather	hrs/days to weeks.
Land surface	hrs/days to months.
Ocean mixed layer	hrs/days to months.
Sea ice	weeks/months to years
Volcanoes	weeks/months to years
Vegetation	hrs/days to millions of years.
Thermocline	yrs/decades to centuries
Mountain glaciers	decades to centuries
Deep ocean processes	centuries (10^2) to 10^4 years
Ice sheets	10^2 to 10^5 years
Orbital forcing	10^3 to 10^5 years
Tectonics	10^6 to 10^9 years
Weathering	10^5 to 10^9 years
Solar constant	$10 - 10^2 - 10^3$ to 10^9 yrs 1-10^9 years
Natural CO_2 cycle	1k to 10k y (k = thousand)
Anthropogenic CO_2	decades to centuries (10 – 10k years)

20.1 Worlds Land Utilization (1990)

Arable land area about 10% worlds land surface (WLS).

Meadows and pasture area about 20% of WLS

Forest area is about 27% of WLS (4035 million hectares)

Deserts area is about 20% of WLS. They occur mostly in high air pressure belts between lat 10° and 35° N/S. Cryosphere or Ice cover area is about 10% of WLS. It was occupied about 30% WLS during the peak of Pleistocene ice-age. If all ice melted the sea level would rise about 60 – 90 meters (200 – 300 ft) flooding many densely populated low lands, and great cities.

20.2 Role Factors (Importance) of Ocean in Climate Variability

Ocean is a pivotal component of the climate system because heat, water, momentum, GHG and many other substances cross the sea surface. Ocean releases heat and water vapour to the atmosphere. More than 80% of water vapour of the atmosphere comes from the oceans. Ocean currents transport heat and salt around the globe. Buffering of atmospheric temperature changes is by oceans huge heat capacity. An Ocean current has immense effect on climate. Example, the confluence of two main ocean currents, the Warm Gulf Stream and the cold Labrador Current over north & west of New founded land. The main fluxes or forces that determine ocean currents are – prevailing wind, earth's rotation and buoyancy (or salinity differences). Ocean currents generally called drifts.

Large – scale circulation is setup whereby the warm sea affects western coasts and cold seas affect eastern coasts.

Cold water coasts have low rate of evaporation hence feeds low water vapour content. This factor aids desertification aridity.

Warm water coasts have high humidity which favours thunderstorm activity and high precipitation.

Important Ocean Currents: Cold water current Oyashio current (that affects Japan), Warm Kuroshio current, cold Labrador current, warm Gulf Stream and Cold Falkland current and Antarctic circumpolar (or west wind drift).

Almost all dynamical process in the ocean are driven by the sun and the atmospheric winds drive the ocean surface circulation up to a depth of top one kilometer, while winds and tidal mixing drive the deeper currents in the ocean.

The principal ocean's sources and sinks of energy are insulation, evaporation; I R emission from sea surface and sensible heating of the sea by warm and cold winds blowing over it.

Similarly, the heat sources and sinks of energy for atmosphere – insulation, uneven heat loss & gain by ocean causes winds, evaporation at sea surface and then condensation.

Regional climate is strongly influenced by thermal properties of the earth's surface and neighbouring ocean. In general ocean moderates extreme conditions. Similar influence exerted on global climate, because of ocean's large heat capacity and its capacity to hold substances in solution. The ocean surves as a reservoir of energy and carbon. As a consequence it (ocean) gives thermal inertia to the climate system mainly in exchanges with the atmospheric heat and CO_2.

Ocean circulations are driven by atmospheric wind stress, which transfers momentum to the ocean. Ocean circulations are also driven by transfer of heat and moisture (by way of density which influence buoyancy of water). The poleward transfer of heat achieved by the general circulation of earth – atmosphere system, of this 60% is contributed by the atmospheric circulation and 40% is contributed ocean circulation.

The spacial distribution of the ocean's heat content and SST (sea surface temperature) trend over the last three decades is not uniform. Climate change models predict non-uniform changes when projected over the next century (2100). It predict greatest warming in Arctic, little change in the sub-polar North Atlantic and Antarctic circumpolar current

20.3 Present and Past Radioactivity of the Earth

Radioactivity is the most important property of our earth. Modern Radioactivity of the earth is mainly associated with radioactive isotopes $\overset{238}{U}, \overset{235}{U}, \overset{232}{Th}$ and $\overset{40}{K}$ which decays as below

$$\overset{238}{U} \rightarrow \overset{206}{p_b} + 8\alpha, \qquad \overset{235}{U} \rightarrow \overset{207}{P_b} + 7\alpha, \qquad \overset{232}{Th} \rightarrow \overset{207}{P_b} + 6\alpha,$$

$$\overset{40}{K} + e \left\langle \begin{array}{l} \nearrow \overset{40}{C_a} + \beta \\ \searrow \overset{40}{A_r} \end{array} \right.$$

During radioactive decay heat energy is liberated. Models of the radioactive earth generates (2.3 to 10) × 10^{20} calories of Radiogenic heat per year. Modern geothermal evidence indicate that by thermal conductivity the earth loses $(1.9 \pm 0.1)\ 10^{20}$ cal of heat annually, which is less than the amount of heat produced by the radioactive model of the earth.

4.5 billion years ago there was twice more ^{238}U on earth than now, which released twice more energy. In addition, some radioactive elements which were present initially are now extinct (some transuranium elememts).

The radioactivity of earth is an important source of its internal heat and can cause melting of material in the earth's interior. Relatively high radioactivity of the young Earth contributed to the rise of its temperature and melting of material. The isotopes of the most long lived radio-active elements including the transuranium onces $\left(^{244}P_u, {}^{247}C_m, z=112-116\right)$ existed for sometime in the early history of the earth.

Temperature records of past few centuries show that the average temperature of the earth remained nearly constant at about 15 °C. This implies that the earth is radiating back (emitting) the same amount of energy as it is receiving from the sun. Hence the earth is in radiative balance with its surroundings.

The Paleo (fossil) records show that the state of radioactive balance as at present was not existing in earlier eras. In the geological history of the earth, there had been warm and cold epochs or ice-ages. This indicate that there were periods in which solar radiation received was above normal during warm epochs and below normal during cold epochs. These warm and cold epochs tell us the important of radioactive balance of the earth for life on earth. The most recent major glaciation, called the Great Ice Age, lasted for a period of between 6×10^5 and 10^6 years until about 2×10^4 years ago.

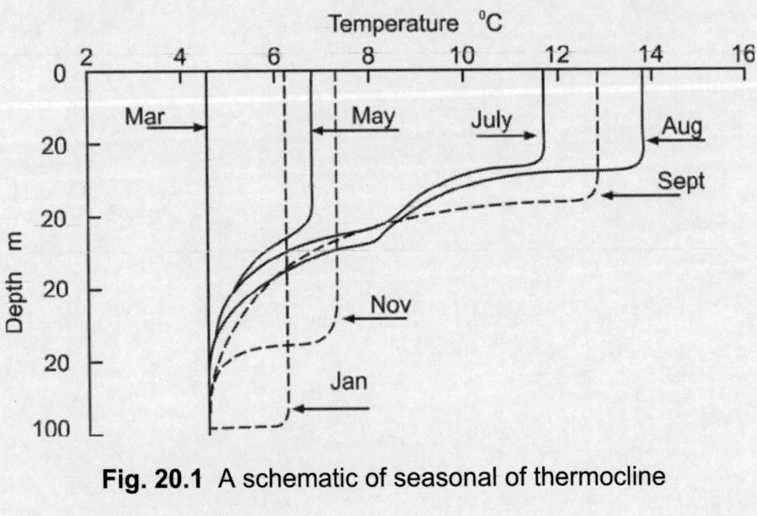

Fig. 20.1 A schematic of seasonal of thermocline

CHAPTER 21
THE SUN

The Sun

The central core of the sun has temperature 15 – 20 million °K. Above this convective core there is a radiative equilibrium layer with temperature ranging 10 million °K to 6000 °K. The convective core and radiative equilibrium region is characterized by nuclear reactions, which is the source region of solar radiative energy.

Above this lies the visible surface of the sun called photosphere. Its temperature is about 6000 °K . It consists of hot gases under high pressure. Photosphere emits radiations of the visible part of the solar spectrum, but the energy is produced in the core region & neighbouring region. Photosphere shows bright regions called faculae and dark regions called sun spots. Sun spots are regions of strong magnetic field.

Solar Activity: when the frequency of observable features of the sun (sun spots, solar wind, solar flares) are more than normal, the sun is said to be Active or disturbed, but when they are absent the sun is said to be "Quite." The solar activity has about 11 year cycle. The magnetic field in sunspots observed to exceed thousand times as compared to the magnetic field outside the sunspots. The polarity of the magnetic field in sunspot region (area) changes in 22 years, which is twice the period of sunspot cycle. No sunspots observed during 1645 to 1715 (70 years), and there does not appear 11 year sunspot cycle before 1645. These facts were brought to the light by

E. Maunder (superintendent of greenwich observatory) in 1890. The period 1645 – 1715 of sunspot minima or complete absence of it is called Maunder minima. There appears a relation between sunspot cycles and climate change. The solar flares are the result of gigantic nuclear explosions in the central core. There may be hundred odd flares daily but a few large flares may occur in a year. The large flares emit electromagnetic radiation and high energy charged particles of plasma, which interact with atmosphere and earth's magnetic field, cause ionospheric storm (high frequency radio blackout) and magnetic disturbances. The impact of solar flares on climatic change is not clearly understood.

The sun's hot gases produce electrically charged particles, which produce electromagnetic waves. These waves constitute electromagnetic spectrum of the sun, which extends from gamma rays. ($\lambda = 10^{-13}$ m) to radio waves ($\lambda = 10^{3}$ m). The total range of λ (wave length) is 10^{16} m.

About 99% of the solar radiation lies between ($\lambda = 0.15$ μm to 4.0 μm). About 9% of this lies in UV-radiation ($\lambda = 0.15$ μm to 0.38 μm). This causes photochemical effects, bleaching and sunburn. About 44% of this lies in visible portion, $\lambda = 0.38$ μm to 0.7 μm (i.e., violet to red). About 46% of this lies in IR region, $\lambda = 0.7$ μm to 2.3 μm. This causes radiant heat with some chemical effects. The sun emits maximum radiation energy at wave length $\lambda_m = 0.5$ μm, while the earth emits maximum radiant energy at wavelength $\lambda = 10$ μm (in IR).

Solar Constant: The amount of solar radiation (energy) incident on unit area in unit time on a surface held at right angle to the solar beam at the outer boundary of the atmosphere is called solar constant and is given by S = 2.0 ly/min or 1359 W/m^2, where 1 ly (one langely) = 1 cal/cm^2.

The energy received from the sun is not remained constant over the 4500 million years of the earth's existence. The sun is growing slowly hotter over the millennia and emitted more and more energy. It is now emits about 30% more heat than it did when the earth was first formed. With all these variations, during the present millennia the temperature of the earth's surface has remained remarkably stable with all interactions of plants, animals and by changes in the heat absorbing capacity of the earth's surface – the extent of its water, lands, vegetation and ice.

The fluctuations in composition of the atmosphere and changes in temperature over the past 200,000 years can be determined by analysis of tiny air-bubbles trapped in ice that has accreted year by year in the Antarctic and Arctic. The cores drilled in ice caps, a record of changes can be measured and deducted. The Vostok core in Antarctica (in Antarctic station Vostok) by USSR, the CO_2 concentration and temperature were estimated for the past 160000 years. The analysis shows CO_2 fluctuated from about 180 to 220 ppm in the atmosphere and temperature rose and fell over the range of 9 to 10 °C, in more or less perfect rhythem with the rise and fall of CO_2 concentrations (or GHG concentrations).

According to prof. George Voitkevich, the mobile gas components of the atmosphere and water experience the fastest cycle, while a very slow cycle in the continental matter. The time required for a full cycle of matter is 7 years for atmospheric carbondioxide, 4000 years through photosynthesis for atmospheric oxygen; nearly one million years for ocean water by way of evaporation; and 80-100 million years for continental matter by way of weathering and removal from land surfaces.

According to WMO, the life time of GHGs in atmosphere vary from hours/weeks to more than 100 years; tropheric ozone has a life period a few hours/ days; CFCs have about 75 to 110 years; N_2O (nitrous oxide)

150 years; CH_4 (methane) 7-10 years, CO_2 (carbondioxide) 50 – 200 years.

CHAPTER 22

THE EARTH-ATMOSPHERIC SYSTEM

Annual Carbon Fluxes

According WMO No. 735 (1990) the atmosphere of the living planet earth the annual carbon fluxes were in 20th Century end.

(i) Photosynthesis on land removes about 100 Gt of carbon (in the form of CO_2) from the atmosphere annually.

(ii) Plant and soil respiration each returns (C) about 50 Gt (total 100 Gt).

(iii) Fossil fuels burning (5 Gt) and deforestation (2 Gt) release (C) into atmosphere 7 Gt.

(iv) Physico-chemical process at sea surface release (C) into atmosphere about 100 Gt.

(v) Physicochemical process at sea absorb (C) about 104 Gt.

Net carbon added to the atmosphere 3 Gt annually $(-100 + 100 + 7 + 100 - 104 = 3)$ in the form of CO_2 $1 \text{Gt} = 10^{12}$kg.

Evolution of Atmosphere and Life on Earth

The primordial substance of the solar system consisted mainly water and carbondioxide (CO_2).

The mass of the earth is about 6×10^{24} kg,

The mass of the atmosphere is about 5.6×10^{18} kg.

The mass of ocean water is about 1.4×10^{21} kg.

That is the mass of ocean is more than 250 times the mass of atmosphere.

The mass of the hydrosphere is about 0.023% of the mass of the earth, while the mass of the atmosphere is about 0.00009% of the mass of the earth. The mass of the O_2 (oxygen) in the atmosphere is about 10^{18} kg, that is one-fifth of the mass of atmosphere.

The plant–kingdom over the globe produces O_2 about 3×10^6 kg and it provides about 10^{17} kg of biomass annually. An average size of tree supplies about 3500 kg of O_2 per year which is sufficient for three people.

The present atmosphere of the earth resulted from the evolution of life on earth. All living organisms are primarily composed of C, O_2, H and N. These are also the basic chemical elements of water and air shells of the earth. The biosphere constitutes about 1440×10^{15} tons of water, 233×10^{10} tons of CO_2 and 11.8×10^{14} tons of O_2.

In summary, at the beginning 4500 million years ago, the surface of the earth was depleted in free oxygen, and UV- solar radiation could have penetrated into the top ocean surface to a depth of about 10 m. However the illuminated top layers of sea water were favourable for development of living matter. This living matter gave rise to uni-cellular photosynthesizing Blue- green algae or their ancestors might have evolved in the zone depth (below 10 m) in ocean where they were protected from the lethal UV- solar radiation, but visible radiation reached into the sea. These ancient life developed at constant density zone, which was responsible for water decomposition to form free oxygen. As a result the biosphere became oxidizing in nature. In this state all CO_2 from the atmosphere used by photosynthesis and carbonatiz. Free oxygen released into atmosphere. The free oxygen in the atmosphere formed an ozone layer, which absorbed the UV-solar radiation. This helped algae spread over the land & sea. Subsequently animals developed as superstructure relative to the photosynthesis and then developed species of animals which consumed plants for food. These species breathed in oxygen and breathed out CO_2. With oxygen fixation in the pigments blood and thus animals spread over the land side by side plants. It must be noted, initially animals have long evolved in marine water in the zone of sea saturated with oxygen. Thus hydrosphere proved to be the home for ancestral plants and animals. The top layer (surficial layer) of the ocean water for long period was the principal zone of intense cycle of C, H, O_2 and other biophile elements. Life will cease when the energy resources of the planet earth are exhausted. It is now beyond doubt the dynamic ocean had played a tremendous role in moderating the climate, that is suitable for life on earth. It is now essential to study ocean dynamics along with studies of Atmosphere, Environment, Biosphere and other basic sciences to understand the climate change in the near future and in the long run.

In order to understand the intricacies of climate change it is required to know the past history of the earth's climate. For billions of years earth has supported life which implies climate remained within narrow limits with all variations in the climate from warm epochs to cold epochs of ice ages.

During past 4.5 billion years the insolation increased by about 30%. Climate models suggest that the input rate of CO_2 from volcanic activity into atmosphere is balanced by the rate of removal by chemical weathering (sink).

A continous record of atmospheric conditions (climate) over Greenland and Antartica dating back to about 4 – lakh years is traced with the help of several ice-cores through the Greenland ice sheet and three through the Antarctic ice sheets. Annual layers in the ice core (count) gives the age. The fall out of volcanic ash also provide common markers in ice cores. Oxygen isotope ratios of the ice give temperature over the ocean upwind of the glaciers. Bubbles in the ice give atmospheric CO_2 and methane concentration. Pollen chemical composition and lava particles provide information of volcanic eruptions, wind speed and direction. Thickness of annual layers gives rate of snow accumulation. Isotopes of some elements give solar activity and cosmic activity. Deep sea sediment cores in the North Atlantic (made by ocean drilling program) give information about:

(i) Sea surface temperature and salinity above the core.

(ii) Ice volume in glaciers,

(iii) Production of ice bergs.

Findings

1. The oxygen-isotope record in the ice cores indicate that there were abrupt temperature changes during the past one lakh years (10^5 years).

   During the last ice age (on many occasions) Greenland temperature warmed over the periods 1-100 years. Subsequently, followed by gradual cooling over long periods.

   About 11500 years ago Greenland temperature warmed by about 8 °C in 40 years in three steps of 5 years. Such abrupt warming is called Dansgaard or Oeschger event.

   The ice core studies show that much of northern hemisphere warmed and cooled in phase (with temperatures).

2. During the past 8000 years, the climate was roughly constant.

   During all of recorded history, our perception of climate change is speculative (as it was during warm and stable climate).

3. Henrich events: North Atlantic sediment studies show coarse material was deposited on the bottom and mid–ocean (Atlantic). Only ice-bergs carry such material out to sea. This shows times during which large numbers of ice-bergs were moved into the north Atlantic. These are called Henrich events.
4. The correlation of Greenland temperature with ice-berg production is related to the deep circulation.

CHAPTER 23

IMPACT OF THE OCEAN ON CLIMATE

23.1 Distribution of Physical Properties in Oceans

Weather: The physical state of atmosphere at any particular location rarely exhibits a steady state even during short intervals of time. The ever changing physical state of atmosphere constitutes the weather. Weather is described in terms of instantaneous values or short period mean values of meteorological elements on surface of the earth and in atmosphere, such as temperature, pressure, humidity, wind, state of sky (clouds), precipitation etc.

Climate: The average condition (not only numerical values but also available information) of weather at a place (location) over a long period (say, more than 30 years), together with its extremes and its probabilities constitutes the climate of the place. Weather and climate have been changing right from the origin of atmosphere and life on earth and reached a balanced state that is suitable for life on earth.

Climate never fits into rigid demarcation. It changes from one type to another, one generation to the next, from one century to the next and from one ice age to the next. Climatic normals are average values of weather elements over a place (or region) for about 10 days, or one month and averaged over a long period about 30 to 100 years of record. These normals are used as yardstick for describing its behaviour or variation of weather & climate.

23.2 Climatic Controls

The factors that control the climate broadly are:

1. Latitude or sun's inclination

2. Altitude
3. Topography
4. Land and sea distribution and
5. Ocean currents

In addition to the above factors, climate is affected by semi – permanent pressure Highs and Lows.

Climate is thus the average condition of the atmosphere, ocean, land surfaces and the ecosystems that dwell in them.

Recently the importance of climate has increased due to rapid changes after industrialization. It is realized that climate change is not restricted to past-eons but it is occurring on time scales that affect human activities.

The surface area of oceans covered over the globe is about 71% and about 80% of the atmospheric water vapour is pumped from the ocean surface. Regional climate is highly influenced by thermal properties of the earth surface, particularly by neighbouring ocean. In general ocean moderates extreme conditions. Similar influence is exerted on global mean climate because of oceans large heat capacity and oceans capacity to hold substances in solution. The ocean serves as a reservoir of energy and carbon. As a result ocean gives thermal inertia to the climate system, mainly in exchanges with atmospheric heat and CO_2.

According to IPCC studies the heat content of entire global system has increased since 1950s. The oceans have absorbed 90% of the heat increased, because there is large heat storage capacity in sea water as compared to the atmosphere, land-ice or the continents.

Major oceans are interconnected by currents. Circulation system exchange mass between ocean basins. Ocean circulations are driven by atmospheric wind stress which transfers momentum to the oceans. Ocean circulations are also driven by transfer of heat and moisture (by way of density which influence buoyancy of sea water). The poleward transfer of heat achieved by the general circulation of earth-atmosphere system; of this 60% is contributed by the atmospheric circulation and 40% by ocean circulation.

The special distribution of ocean's heat content and SST trend over the past 50 years is not uniform. Climate change models predict non–uniform changes when projected over the next century (2100). The model predict greatest warming in Arctic, little change in the sub - polar North Atlantic and Antarctic circumpolar current.

Table 23.1 Global monthly surface temperatures for the period 1901-2000 (*Source:* NCDC website)

Month	Mean temperature °C		
	Land (area 29%)	**Sea (area 71%)**	**Global (land-sea combined)**
Jan	2.8	15.8	12.0
Feb	3.2	15.9	12.1
March	5.0	15.9	12.7
April	8.1	16.0	13.7
May	11.1	16.3	14.8
June	13.3	16.4	15.5
July	14.3	16.4	15.8
Aug	13.8	16.4	15.6
Sept	12.0	16.2	15.0
Oct	9.3	15.9	14.0
Nov	5.9	15.8	12.9
Dec	3.7	15.7	12.2
Annual	8.5	16.1	13.9
Range	11.5	0.7	3.8

The above data shows that global land area (29%) average temperature 8.5 °C, range 11.5 °C, while global ocean area (71%) average temperature 16.1 °C, range 0.7 °C. Combined global land and ocean average temperature 13.9 °C and range 3.8 °C.

The following points are important in respect of the role of ocean on climate variability

1. Heat, water, momentum, GHGs and many other substances cross the sea surface. This shows that oceans play a central component role in climate system.
2. Oceans/seas release heat and water vapour into the atmosphere.
3. Ocean currents transport heat and salt around the globe.
4. In global hydrological cycle, oceans maintain pole to equator temperature gradient and fresh water transport.
5. Buffering of atmospheric temperature changes by oceans enormous heat capacity.
6. In middle latitudes, atmospheric changes tend to precede oceanic changes through air-sea interactions.
7. In tropical latitudes, variation in SST and tropical air temperature and winds are in phase with one another that is they are interdependent.

8. In tropical oceans (particularly in Pacific) EINino – Southern Oscillation phenomena are closely related.
9. According to Houghton and Woodwell (WMO-No. 735), the physical and chemical processes at sea surface release carbon fluxes about 100 Gt into atmosphere, while it absorb about 104 Gt. (1 Gt = 10^9 metric tons)
10. The global ocean carbon reservoir 36000 metric tons while the atmospheric carbon reservoir 735 mt.
11. The specific heat of ocean waters is about 4 times that of atmospheric air.
12. According to Lvovich MI, over periphery land area average annual precipitation depth 910 mm, evaporation 560 mm, over inland area precipitation depth 238 mm, evaporation 238 mm while on world ocean precipitation depth 1140 mm and evaporation 1251 mm.
13. Global land surface annual mean temperature 8.5 °C, range 11.5 °C while global sea surface annual mean temperature 16.1 °C and range 0.7 °C.

23.3 Components of Climate System

Climate variation and climate change: The changes that take place in climate over a period due to natural causes is called climate variation. However the changes that take place in climate over a period of time due to man-made activities is called anthropogenic climate change or simply climate change.

Climate variation includes ice-ages, long term warm climate enjoyed by dinosaurs and the event of prolonged drought over Sahel region in Africa. Climate change includes ozone hole over Antarctica and present global warming.

Climate models are mathematical representations of climate system by equation containing elements like temperature, wind, ocean currents, precipitation, evaporation, snow melt and other climate variables. Climate modeling includes interaction of fields or interlocking systems of atmosphere, ocean, land surface, sea-ice, land-ice, part of biosphere, hydrosphere. Earth system models also include physical, chemical and biological aspects over the same period.

Global warming is associated with changes in climate which are attributed to increased GHGs pumped into atmosphere by human activities.

Environmental changes include air pollution, water pollution, deforestation, soil erosion, endangerment of species or ecosystem (by loss or pollution of habitat). Climate prediction models largely depend on statistical theories (time series, probability occurrence of extreme natural events).

In order to arrive at a decisive prediction of future climate the study and clear understanding of climate components is essential. Regional weather, ELNinos, North Atlantic oscillation, Asian monsoon variations, North American monsoon variations, droughts, floods, ocean circulation processes, ice-ages, must be understood. In addition to these environmental chemistry, biosphere evolution and linkages are required. The fig.23.1 shows schematic climate system components.

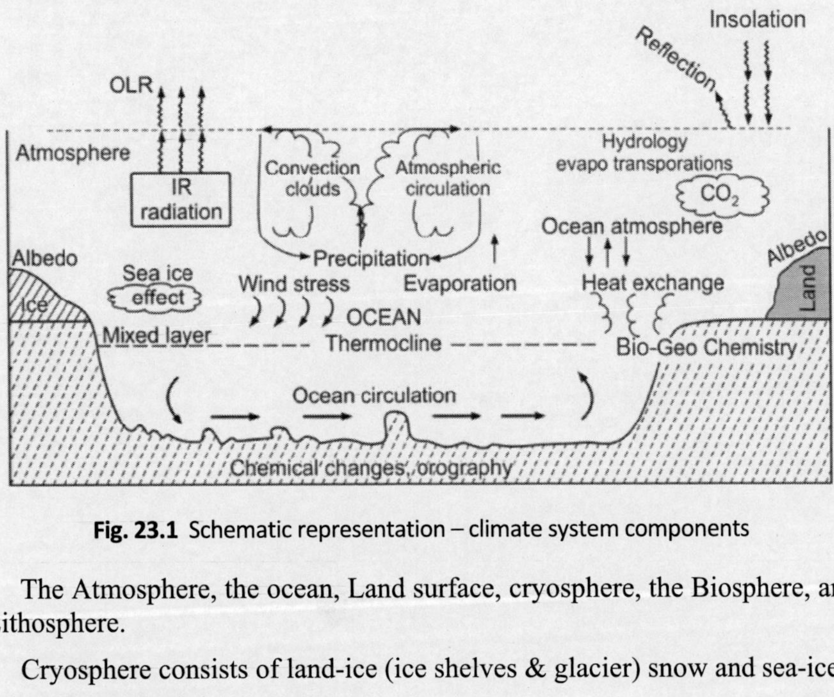

Fig. 23.1 Schematic representation – climate system components

The Atmosphere, the ocean, Land surface, cryosphere, the Biosphere, and Lithosphere.

Cryosphere consists of land-ice (ice shelves & glacier) snow and sea-ice.

The Biosphere is all living beings & vegetation on the earth & oceans.

The Lithosphere-solid earth (volcanoes, earthquakes).

23.4 Climate System or Earth System

The global interlocking system of atmosphere, ocean, land surfaces, sea ice and land ice and parts of biosphere and also solid earth. Thus earth system emphasizes the simultaneous study of all parts in this system, that contribute due to chemical reaction and biological contribution. Thus in a nut shell earth system models include physical, chemical, biological aspects.

Observed Impact of Human activities during last 150 years on climate change and variation.

Overall, man has achieved to add green house gases and reduced the sinks. Of the green housegases increased by human activity, 70% account to energy sector.

CHAPTER 24

GEOLOGICAL TIME SCALE

According to James Hutton (1726-1797), a Scottish scientist, the following modern geological time scale came into use since late eighteenth century. Based on sediment deposits of sequence of layers, he propounded principle of super–position and includes principles of uniformity of process.

William Smith (1764-1839) used rock layers identified by fossils they contained or possessed. This classification is called fossil correlation. Fossils are the remains of animals and plants found preserved in rocks.

Geological time scale divided into four main Eras, built on rock strata. Periods are based on fossils discovery.

Eras derived from Greek words "palaes" (old) "Mesos" (middle) and "Kainos" (recent), combined with "zos" (life).

Discovery of radioactivity helped to determine actual ages of rocks. Isotopes of some elements enables scientists to know about solar activity and cosmic activity. Radiocarbon dating used to fix the age of wood, bones, shells, peats etc. The Table 24.1 gives geological time scale.

Table 24.1 Geological Time Scale

Era	Period	Epoch	Time scale in year
Cenozoic	Quarternary	Recent	10^4
		Pleistocene	10^6
	Tertiary	Pliocene	10^7
		Miocene	2.5×10^7
		Oligocene	4.0×10^7
		Eocene	6.0×10^7
		Paleocene	7.0×10^7
Mesozoic	Cretaceous		1.35×10^8
	Jurassic		1.80×10^8
	Triassic		2.25×10^8

Table 24.1 contd...

Era	Period	Epoch	Time scale in year
Palaeozoic	Permian Carboniferous Devonian Silurian		2.70×10^8 3.50×10^8 4.00×10^8 4.40×10^8
	Ordovician Cambrian		5.00×10^8 6.00×10^8
Pre – Cambrian			4.50×10^9

MY = millions years.

1. **The pre-Cambrian Era (700 – 4500 MY ago):** It was the oldest and longest of the main divisions. During this period mountain–building took place and oldest rocks were formed.
2. **Cambrian Period (600 – 700 MY ago)**: In this period shallow seas covered over large part of surface of the earth. Life existed in sea as sea weeds, sponges, marine invertebrates, however no life existed on land.
3. **Ordovician Period (500 – 600 MY ago):** Volcanic eruptions. Seas continue to expand. All life co
4. nfined to sea (no life on land). Fishes developed, invertabrates.
5. **Silurian Period (440 – 500 MY ago):** Periodic rise and fall of sea levels. Continous change on land. Plants began to adopt on land. New species of vertebrate animals formed. First air–breathing animals evolved.
6. **Devonian Period (400 – 440 MY ago):** In this period volcanic activity increased. Land areas expand. Mountains began to form. Vertebrate animals develop. A variety of fish appears in sea water. Primitive amphibians. Invertabrates slowly move over to land.
7. **Carboniferous Period (350 – 400 MY ago):** Seas spread. Most of present Europe and Russia lie under water, which slowly emerge as swampy areas.

   Coal began to form in swamp vegetation. Large trees over tropical swamps. Amphibian creatures develop. Various marine life develop and spread. Reptiles breed on land.
8. **Permian Period (270 – 350 MY ago):** Warping of earth crust. Northern Hemisphere covered with ice. Deciduous plants appear. End of marine creature domination, creatures increase on land. Insects emerge. Spiders and primitive reptiles.
9. **Triassic Period (225 – 270 MY ago):** Shrubs cover mountains and deserts. More development of mountain ranges (in NH). Fish shaped

reptiles, flying fishes and Lobster like creatures formed. Rise of dinosaurs. Primitive mammals appear.

10. **Jurassic Period (180 – 225 MY ago):** Rockies, Endies and Panama develop. High mountains of previous period suffer from erosion. Lime stone forms. Coniferous forms develop, flowers begin to bloom. Aquatic reptiles dominate in sea. On land birds evolve. Giant dinosaurs lived in swamps Turtles, egg laying mammals appear.

11. **Cretaceous Period (135 – 180 MY ago):** Swamps deltas appear. Rivers flow, chalk deposits appear. Major mountains build up. Deciduous trees and flowering plants spread. Flying reptiles dominate in sea. On land birds evolves. Giant dinosaurs become extinct.

Epochs: 11. Paleocene (70 – 135 MY ago):

11. **Eocene (60 – 70 MY ago):** Severe volcanic activity, warming of climate, flowering plants dominate. Present species of fishes developed in sea. Big whales and sea-cows appear. On land modern animals and giant reptiles and primitive monkeys appear.

12. **Epoch of Oligocene (40 – 60 MY ago):** Alps formed. In seas crabs, snails evolve. On land animals and plants found in abundant. Primitive anthropoids (ape like man) and Mesohippus appeared.

13. **Epoch of Miocene (25 – 40 MY ago):** Earth's crust completely formed. Alps and Himalayas formed. Boney fish, sharks, protohippus, whales develop in sea. On land mammals, water birds, penguins in Antarctica appeared.

    Equatorial regions cooled by 3 to 4 °C, while Europe, America warmed up by 7 to 10 °C.

14. **Epoch of Pliocene (10 – 25 MY ago):** Continents and oceans develop into present form. Europe and Asia land masses join. Vegetation on land limited. Rise of Man and Pliohippus.

15 **Epoch of Pleistocene (1- 10 MY ago):** A great ice-age, ice sheets and glaciers cover most of Europe and America. On land rise modern horse. Emergence of Homosapince (man).

16 **Recent epoch or Holocene (10^4 to 10^6 years ago):** Ice sheets retreat. Sea level rises. Man's dominance grows. Domestication of animals.

KY = Thousand years.

LGM (Last Glacial Maximum) (18 – 23 KY ago).

CLIMAP (Climate Long range Investigation, Mapping and Prediction).

This is based on proxy data from ocean sediments. Thick ice covered over Canada, Northern United States of America, Northern Europe and parts of Euresia.

During this period 1km thickness ice layer covered over Chicago, Glasgow & Stockholm. Antarctica sea level was about 120 – 130 meters lower than the present.

The average SST was 4 °C colder than the present.

North Atlantic SST was colder by more than 8 °C. Low latitude temperatures were about 2 °C lower than the present.

24.1 Milankovitch Cycles

Climate variation occurs due to earth's position (elliptical orbit) relative to the sun, orientation of earth's axis with respect to the orbital plane and precession of axis. These variation of climate is on the scales of 10 KY to 100 KY.

(i) Eccentricity (e) of elliptical orbit, time scale 100 KY to 400 KY.

If e = 0 (circle) to e = 0.07.

(ii) Obliquity-spin axis of earth varies 22.1° to 24.5° on time scale 41 KY.

Present obliquity 23.5°. At position 22.1° it will be ice age.

(iii) Precession-the earth's spin vector precesses with period 23 KY.

The earth is 3% closer to the sun in January perihelion than in July – aphelion. At perihelion earth receives 7% more solar energy than when the earth is at aphelion.

1. When the 'e' is small there is little change in earth-sun distance.
2. When the 'e' large, earth will receive about 20% more solar energy at perihelion than when at aphelion.

When the earth's axis tilt is more, earth experience stronger summer sun and weaker winter sun (particularly at high latitudes).

When the tilt is small ice-ages set in. Cooler summer will not melt winter ice/snow. The precession scale 23 KY. Earth will be closest to the sun in July instead of January. This causes stronger summer.

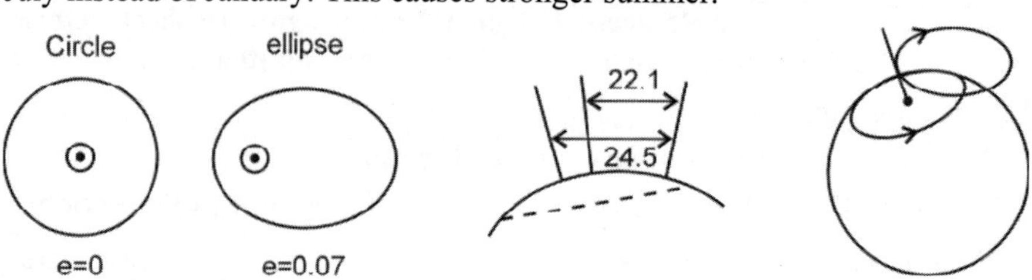

eccentricity Obliquity Precession

CHAPTER 25

OCEAN WAVES

The physical properties of sea water is very important in marine studies. Physical oceanography mainly deals with wave studies and Tidal studies. Initially sea water was assumed to be ideal fluid. The physical properties of sea water depends on suspended particles of organic, inorganic matter and includes density, specific heat, eddy viscosity, conductivity and diffusion.

Motion of ocean water deals with:

(i) Transport of water masses in particular direction over a period of time.

(ii) Random motion (turbulence), which is superimposed upon general flow, and

(iii) Oscillations or oscillating characteristic of waves.

In general motion of ocean water shows rise and fall of the sea surface instead of motion of individual water particles (or parcels).

Most of the ocean waves are generated by the action of wind on sea surface. Wave forms depend on wind strength and period of action. Stronger the wind and that blows over a longer period develops larger wave forms. Depending on the energy transferred to water, ocean waves continue to travel even after wind stopped.

Def. Sea and a Swell

A system of waves that generated by the local wind blowing at the place and time of observation is called **sea**.

A wave system that is not generated by the local wind, but may be developed either by winds blowing in another place (region/location) or by the winds that have ceased to blow is called a **swell**.

The wave heights change at random both in time and space. The statistical wave properties like average height (averaged over a few hundred waves) also changes from day to day. Offshore waves are generated by wind.

Sea level changes from hour to hour, increases and decreases by about a meter with some fixed reference point on the shore. The slow rise and fall of sea level is due to the tides. Tidal waves lengths – thousands of kilometers, which are generated by variations (small changes) in gravity. Gravity changes occur due to the motions of the sun and the moon relative to the earth.

Ocean surface waves are non-linear. The solution of the equations of motion, depend on surface boundary conditions. For simplicity, we assume the flow is two dimensional and suppose the waves are travelling in x – direction. Since we are considering flow locally, the coriolis force and viscosity can be neglected.

Wave Characteristics

With reference to the Fig, 25.1 explanation of terms.

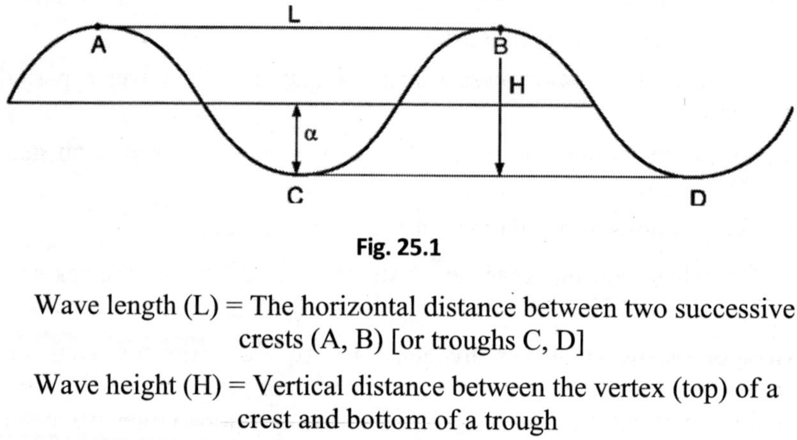

Fig. 25.1

Wave length (L) = The horizontal distance between two successive crests (A, B) [or troughs C, D]

Wave height (H) = Vertical distance between the vertex (top) of a crest and bottom of a trough

Wave speed (c) = The distance travelled by a wave in unit time

$$c = \text{frequency} \times \text{wave length}$$

$$c = n\,L$$

where $n = \text{frequency} = \dfrac{1}{T}$

T = wave period.

n = number of crests (or troughs) passing at a given point in unit time

wave period (T) = It is the time interval between the passage of successive crests (or troughs) past a given point

$$T = \frac{1}{n}$$

The height of free surface (η) or sea surface elevation is given by simple harmonic function

$$\eta = a \sin\left(\frac{2\pi t}{T}\right) = a \sin \sigma t \qquad .....(25.1)$$

where $\sigma = \frac{2\pi}{T}$

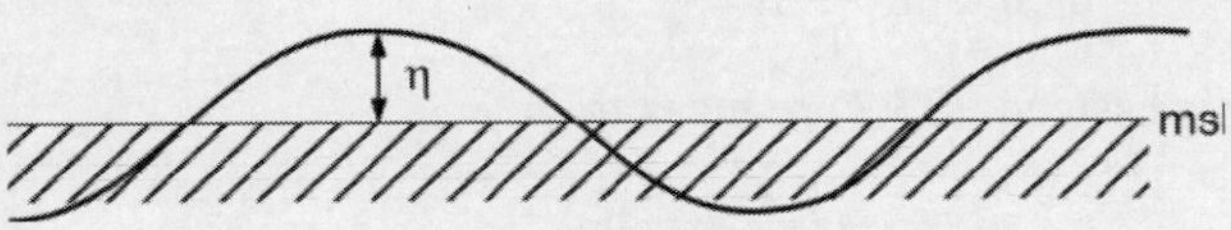

Fig. 25.2

For a progressive wave of small amplitude, the variations in time and space of the free surface travelling in x–direction can be written as

$$\eta = a \sin 2\pi\left(\frac{t}{T} - \frac{x}{L}\right)$$

$$\eta = a \sin\left(\sigma t - \hat{k}\, x\right) \qquad .....(25.2)$$

where $\sigma = \frac{2\pi}{T}, \hat{k} = \frac{2\pi}{L}$

σ is wave frequency in radiance per second

It is assumed that k a = O (o)

Note: A standing wave can be considered as a contribution of two progressive waves travelling in opposite direction.

From classical hydrodynamics, the wave velocity (c) is given by the equation

$$c^2 = \frac{g\,L}{2\pi} \tanh = \frac{2\pi}{L} h \qquad .....(25.3)$$

where h = water depth to the bottom.

Eq. (25.3) can also be written as

$$\sigma^2 = g\,\hat{k} \tanh\left(\hat{k}h\right) \qquad .....(25.4)$$

The following two approximations are very useful.

(i) *Short-waves (or surface waves), also called deep-water approximation.*

If $h > \frac{L}{2}$, then $\tanh\frac{2\pi}{L}h \simeq 1$

In this case eq. (25.3) reduces to

$$c^2 = \frac{gL}{2\pi} \quad \text{or} \quad c=\sqrt{\frac{gL}{2\pi}} \qquad .....(25.5)$$

Eq. (25.5) is deep water phase velocity.

(ii) Long waves, also called shallow water approximation

If $h \ll L$, then $\tanh\frac{2\pi}{L}h \simeq \frac{2\pi}{L}h$

In this case eq. (25.3) reduces to

$$c^2 = \frac{g\,L}{2\pi}\cdot\frac{2\pi}{L}h$$

$$c^2 = gh \quad \text{or} \quad c=\sqrt{gh} \qquad .....(25.6)$$

Equation (25.6) is shallow-water phase velocity.

If $h \ll L$ (from eq. (25.6)), c is dependent on h, and the waves are called long waves.

If $h > L$ [from eq. (25.5)], c is independent of h, and the waves are called short waves or surface waves.

In case of surface waves, individual water particles near surface move in circular orbits, but radii of these orbits or velocities decrease rapidly with depth.

In shallow water, different physical processes occur. The depth of the water may be insufficient for the lower particles to complete circular path. The waves will drag on the ocean floor as shown in the Fig. 25.3.

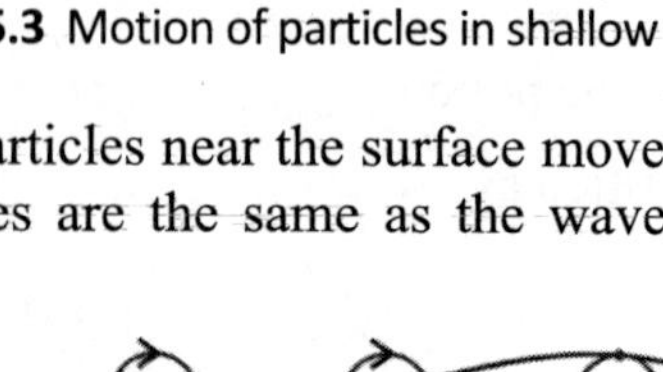

Fig. 25.3 Motion of particles in shallow water

In deep water, the particles near the surface move in a circular paths. The diameter of these circles are the same as the wave height H as shown in Fig.25.4.

Fig. 25.4 Motion of surface particles in deep water

Below the surface, the water particles move in circles of smaller and smaller diameter within half of wave length of the surface, the diameter of the circular paths reduces to less than $\frac{1}{20}$ (one twentieth) of the surface value. The Fig. 25.5 shows this effect.

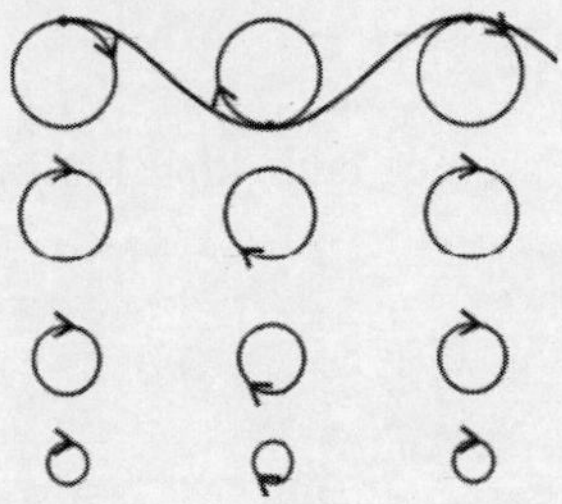

Fig. 25.5 Variation of circular orbits with depth

In shallow water, when $h < \frac{L}{2}$, the lower part of the wave moves faster than the top surface leading to break off of crest, which is called "break".

Note: There is no difference between the speeds of short waves and long waves in shallow-water. However all shallow-water waves slow down as the depth of the water decreases. Hence faster deep water waves begin to approach the slower shallow-water waves. They do not actually overtake the waves ahead, but the wavelength becomes shorter.

Group Velocity (C_g)

The velocity at which a group of waves (in two dimensional, propagation, linear or non- linear waves) travel across the ocean is defined by the relation

$$C_g = \frac{\partial \sigma}{\partial \hat{k}}$$

(i) For Deep Water

We know $\sigma^2 = g\hat{k}\tanh(\hat{k}h)$ .....(25.7)

Which is approximated as $\sigma^2 \simeq g\hat{k}$, (Since tanh kh = 1; for deep water) .....(25.8)

$$\therefore \quad 2\sigma \frac{\partial \sigma}{\partial \hat{k}} = g \qquad \text{(differentiating (25.8) w. r. t '}\hat{k}\text{')}$$

or $\frac{\partial \sigma}{\partial \hat{k}} = \frac{g}{2\sigma}$ ie., $C_g = \frac{g}{2\sigma}$ (by def) .....(25.9)

Since for deep water $C = \sqrt{\frac{g\,L}{2\pi}} = \sqrt{\frac{g}{k}} = \frac{g}{\sigma}$ .....(25.10)

Substituting this (25.10) in eq (25.9) we have

$$C_g = \frac{c}{2} \qquad .....(25.11)$$

Eq. (25.11) is for deep water group velocity.

(ii) For shallow water

We know $\sigma^2 = g\,\hat{k}\tanh(\hat{k}h)$

$$\sigma^2 \simeq g\,\hat{k}\cdot\hat{k}h \qquad \text{since } \tanh\hat{k}h \simeq \hat{k}h \text{ for shallow water}$$

$$\sigma^2 = gh\,\hat{k}^2$$

or $\quad \sigma = \sqrt{gh}\,\hat{k}$

$$\therefore \quad \frac{\partial\sigma}{\partial\hat{k}} = \sqrt{gh} \qquad (\text{differentiating w. r. t } \hat{k})$$

$$c_g = \frac{\partial\sigma}{\partial\hat{k}} = \sqrt{gh} \qquad .....(25.12)$$

But for shallow-water approximation

$$c = \sqrt{gh} \qquad .....(25.13)$$

From (25.12) & (25.13) we have

$$C_g = c \qquad .....(25.14)$$

Equation (25.14) is shallow water group velocity

The lowest frequency waves travel the faster $C_g = c$.

while the higher frequency waves travel slower $C_g = \frac{c}{2}$

Comparative properties of surface waves and long waves.

Surface waves	Long waves
1. Waves are progressive, they are forced or free	1. Waves are progressive standing and they may be forced or free
2. Velocity of progress depends on wavelength (L), but independent of depth (h)	2. Velocity of progress depends on depth (h), but independent of wavelength (L)
3. Movement of water particles in vertical plane are circle; radii of circles decrease rapidly with increasing depth from surface. Motion (particles) at a depth of wave length imperceptible. In some cases the motion may be wide ellipses.	3. The movement of water particles in vertical plane depends on depth (h), but independent of wavelength (L). In ellipses which are flat, the water particles will be oscillating back and forth in horizontal plane. Horizontal plane. Horizontal motion is independent of depth (h)

Cont...

Surface waves	Long waves
4. Vertical displacement of water particles decrease rapidly with increasing depth from surface and become imperceptible at depth h = L	4. Vertical displacement of particles decrease linearly from the surface to the bottom.
5. Pressure distribution below the depth of perceptible motion is not influenced by the waves	5. The wave influences the pressure distribution in the same manner at all depths
6. Influence of the earth's rotation (coriolis force) is negligible	6. The influence of earth's rotation cannot be neglected (if the period of wave $\simeq$ to the period of earth's rotation.

25.1 Ocean Surface Waves

Over a sequence of ocean waves, when wind blows eddies will be formed on the Lee side of the waves. This is because the pressure of wind will be greater on the wind ward slopes (as compared to the Lee ward slope which are sheltered by the crests). This happens only when ocean waves travel at a lesser velocity than the wind speed.

Under this condition, waves may increase only if

$$c(w-c)^2 \geq \frac{4\upsilon g(\rho-\rho')}{s\,\rho'} \qquad \text{.....(25.15)}$$

where w = wind velocity

c = velocity of the waves

υ = kinetic viscosity of the water

g = gravitational attraction of the earth

ρ = density of water

ρ' = density of air

s = sheltering coefficient.

For a given velocity w, the L H S of eq (25.15) is maximum

when $c = \frac{w}{3}$

That is, $\frac{w}{3}\left(w-\frac{w}{3}\right)^2 \geq \frac{4\upsilon g(\rho-\rho')}{s\,\rho'}$

or $\frac{4}{27}w^3 \geq \frac{4\upsilon g(\rho-\rho')}{s\,\rho'}$

or $$w^3 \geq \frac{27\upsilon g(\rho-\rho')}{s\,\rho'}$$

For general surface waves of the sea, we have simple formulae

$$c = \frac{L}{T} = \sqrt{\frac{g\,L}{2\,\pi}} = g\frac{T}{2\pi}$$

and the following relations

$$L = \frac{2\pi}{g}c^2 = \frac{g}{2\pi}T^2$$

$$\hat{k} = \frac{2\pi}{L} = \frac{g}{c^2}$$

$$T = \sqrt{\frac{2\pi L}{g}} = \frac{2\pi c}{g}$$

$$\sigma = \frac{2\pi}{T} = \frac{g}{c}$$

1. Two surface waves that travel in the same direction represented by

$$\eta_1 = a_1 \sin(\sigma_1 t - k_1 x)$$

$$\eta_2 = a_2 \sin(\sigma_2 t - k_2 x)$$

The actual surface waves that travel in the same direction is obtained by adding the displacements (of two individual waves).

If amplitudes are equal, then

$$\eta = 2a\ \cos\left[\frac{(\sigma_1-\sigma_2)}{2}t - \frac{(k_1-k_2)x}{2}\right]\sin\left[\left(\frac{\sigma_1+\sigma_2}{2}\right)t - \frac{1}{2}\left(\frac{k_1+k_2}{2}\right)x\right]$$

$$\text{Velocity } c = \frac{\sigma_1-\sigma_2}{k_1-k_2} = \frac{c_1 c_2}{c_1+c_2}$$

2. Standing Waves in Bays – Seiches

In Bays, standing waves may develop that are similar to the oscillations in the Lakes. These waves are called Seiches.

Seiches are free oscillations of a period which depend on horizontal dimensions, depth of the lake and the number of nodes of the standing wave. The wavelength will be the order of magnitude as that of the Lake.

(The length of the lake >> depth) and the waves will have the character of long waves.

A long wave that proceeds in water of constant depth (in positive or negative direction) must satisfy the equations of motion and continuity. Neglecting coriolis force and friction force,

We have

$$\left.\begin{aligned}&\frac{du}{dt} \cong \frac{\partial u}{\partial t} = -g\frac{\partial \eta}{\partial x}\\ \text{and}\quad &\frac{\partial \eta}{\partial t} = -h\frac{\partial u}{\partial x}\end{aligned}\right\} \qquad \text{.....(25.16)}$$

$\frac{du}{dt} \simeq \frac{\partial u}{\partial t}$ in case of long-waves because $\frac{\partial u}{\partial x}$ is small in quantity.

The surface vertical displacement is called η and $\frac{\partial \eta}{\partial X}$ is the slope of the free surface.

Let ξ = horizontal displacement and $u = \frac{\partial \xi}{\partial t}$

The equations (25.16) become

$$\left.\begin{aligned}&\frac{\partial}{\partial t}\left(\frac{\partial \xi}{\partial t}\right) = -g\frac{\partial \eta}{\partial x}\\ &\frac{\partial \eta}{\partial t} = -h\frac{\partial}{\partial x}\left(\frac{\partial \xi}{\partial t}\right) \quad \text{or} \quad \eta = -h\frac{\partial \xi}{\partial x}\end{aligned}\right\} \text{(by integrating)} \qquad \text{.....(25.17)}$$

(The solution of) these equations are satisfied if

$$\left.\begin{aligned}&\xi = a \sin\left[\sigma t \pm \frac{\sigma x}{c} + \varepsilon\right]\\ &\eta = \mp a\, h\frac{\sigma}{c}\cos\left[\sigma t \pm \frac{\sigma}{c}x + \varepsilon\right]\end{aligned}\right\} \qquad \text{.....(25.18)}$$

where c = the velocity of progress of the wave

$$c = \frac{\sigma}{k} = \sqrt{gh}$$

Eq. (25.18) define two waves that progress in opposite directions. The equations of motion are also satisfied by a superposition of two waves that proceed in opposite directions.

The velocity of progress is

$$\xi = \xi_1 + \xi_2 = 2a \sin\left(\frac{\sigma}{c}x\right)\cos\,(\sigma t + \varepsilon) \qquad \text{.....(25.19)}$$

and $$\eta = \eta_1 + \eta_2 = -2a\ h\frac{\sigma}{c}\cos\left(\frac{\sigma}{c}x\right)\cos\left(\sigma t + c\right) \quad .....(25.20)$$

Boundary Conditions

Let L = length of the basin, measure from x = 0

The boundary conditions would be

$$x = 0,\ \xi = 0 \quad \text{and} \quad x = l,\ \xi = 0$$

The first condition is satisfied if the period of oscillation is

$$\frac{\sigma}{c}l = n\pi, \quad \text{where } n > 0$$

Since $\sigma = \dfrac{2\pi}{T}$ and $c = \sqrt{gh}$

$$T_n = \frac{1}{n}\frac{2l}{\sqrt{gh}} \quad .....(25.21)$$

25.2 Plane Waves in Three Dimensions

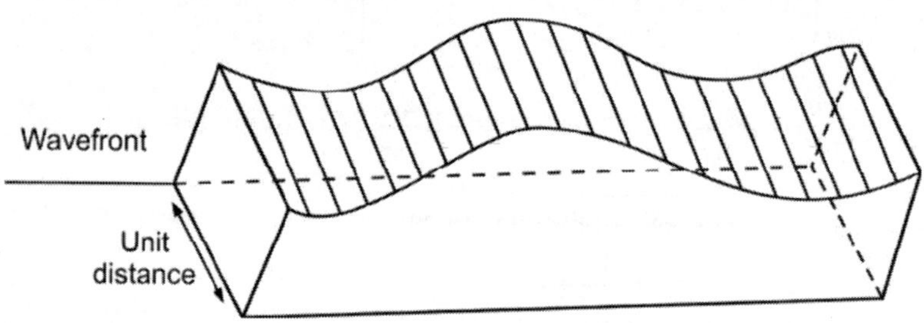

Fig. 25.6

If, at all points of a plane drawn perpendicular to the direction of propagation, the disturbance is constant, then the wave is called a plane wave and the plane, a wavefront (see Fig.25.6)

Since wave front moves at right angles to itself, with velocity of propagation 'c', its equation must be of the form

$$lx + my + nz = \text{constant} \quad .....(25.22)$$

where l, m, n are direction cosines to each wave front.

At any time 't', the distance ϕ is to be constant for all x, y, z, satisfying the equation lx + my + nz = constant.

It follows that $\phi = \phi\ [lx + my + nz - ct]$ .....(25.23) satisfies all the requirements.

Thus eq. (25.23) represents a plane wave moving with velocity 'c' in the direction of the normal without any change of shape. It is easily seen that eq. (25.23) is a particular solution of the wave equation.

$$\nabla^2\phi = \frac{\partial^2\phi}{\partial x^2} + \frac{\partial^2\phi}{\partial y^2} + \frac{\partial^2\phi}{\partial z^2} = \frac{1}{c^2}\frac{\partial^2\phi}{\partial t^2} \quad .....(25.24)$$

To verify, we put

$$lx + my + nz - ct = p,$$

So that $\phi = \phi\ (p)$

$$\therefore \frac{\partial\phi}{\partial x} = \frac{\partial\phi}{\partial p}\frac{\partial p}{\partial x}$$

$$= 1\frac{\partial\phi}{\partial p} \qquad \left[\text{using eq. (25.23) } \frac{\partial\phi}{\partial x} = 1\right]$$

and $$\frac{\partial^2\phi}{\partial x^2} = 1\frac{\partial^2\phi}{\partial p^2}\frac{\partial p}{\partial x}$$

$$\frac{\partial^2\phi}{\partial x^2} = l^2\,\frac{\partial^2\phi}{\partial p^2} \quad .....(25.25)$$

$$\text{Similarly}\ \frac{\partial^2\phi}{\partial y^2} = m^2\,\frac{\partial^2\phi}{\partial p^2} \quad .....(25.26)$$

$$\frac{\partial^2\phi}{\partial z^2} = n^2\,\frac{\partial^2\phi}{\partial p^2} \quad .....(25.27)$$

Substituting eq. (25.25), (25.26), (25.27) in eq. (25.24) we have

$$\nabla^2\phi = \left(l^2 + m^2 + n^2\right)\frac{\partial^2\phi}{\partial p^2}$$

or $$\nabla^2\phi = \frac{\partial^2\phi}{\partial p^2} \qquad \text{since } l^2 + m^2 + n^2 = 1 \quad .....(25.28)$$

$$\text{Also}\quad \frac{\partial\phi}{\partial t} = \frac{\partial\phi}{\partial p}\frac{\partial p}{\partial t}$$

$$= -\,c\frac{\partial\phi}{\partial p} \qquad \left[\text{using eq. (25.23) } \frac{\partial\phi}{\partial t} = -c\right]$$

$$\frac{\partial^2\phi}{\partial t^2} = -c\frac{\partial^2\phi}{\partial p^2}\frac{\partial p}{\partial t}$$

$$\frac{\partial^2\phi}{dt^2} = c^2\frac{\partial^2\phi}{\partial p^2}$$

or
$$\left.\begin{aligned}\frac{\partial^2\phi}{\partial p^2} = \frac{1}{c^2}\frac{\partial^2\phi}{\partial t^2}\end{aligned}\right\}\text{[using eq. (25.23)]} \qquad ...(25.29)$$

substituting eq. (25.29) in eq. (25.28) we get

$$\nabla^2\phi = \frac{1}{c^2}\frac{\partial^2\phi}{\partial t^2} \qquad .....(25.30)$$

eq. (25.30) is the required wave equation.

Solutions of wave equation

(A) One dimensional wave equation.

viz $$\frac{\partial^2\phi}{\partial x^2} = \frac{1}{c^2}\frac{\partial^2\phi}{\partial t^2} \qquad .....(25.31)$$

Put $u = x - ct, \qquad v = x + ct$ .....(25.32)

$$\frac{\partial\phi}{\partial x} = \frac{\partial\phi}{\partial u}\frac{\partial u}{\partial x} + \frac{\partial\phi}{\partial v}\frac{\partial v}{\partial x}$$

$$= \frac{\partial\phi}{\partial u} + \frac{\partial\phi}{\partial v}, \quad \frac{\partial u}{\partial x} = 1, \quad \frac{\partial v}{\partial x} = 1 \quad \text{[using eq. (25.32)]}$$

$$\frac{\partial\phi}{\partial x} = P + Q \quad \text{say} \qquad .....(25.33)$$

where $P = \frac{\partial\phi}{\partial u}, \quad Q = \frac{\partial\phi}{\partial v}$

$$\frac{\partial^2\phi}{\partial x^2} = \frac{\partial P}{\partial x} + \frac{\partial Q}{\partial x} \quad \text{[Differentiating (25.2) w.r.t x]}$$

$$= \left(\frac{\partial p}{\partial u}\frac{\partial u}{\partial x} + \frac{\partial p}{\partial v}\frac{\partial v}{\partial x}\right) + \left(\frac{\partial Q}{\partial u}\frac{\partial u}{\partial x} + \frac{\partial Q}{\partial v}\frac{\partial v}{\partial x}\right)$$

$$= \left(\frac{\partial^2\phi}{\partial u^2} + \frac{\partial^2\phi}{\partial u\partial v}\right) + \left(\frac{\partial^2\phi}{\partial v\partial u} + \frac{\partial^2\phi}{\partial v^2}\right) \quad \text{[using eq. (25.33)]}$$

$$\frac{\partial^2\phi}{\partial x^2} = \frac{\partial^2\phi}{\partial u^2} + 2\frac{\partial^2\phi}{\partial u\partial v} + \frac{\partial^2\phi}{\partial v^2} \qquad .....(25.34)$$

Also $\dfrac{\partial\phi}{\partial t}=\dfrac{\partial\phi}{\partial u}\dfrac{\partial u}{\partial t}+\dfrac{\partial\phi}{\partial v}\dfrac{\partial v}{\partial t}$ $\qquad\left[\because \dfrac{\partial u}{\partial t}=-c \quad \dfrac{\partial v}{\partial t}=+c\right]$

$$=-c\frac{\partial\phi}{\partial u}+c\frac{\partial\phi}{\partial v}$$

$$\left.\begin{aligned}&\frac{\partial\phi}{\partial t}=L+M \qquad \text{say}\\ \text{where}\quad &L=-c\frac{\partial\phi}{\partial u}\quad M=c\frac{\partial\phi}{\partial v}\end{aligned}\right\} \qquad \text{.....(25.35)}$$

$$\therefore\quad \frac{\partial^2\phi}{\partial t^2}=\frac{\partial L}{\partial t}+\frac{\partial M}{\partial t}=\left(\frac{\partial L}{\partial u}\frac{\partial u}{\partial t}+\frac{\partial L}{\partial v}\frac{\partial v}{\partial t}\right)+\left(\frac{\partial M}{\partial u}\frac{\partial u}{\partial t}+\frac{\partial M}{\partial v}\frac{\partial v}{\partial t}\right)$$

$$=\left(-c\frac{\partial L}{\partial u}+c\frac{\partial L}{\partial v}\right)+\left(-c\frac{\partial M}{\partial u}+c\frac{\partial M}{\partial v}\right)$$

$$=\left(c^2\frac{\partial^2\phi}{\partial u^2}-c^2\frac{\partial^2\phi}{\partial u\partial v}\right)+\left(-c^2\frac{\partial^2\phi}{\partial u\partial v}+c^2\frac{\partial^2\phi}{\partial v^2}\right)$$

$$\frac{\partial^2\phi}{\partial t^2}=c^2\left(\frac{\partial^2\phi}{\partial u^2}-2\frac{\partial^2\phi}{\partial u\partial v}+\frac{\partial^2\phi}{\partial v^2}\right) \qquad \text{.....(25.36)}$$

Eq. (25.31) is $\dfrac{\partial^2\phi}{\partial x^2}=\dfrac{1}{c^2}\dfrac{\partial^2\phi}{\partial t^2}$

Substituting eq. (25.35) & (25.36) in this equation, we have

$$\frac{\partial^2\phi}{\partial u^2}+2\frac{\partial^2\phi}{\partial u\partial v}+\frac{\partial^2\phi}{\partial v^2}=\frac{1}{c^2}\times c^2\left(\frac{\partial^2\phi}{\partial u^2}-2\frac{\partial^2\phi}{\partial u\partial v}+\frac{\partial^2\phi}{\partial v^2}\right)$$

This gives $\dfrac{\partial^2\phi}{\partial u\partial v}=0$ .....(25.37)

The most general solution of eq. (25.37) is given by

$$\phi=f(u)+g(v)$$

$$=f(x-ct)+g(x+ct)$$

where f and g are arbitrary functions.

(B) Solution of wave equation in two dimensions.

viz $\dfrac{\partial^2\phi}{\partial x^2}+\dfrac{\partial^2\phi}{\partial y^2}=\dfrac{1}{c^2}\dfrac{\partial^2\phi}{\partial t^2}$ .....(25.38)

$$\text{put}\quad \left.\begin{aligned}u&=lx+m\,y-ct\\ v&=lx+m\,y+ct\end{aligned}\right\} \qquad \text{.....(25.39)}$$

where $l^2 + m^2 = 1$

Now $\dfrac{\partial\phi}{\partial x} = \dfrac{\partial\phi}{\partial u}\dfrac{\partial u}{\partial x} + \dfrac{\partial\phi}{\partial v}\dfrac{\partial v}{\partial x}$

$$= l\frac{\partial\phi}{\partial u} + m\frac{\partial\phi}{\partial v} \qquad \because \frac{\partial u}{\partial x} = l,\ \frac{\partial v}{\partial y} = m$$

$\dfrac{\partial\phi}{\partial x} = P + Q$ say,

where $\quad P = l\dfrac{\partial\phi}{\partial u}, \quad Q = m\dfrac{\partial\phi}{\partial v}$ .....(25.40)

$$\frac{\partial^2\phi}{\partial x^2} = \frac{\partial P}{\partial x} + \frac{\partial Q}{\partial x} = \left(\frac{\partial P}{\partial u}\frac{\partial u}{\partial x} + \frac{\partial P}{\partial v}\frac{\partial v}{\partial x}\right) + \left(\frac{\partial Q}{\partial u}\frac{\partial u}{\partial x} + \frac{\partial Q}{\partial v}\frac{\partial v}{\partial x}\right)$$

$$= l\left(\frac{\partial P}{\partial u} + \frac{\partial P}{\partial v} + \frac{\partial Q}{\partial u} + \frac{\partial Q}{\partial v}\right) \qquad \because \frac{\partial u}{\partial x} = l,\ \frac{\partial v}{\partial x} = l$$

$$\frac{\partial^2\phi}{\partial x^2} = l^2\left(\frac{\partial^2\phi}{\partial u^2} + 2\frac{\partial^2\phi}{\partial u\partial v} + \frac{\partial^2\phi}{\partial v^2}\right) \quad \text{[using 25.40]} \qquad .....(25.41)$$

Similarly $\dfrac{\partial^2\phi}{\partial y^2} = m^2\left(\dfrac{\partial^2\phi}{\partial u^2} + 2\dfrac{\partial^2\phi}{\partial u\partial v} + \dfrac{\partial^2\phi}{\partial v^2}\right)$ .....(25.42)

Adding (25.41) & (25.42) we get

$$\frac{\partial^2\phi}{\partial x^2} + \frac{\partial^2\phi}{\partial y^2} = \left(l^2 + m^2\right)\left(\frac{\partial^2\phi}{\partial u^2} + 2\frac{\partial^2\phi}{\partial u\partial v} + \frac{\partial^2\phi}{\partial v^2}\right)$$

$$\frac{\partial^2\phi}{\partial x^2} + \frac{\partial^2\phi}{\partial y^2} = \left(\frac{\partial^2\phi}{\partial u^2} + 2\frac{\partial^2\phi}{\partial u\partial v} + \frac{\partial^2\phi}{\partial v^2}\right) \quad \left(\because l^2 + m^2 = 1\right) \qquad ..... (25.43)$$

Again $\dfrac{\partial\phi}{\partial t} = \dfrac{\partial\phi}{\partial u}\dfrac{\partial u}{\partial t} + \dfrac{\partial\phi}{\partial v}\dfrac{\partial v}{\partial t}$, [since using (25.39) $\dfrac{\partial u}{\partial t} = -c$, $\dfrac{\partial v}{\partial t} = +c$]

$$= -c\frac{\partial\phi}{\partial u} + c\frac{\partial\phi}{\partial v}$$

As before

$$\therefore \quad \frac{\partial^2\phi}{\partial t^2} = c^2\left[\frac{\partial^2\phi}{\partial u^2} - 2\frac{\partial^2\phi}{\partial u\partial v} + \frac{\partial^2\phi}{\partial v^2}\right] \qquad ..... (25.44)$$

Substituting (25.43) & (25.44) in eq. (25.36)

Viz. $\dfrac{\partial^2\phi}{\partial x^2} + \dfrac{\partial^2\phi}{\partial y^2} = \dfrac{1}{c^2}\dfrac{\partial^2\phi}{\partial t^2}$

we have $\frac{\partial^2\phi}{\partial u^2}+2\frac{\partial^2\phi}{\partial u\partial v}+\frac{\partial^2\phi}{\partial v^2}=c^2\times\frac{1}{c^2}\left(\frac{\partial^2\phi}{\partial u^2}-2\frac{\partial^2\phi}{\partial u\partial v}+\frac{\partial^2\phi}{\partial v^2}\right)$

This gives $\frac{\partial^2\phi}{\partial u\partial v^2}=0$ .....(25.45)

This most general solution of eq. (25.45) is

$$\phi=f(u)+g(v)$$

$$=f(lx+my-ct)+g(lx+my+ct)$$

where f, g are arbitrary functions and $l^2 + m^2 = 1$

(C) Solution of three dimensional wave equation.

viz $$\frac{\partial^2\phi}{\partial x^2}+\frac{\partial^2\phi}{\partial y^2}+\frac{\partial^2\phi}{\partial z^2}=\frac{1}{c^2}\frac{\partial^2\phi}{zt^2}$$

or $$\nabla^2\phi=\frac{1}{c^2}\frac{\partial^2\phi}{\partial t^2}$$

As before put $u = lx + m\,y + n\,z - ct$ etc.

Proceeding as above case we can show that its solution is given by

$$\phi=f(lx+m\,y+nz-ct)+g(lx+my+nz+ct)$$

where f and g are arbitrary functions and that $l^2+m^2+n^2=1$

CHAPTER 26

TIDES

Def: Tides: The rise and fall of the ocean water is termed "Tide" and the associated currents as "tidal currents".

Tidal currents do not transport water to large distances. Tidal currents change from place to place. They depend on the character of the tide, depth (h) to the bottom and configuration of the coast. In any given locality tidal currents repeat themselves as regularly as the tide (to which they are associated)

In samaveda of Indian Vedic period (2000 to 1400 Bc), the relation between tides and the moon and the sun described, indicating the tidal phenomena was known in the Vedic period. Polynesian navigators traded over long distance in the Pacific ocean (4000 BC). They understood the ocean currents, winds, waves and tides.

The Physical Features of the Tides: The water near a sea coast rises twice a day and covers (spreads) a portion of the beach. This is called High tide. About six (6) hours later the water level goes down below normal (recedes) and a large area of the beach is uncovered. This is called Low tide.

The rise and fall of the oceans water is due to the gravitational attraction of the moon and to a smaller extent due to the attraction of the sun.

During a period of a day (about 24 hours 50 minutes) two High tides and two Low tides occur at any coastal location. During this period moon orbits around the earth. During the orbital motion of the moon around the earth, the gravitational attraction of the moon causes the earth's ocean on the sides of the earth facing the moon (position A in Fig 26.1) water is pulled towards the moon. This causes High tide. At the same time, a High tide occurs on the opposite side of the earth (position B in Fig. 26.1).

Let G be the common centre of mass of Earth and Moon. Mass of the earth is about 81 times the mass of the moon. Hence their common centre of mass 'G' is located close the centre of earth (is about 1700 km below the earth's surface). However, the position of G will be constantly changing as the Earth's & Moon's (distance) direction changes in space. The common centre of mass (G) moves in an elliptical orbit around the sun. In general, the gravitational attraction between the earth and moon tend to balance the centrifugal force arising because of their rotation about their common centre of mass (G). Some mass particles of the earth are closer to the moon than others. The gravitational attraction of the moon on the closer particles will be greater than the average, while the force of attraction on the distant particles will be less than the average. Liquid water on the earth respond freely to these force difference compared to earth's solid particles. Because of this, water mass (on earth) tend to flow towards the moon. As a result the ocean water is piled up (heaped up) near the point A of the earth (directly below the moon), and also at 'B' the far side of the earth, there is High tide at the same time.

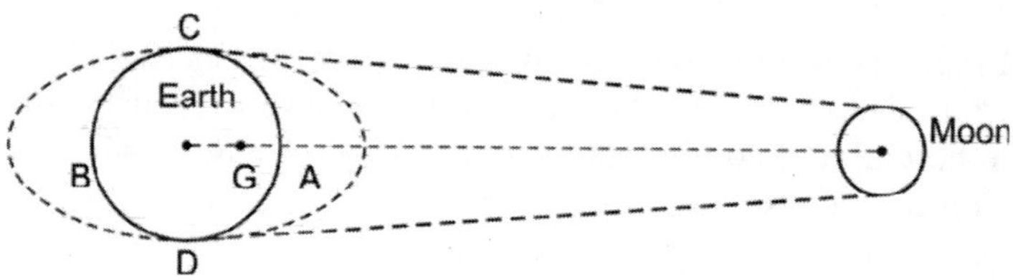

Fig. 26.1

Since water heaps up at A and B positions, there will be a corresponding lowering of water levels at position C & D, that is, there will be Low tides at positions C & D. The two High tides (at A & B) and the two Low tides (at C & D) change their positions as the moon orbits around the earth. Any point on the ocean, below the two High tides there will be two Low tides during 24 hours 50 minutes (that moon orbits around the earth.)

Solar Tides: Explanation of the tides is more complicated when we consider about the effect of sun's attraction. The gravitational attraction of the sun (on earth) is much less than the moon because of its distance (earth-sun distance).

After full moon & new moon, the tidal force of the moon, the tidal force of the sun is added to the moon's effect. The High tides at full moon & new moon are called spring tides.

At half moons, the gravitational attraction of the sun is perpendicular to the moon. The High tides are below average and are called Neap tides. The solar tide is about $5/11^{th}$ of the lunar tide, hence the ratio of the height of spring tide to that of a Neap tide ratio is $(11+5)$: $(11-5)$ or 8:3.

When the moon is closest to the earth it is perigee and when it is farthest to the earth it is apogee.

The variation in the moon's distance from earth causes the tide-raising force to be 1/3 greater at perigee than at apogee. When perigee occurs at time of new or full moon, the High tides are (particularly) high, while at the same time the heights of the Low tide are much below average.

In open ocean, the tidal rise and fall due to the attraction of the moon and the sun is only about one meter. But the shape and location of the coastline develops much larger variation. Consequently, coastal tides may be more than 15 times greater than they are in the open sea.

Diurnal (daily) changes of weather are related to the rotation of the earth on its axis. Days are warmer than nights. Diurnal variation of cloud and wind velocity indirectly related to the rotation of the earth.

Tidal action is very important in erosion of ocean floor, coast line.

26.1 Mathematical Features of Waves

Ocean waves are generated by wind. Largest waves are generated by strong winds blown over a long period (duration). Wavelength and frequency are related through the dispersion relation.

For deep water dispersion, we have approximation

$$\sigma^2 = g\hat{k}\,,\ h > \frac{L}{4}$$

For shallow-water dispersion we have approximation

$$\sigma^2 = g\hat{k}^2 h\,,\ h < \frac{L}{11}$$

Waves in deep water are dispersive. Longer wavelengths travel faster than shorter wavelengths. However waves in shallow-water are not dispersive.

Wave Energy: Wave energy is proportional to variance of surface displacement.

Let E denote the wave energy in Joules/m^2

η = sea surface deviation

Then $E = \rho g \left(\overline{\eta}\right)^2$

where $\overline{\eta}$ = average variance of sea surface displacement

ρ = density of water (sea)

g = gravity.

Shoaling Waves and Coastal Process

In Hydrodynamics dispersion is given by

$$\sigma^2 = g\hat{k} \tan (kh) \qquad \text{......(26.1)}$$

$$\sigma = \frac{2\pi}{T}$$

where $\sigma =$ wave frequency in radiance

$$\hat{k} = \frac{2\pi}{L} = \text{wave number}$$

L = wave length

T = wave period

Since σ is constant, the equation (26.1) gives

$$\frac{L}{L_0} = \frac{c}{c_0} = \frac{\sin\alpha}{\sin\alpha_0} = \tanh\left(\frac{2\pi h}{L}\right) \qquad \text{.....(26.2)}$$

where $L_0 = \frac{gT^2}{2\pi}, \quad c_0 = \frac{gT}{2\pi}$ .....(26.3)

$$c = \frac{L}{T} = \frac{\sigma}{k} = \text{phase speed}$$

$\alpha =$ angle of the crest relative to contours of constant depth

h = sea water depth.

Subscript '0' stands for values in deep water.

$\frac{h}{L}$ is calculated from the solution of

$$\frac{h}{L_0} = \frac{h}{L}\tanh\left(\frac{2\pi h}{L}\right) \qquad \text{.....(26.4)}$$

Since wave velocity (c) is a function of depth 'h', in shallow water, variations in offshore water depth can focus or defocus wave energy reaching the shore.

Wave group velocity is faster in the deep water between the ridges and the wave crests become progressively deformed as the wave propagates toward beach. Wave energy, which propagates perpendicular to wave crests, is refracted out of the region between the headland. As a result, wave energy is focused into headlands (and breakers there are much larger than breakers in the bay).

Breaking waves

When waves move into shallower water the group velocity becomes small, while wave energy per square meter of area of sea surface increases. These processes cause wave to steepen, with short steep crests and broad shallow troughs. When wave slope at the crest becomes very steep, the wave breaks.

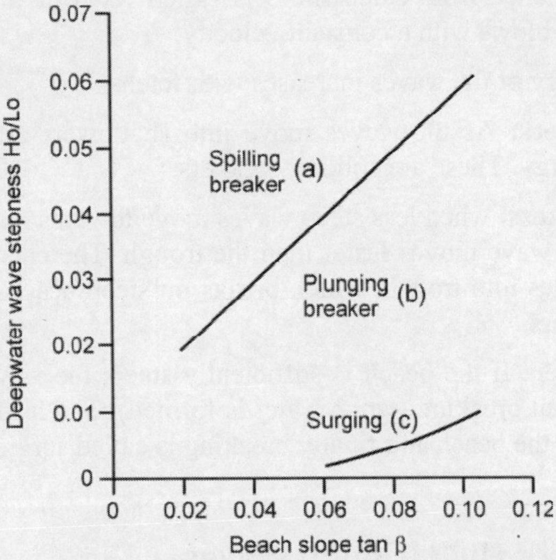

Fig. 26.2(A) Classification of breaking waves as a function of beach steepness and wave steepness offshore

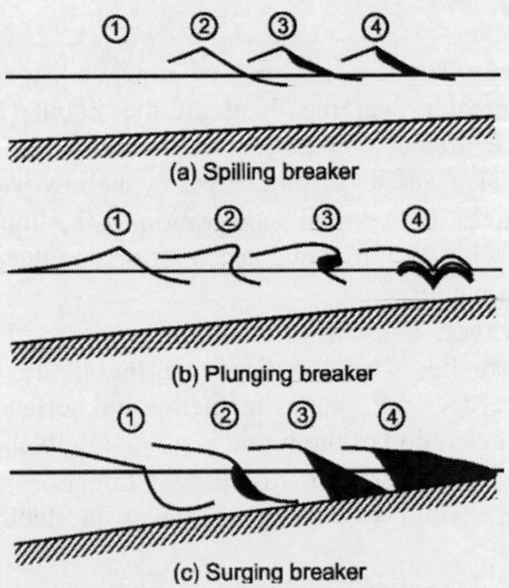

Fig. 26.2(B) Sketch of types of breaking waves , Horikawa (1988, 79, 81)

Note: In general, waves do not have

(i) momentum,

(ii) a mass of flux, but waves have

(iii) a momentum flux.

Fetch: The distance from a lee shore is called fetch or the distance over which the wind blows with a constant velocity.

Note: The energy of the waves increases with fetch.

Spilling breakers: As the waves move into shallower water steep waves slowly lose energy. These are spilling breakers.

Plunging breakers: when less steep waves move towards/ on steep beaches, the crest of the wave moves faster than the trough. The crest moving ahead of trough plunges into trough, which breaks on steep beach. This is termed plunging breakers.

Surging breaker: If the beach is sufficiently steep, the wave can surge up the beach without breaking (white water is formed). The leading edge of the water surges up the beach and finally breaking is called surging breaker.

26.2 Types of Fluid (Liquid) Waves

According to hydrodynamics wave motion in liquids is categorized into two groups.

(i) Tidal waves and

(ii) Surface waves

(i) Tidal waves: These waves arise when wave length of the oscillation is much greater than the depth of the liquid ($L >> h$), and the disturbance effects the motion of the whole of the liquid. Tidal waves are also called as Long waves in shallow-water.

In tidal waves, the vertical acceleration of the liquid is negligible as compared with the horizontal acceleration and the plane of the liquid moves as a whole.

(ii) Surface waves: These arise when the wave length of the oscillation is much smaller than the depth of the liquid ($L << h$) and the disturbance does not extend far below the surface. In this case the vertical acceleration of the liquid is not negligible.

Waves in deep water are dispersive. Longer wave lengths travel faster than shorter wavelengths. Waves in shallow water are not dispersive.

Wave energy is proportional to variance of surface displacement.

Waves are generated by wind stress. Largest waves are generated by strong wind blown over a longer period.

26.3 Wave Profile of Small Elevation Surface Waves

Consider a stream of constant depth 'h' in which wave of height $\eta = \eta(x,t)$ above msl propagated (see Fig.26.3) along x-axis

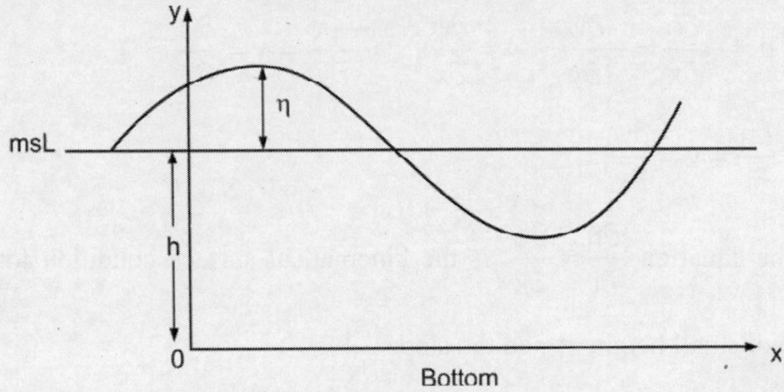

Fig. 26.3

The equation of the free surface is

$$y = h + \eta \qquad \text{.....(26.5)}$$

Since the bottom surface never moves along the fluid,

$$\frac{d}{dt}(y - h - n) = 0$$

or $$\frac{dy}{dt} = \frac{d\eta}{dt} \qquad \because h = \text{constant}$$

$$= \frac{\partial \eta}{\partial t} + \frac{\partial \eta}{\partial x}\frac{\partial x}{\partial t}$$

$\therefore$ $$\frac{dy}{dt} = \frac{\partial \eta}{\partial t} + u\frac{\partial \eta}{\partial x}$$

or $$v = \frac{\partial \eta}{\partial t} + u\frac{\partial \eta}{\partial x} \qquad \frac{dy}{dt} = v = \text{velocity} \qquad \text{.....(26.6)}$$

If we consider very small disturbance, we can neglect square & product terms of small quantities.

$\therefore$ at free surface we get from (26.6)

$$v = \frac{\partial \eta}{\partial t}$$

We know, for stream function ψ,

$$u = -\frac{\partial \phi}{\partial x} = -\frac{\partial \psi}{\partial y}, \quad v = \frac{\partial \psi}{\partial x}, \qquad \frac{-\partial \phi}{\partial y} = v = \frac{\partial \psi}{\partial x}$$

$$\therefore \quad \frac{\partial \eta}{\partial t} = v = \frac{\partial \psi}{\partial x}$$

i.e., The equation $\frac{\partial \eta}{\partial t} = \frac{\partial \psi}{\partial x}$ is the kinematical surface condition for wave profile of small height η and the slope $\frac{\partial \eta}{\partial x}$.

26.4 Tidal Currents

Tides are important for navigation. Tides are produced by a combination of gravitational potential of the moon and sun and the centrifugal force generated by earth's rotation about the common mass centre of the earth-moon-sun system.

Tides develop strong currents over large areas of the ocean. Tidal current speed reaches up to 5 mps in coastal waters. Tidal currents also generate internal waves over sea mounts, continental slopes and at mid-ocean ridges. Tidal currents aid for the deep circulation. It influences climate and abrupt climate change. Even in deep oceans, tidal currents can stop for a while (suspend) bottom sediments.

Earth's crust is elastic and it bends under the impact of tidal potential. The crust also bends under the weight of oceanic tides. Consequently the sea floor and the continents move up and down by about 10 cm in response to the tides.

Oceanic tides lag behind the tide-generating potential. This develops forces that transfer angular momentum between earth and the tide producing body namely the moon and the sun. The coriolis force (earth's rotation) and tidal force slows down the earth's axis of rotation leading to increase of day length and the moon's orbital motion around the earth also slows down. Further this causes the moon to face same side to the earth. Tides also influence the orbits of earth satellites.

It may be noted that it is very very difficult to find or forecast of the amplitude and phase of tides at any place on the ocean or along the coast. It is equally difficult to find the speed and direction of tidal currents. However, predicting tides along coasts and at ports is easier, which is achieved from data of tide gauge and the theory of tidal forcing.

Tides are calculated from the hydrodynamic equation for self-gravitating ocean (on a rotating elastic earth). The driving force is the gradient of the gravity field of the moon and sun. However, hydrodynamic theory of calculation of tides not accurate (because the dissipation of tidal energy is not clear).

26.5 Tsunamis

In Japanese language, Tsunamis meaning harbour waves. In Pacific ocean and Hawaiian islands Tsunamis are prominent. They cause great damage to life and property and also cause coastal climate change.

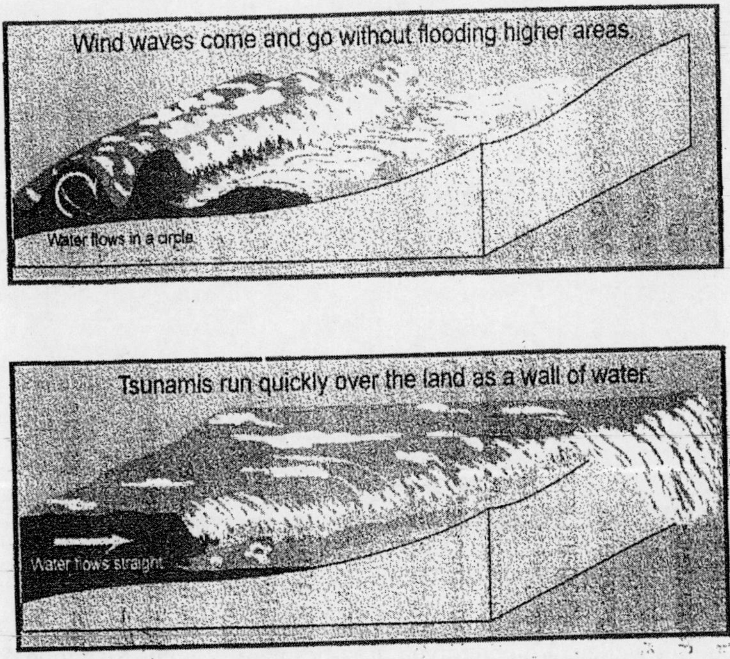

Fig. 26.4 Seawave and Tsunami

Tsunamis are low frequency ocean waves generated by submarine earthquakes/volcanoes/explosions and great landslides into the oceans. The abrupt motion of sea floor over great distance (100-1000 km) generate waves with period 12-15 minutes. These are shallow-water waves propagating in all directions in circular form from the earthquake epicentre with speed ranging 500-1000 kmph, wavelength more than 500 km and wave height over coastal area 2-6 meters. The Tsunamis waves are not noticeable at sea, but after slowing on approach to the coast and after refraction by sub-sea features they come over land shore and waves surge to heights more than 10 m above sea level. Tsunamis can travel around globe like Rossby waves. The first wave of Tsunamis is unlikely to be biggest. Wave amplitudes are relatively large shoreward of submarine ridges. Wave amplitudes decrease in the presence of coral reefs bordering the coasts.

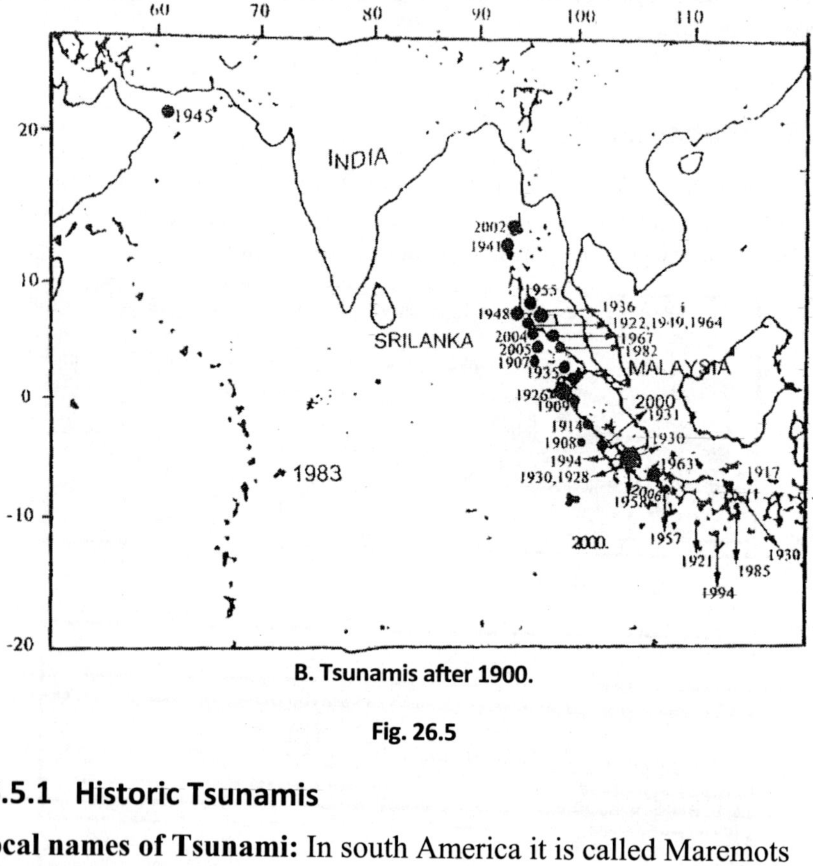

B. Tsunamis after 1900.

Fig. 26.5

26.5.1 Historic Tsunamis

Local names of Tsunami: In south America it is called Maremots

In Tamilnadu (India) it is termed Kadal Kolor Azlupperalai. Other local names – Flut wellen, Veloedgolven, Raz-de-mare, Vagues Sismique, Kaiee.

In 1958 a very huge rockfall dropped into Lituya Bay (Alaska). It created a wave of water that travelled up to the opposite slope of Bay stupping the forest off the slopes at elevation up to 520 m above sea level.

In 1628 BC, a great volcano santorini erruption in the Aagean Sea burried the Bronze age city of Akrotoni on Thera to a depth of 70 m. A tsunami 100-150 m high developed that devastated the north coast of Crete. The Santorini erruption spewed out 30 km explosive magma in four separate phases with tsunami events.

In 1979, in Marare island, an atombomb explosion experiment about 10 lakh m^3 coral reaf sliding in submarine, created a local tsunami wave. Such atombomb explosion tests (about 350) created local tsunamis.

In 1755 (1st Nov) Lisbon rocked by closely spaced earthquakes of magnitude 8.7 on Richter scale, offshore in Atlantic Ocean created. Tsunami wave 10 m high and killed thousands of coastal residents of Spain, Portugal and North Africa.

In 1896, one of the worst Tsunami engulfed entire villages in Sanriktu, Japan. A wave of 20 m high drowned 26000 people.

In 1946, an earthquake in Aleutian island generated a tsunami wave height about 8 m that killed 165 people in Hawaii and Alaska.

In 1948 (1st April) a tsunami hit Hilo (Hawaii), travelled with a speed of about 776 kmph is the deep ocean and slowed down to 56 kmph near shore in Hilo. A first wave was the biggest, height about 30 m near coast swept the Scotch, Gaplighthouse.

In 1948, USA established a Pacific Tsunami warning centre (PTWC) at Hilo in Hawaii.

Succession of Tsunamis: On 22 may 1960, an earthquake of magnitude 9.5 on R-scale triggered a Tsunami in South Chilee (South America) which is the biggest Tsunami on record till that time. This earthquake had downward movement of a subducting oceanic plate (size 1000 km × 300 km). The area moved at a depth was almost as long as California. There were big aftershocks, one of the aftershocks on 22 may 1960 (Sunday) and people ran out to streets from houses. About 15 minutes later sea rose like a tide of 4.5 m above sea level, subsequently sea receded rapidly with hissing noise dragging broken houses and boats into the sea. After about 1hour 15 minutes later a second Tsunami arrived with 8 m wave height with a speed of about 200 kmph, crushed the boats, coastal briddings and men. A third wave arrived at a speed of about 100 kmph with wave height 11 m. The Chilean Tsunami arrived at Hawaii was correctly predicted with one minute delay (9.57 AM was predicted, arrived at 9.58 AM). This tsunami took a human toll of more than 1000 people in Chile, 61 in Hawaii and 199 in Japan. Very high amount of wave pulses bounced back and for the across entire Pacific Ocean.

2004, Indian Ocean Tsunami: On 26 December 2004, the Indian plate slipped below the Burma plate at 6.29AM (IST) releasing phenomenal pressure which forced the sea rise upward. The slipping of plates caused earthquake of magnitude 9.0 on Richter scale. The seismic yield was of the order 32 billion tons of TNT. The epicentre of the quake was in the sea, North of Aceh Province in the Indonesian island of Sumatra. The earthquake was felt in Andaman islands at 8.38 AM (IST) and its magnitude was 6.1 on Richter scale. The pressure wave travelled about 1200 km in 2 hrs 08 minutes (about 600 kmph). The Tsumani waves travelled as far as coast of Africa-Somalia, a distance of about 6500 km from the epicentre. It took a human toll of about 3 lakh people. It hit the Kerala coast (India) and a toll of 162 people there. Earthquakes are highly localised phenomena, but 26 Dec 2004 earthquake/Tsunami caused damage in 13 countries mainly Indonesia, srilanka, India, Thailand. This Tsunami is called the world's worst ever natural disaster because it affected 13 countries.

2011, Japan Tsunami and Fukoshima Nuclear reactor Leakage: On 11 March 2011, a sub-marine earthquake of magnitude 9.0 on Richter scale in Japan generated a ferocious Tsunami on the east coast of Japan. The earthquake was 5th largest in the world since 1900. Walls of water whisked away houses, cars, ships. A ship carrying more than 100 people was swept away by Tsunami, which was swallowed everything in its path as it surged kilometers island before retreating. An explosion at a nuclear plant in Japan destroyed the building, housing the reactor at Fukoshima on next day 12 March, triggering radiation leaks, under the impact of massive earthquake and Tsunami 8 ft wave spewed water, radiation climbed 100 times normal. The earthquake also caused earth's axis to wobble by about 4 inches. The Indonesian 2004 earthquake Tsunami shifted earth's axis by 3 inches, while the Chilean earthquake of 2010 changed axis by 3 inches.

Fukoshima released 25-50% of the amounts of radioactivity released from Chernobyl.

According to Shepard, Pacific Tsunamis have the following features.

(i) Tsunamis appear to be generated by submarine plate movement along a linear fault.

(ii) Tsunamis can travel thousands of kilometers and still cause heavy damages to costal life and property. In general first Tsunami wave may not be the biggest.

(iii) Amplitude of Tsunami waves relatively large shoreward of submarine ridges. They are relatively low shoreward of submarine valleys, provided the features extend into deep sea water.

(iv) Wave amplitudes are decreased by the presence of coral reef bordering the coast.

(v) Tsunami waves can bend and around circular islands with little or no loss of energy. However they are significantly smaller on backsides of elongated angular islands.

(vi) Like Rossby waves Tsunami waves can travel around the globe.

In extreme cases of earthquake intensity scale 10 Richter scale wave surge height may reach to 45-50 m above mean sea level.

Historical records of world Tsunamis up to 2007: Total recorded 1243, associated earthquake magnitudes 6 – 9.5 on Richter scale.

During 1900 – 2001 recorded 790 Tsunamis, of which 82% occurred in Pacific. 80% associated with earthquakes rest with volcanoes & landslides.

80% (624) occurred in Pacific, 12% in Atlantic and 6% in Indian Ocean.

About 17% (134) generated near Japan

15% (118) generated near South America

11% (81) generated near Kuril Islands and Kamchatka

10% (79) generated near Maxico & Central America

9% (78) generated near Philippines

7% (55) generated near Alaska, Westcoast of Canada

7% (55) generated near New Zealand & Tonga

3% (24) generated near Hawai Islands.

Largest number of Tsunamis recorded during 1500 – 2000, about 1139.

26.6 Storm Surges

Storm surge is the rapid rise in the sea level water due to winds, steep pressure falls. Oceans waves and ocean currents associated with a single storm at the time of storm crossing the coast. Storm winds blowing over shallow continental shelves pile up water against the coast. This increase in sea level is called storm surge.

Storm surges inundate coastal low lying areas and cause havoc to life and property. More than 75% loss of life and property in tropical cyclone is attributed to storm surges. All storm surges are associated with severe cyclones but all severe cyclones are not accompanied by storm surges.

The main processes involved in storm surges are:

Winds blowing toward the coast push the water directly toward the coast. Wave interaction (viz wave run- up and other waves) transport water to the coast adding to the above process. Ekman transport by winds parallel to the coast, transport water toward the coast cause rise in sea level. Inverted barometer effect due to pressure fall in the storm centre raises sea level. Storm surge adds to the tides.

The following empirical relation gives slope in surface is proportional to the wind stress

$$\frac{\partial \eta}{\partial x} = \frac{\tau_0}{\rho g h}$$

where

η = sea level (asl)

x = horizontal distance

h = water depth

τ_o = wind stress at the sea surface

ρ = density of sea water

g = acceleration due to gravity.

CHAPTER 27

NUMERICAL MODELS

V. Bjerknes (1904) first suggested that given observed initial fields of mass and velocity, it is possible in principle to determine, the mass and velocity distribution at some future time by solving the hydrodynamical equations including thermodynamic equations [as an initial value problem.] Knowing its initial state.

Analytical solutions to equation of motion practically not possible because of non-linear nature of partial differential equation. Non linear terms exist because of turbulence, realistic orography of seafloor and coast lines. Actual measurements of data both on sea-surface and undersea are sparse, as a consequence numerical models at best provide global view of ocean currents. However models based on observations provide remarkably detailed realistic view of the ocean. Numerical models are algebraic approximations of hydrodynamic equations of motion, continuity equation, hydrostatic equation and equation of state, called primitive equations. The primitive equations are discretized (with the help of numerical interpolations) in terms of weather/ocean governing variables at a number of finite discrete points both in time and space at a local place/region. In this way the continuous equations approximated by equivalent set of algebraic equations. By numerical interpolation techniques the values of variables at some nearby time for that location is found (predicted). This process of prediction (variable values) at a local place is repeated to generate forecast at some arbitrary future time.

The concept is well explained in terms of linear advection equation in one dimension.

$$\frac{\partial F}{\partial t} + c\frac{\partial F}{\partial x} = 0, \quad c > 0, \qquad .....(27.1)$$

where F = F(x, t) any variable is a function of *x* (horizontal distance) and t (time), C is adverting velocity/spead.

The equation (27.1) is a similar to a linear form of first two terms in the equation of motion, the thermodynamic or vorticity equations etc. C is interpreted, in general, as phase velocity (like gravity wave propagation in x-direction).

The analytical solution to equation (27.1) and corresponding numerical difference equations are found which are comparable accuracy.

Let the initial distribution of F as F(x), the solution of this equation is

$$F\,(x, t) = F_{\partial}(x - et) \qquad .....(27.2)$$

i.e., any arbitrary initial wave distribution of F moves at a spead of C along x – axis without change in shape.

The equation $\frac{\partial F}{\partial t} + c\frac{\delta F}{\partial x} = 0$, is a prototype of several important terms that are found in the primitive equations.

Numerical solutions: The equation (27.1) may be discritised by defining values of F on a grid of points in the (x, t) plane.

Let $\quad x = m\Delta x, \qquad m = 0 \pm 1, \pm 2, \pm 3, \qquad .....(27.3)$

$$t = n\Delta t,\ n = 0, 1, 2, 3, \qquad .....(27.4)$$

Which replace the continuous (x, t) space by a mesh or gird of discrete points.

$$F(\,x_0 + m\Delta x, t + n\Delta t) - F(m, n)$$

Using second order finite difference approximation of

$$f'(x) = \frac{f(x+\Delta x) - f(x-\Delta x)}{2\Delta x} + O(\Delta x^2)$$

$$f''(x) = \frac{f(x+\Delta x) - 2f(x) + f(x-\Delta x)}{\Delta x^2} + O(\Delta x^2)$$

Similarly f'(t), f''(t) be written where $O(\Delta x^2)$ and $O(\Delta t^2)$ are truncated errors in finite difference approximation.

With finite difference approximation, the advection equation

$$\frac{\partial F}{\partial t} + C\frac{\partial F}{\partial x} = 0, \qquad e>0$$

is transformed to

$$\frac{F_{m,n+1} - F_{m,n-1}}{2\Delta t} + C\frac{F_{m+1,n} - F_{m-1,n}}{2\Delta x} = 0$$

or $$F_{m,n+1} = F_{m,n-1} - \frac{C\Delta t}{\Delta x}\left(F_{m+1,n} - F_{m-1,n}\right) \qquad .....(27.5)$$

$$F_{m,n+1} = F_{m,n-1} - \frac{C\Delta t}{\Delta x}\left(F_{m+1,n} - F_{m-1,n}\right)$$

Equation (27.5) is called as a three-level scheme, because three time levels are involved, $(m+1)^{th}$ time level, m^{th} time level and $(m-1)^{th}$ time level.

Equation (27.5) represents a simple marching procedure, which is termed leapfrog scheme. In this method the value of F at some location point $m\,\Delta x$ time $(n+1)\Delta t$ is extrapolated from the known values at earlier times n and n–1.

However equation (27.5) cannot be used to calculate $F_{m,1}$ because the values are not known at n = 1 (that is earlier to t = 0, the first time step).

Equation (27.5) can be solved analytically using the method of separating variables.

Exponential solution:

Let $$F_{m,n} = B^{n\Delta t} e^{i\mu\, m\Delta x} \qquad .....(27.6)$$

where B may be complex nuclear.

From equation (27.6) we have

$$F_{m,n} = B^{n\Delta t} e^{i\mu\, m\Delta x}$$

$$F_{m,n+1} = B^{(n+1)\Delta t} e^{i\mu\, m\Delta x}$$

and $$F_{m,n-1} = B^{(n-1)\Delta t} e^{i\mu\, m\Delta x}$$

Substituting the above values in equation (27.5) we have

$$\left[B^{(n+1)\Delta t} - B^{(n-1)\Delta t}\right] e^{i\mu m\Delta x} = -\frac{C\Delta t}{\Delta x}\left[e^{i\mu(m+1)\Delta x} - e^{i\mu(m-1)\Delta x}\right]$$

Cancelling the common factor $F_{m,n}$ we get

$$\left[B^{\Delta t} - B^{-\Delta t}\right] = -\frac{C\Delta t}{\Delta x}\left[e^{i\mu\Delta x} - e^{-i\mu\Delta x}\right]$$

$$\left[\begin{array}{ll} \text{Eulers formula } e^{\pm i\theta} & = \cos\theta \pm i\sin\theta \\ e^{i\mu\Delta x} - e^{-i\mu\Delta x} & = -2i\,\sin\mu\Delta x \end{array}\right]$$

Multiplying the above equation by $B^{\Delta t}$, we have

$$B^{2\Delta t} - 1 = -C\frac{\Delta t}{\Delta x}B^{\Delta t}(2\,i\sin\mu\Delta x)$$

or $$B^{2\Delta t} + 2\,i\,\sigma\,B^{\Delta t} - 1 = 0$$

where $$\sigma = C\frac{\Delta t}{\Delta x}\sin\mu\Delta x$$

which is a quadratic equation in $B^{\Delta t}$

$\therefore$ $$B^{\Delta t} = -i\sigma \pm \sqrt{1-\sigma^2}$$

Two cases considered $|\sigma| \lesssim 1$

(i) Stable case $|\sigma| \leq 1$, no exponential growth

(ii) Unstable case $|\sigma| > 1$, there will be exponential implication

Note:

1. Advection process is that in which a wind or current carries properties like temperature, moisture, density, concentration, etc. between regions/locations that involves gradients of that property (temperature gradient etc) time and the velocity in the direction of the gradient.
2. Equation of state gives the relation between temperature, pressure, salinity and density.

Since analytical solutions are not possible for the primitive hydrodynamic and thermodynamic equations, a recourse is sought NWP (numerical weather prediction) with real initial atmospheric state. NWP models developed are

(i) Quasi Geostrophic limited Area Barotropic model

(ii) Quasi Geostrophic multilevel omega equation

(iii) Two level and four level Quasi Geostrophic models

(iv) Non-linear Balance, using Cressman Technique

(v) Objective analysis using Cressman Technique

(vi) Assimilating data received through Global Telecommunication system (GTS).

National Meteorological Centre Washington, the Florida State University and other institutions and India meteorological Department developed data assimilation systems developed in NWS (National Weather Service's)

27.1 Ocean Models

Oceanographic numerical models: (i) Mechanistic models (ii) simulation models

Mechanistic models developed based on:

1. Planetary waves 2. Interaction of the flow with sea-floor features 3. Wind effect on surface of sea.

These models provide insight into physical mechanisms affecting the ocean.

Simulation models developed for forecasting circulation of oceanic regions. These models make use of primitive equations of motion, continuity equation, hydrostatic equation, sample of equation of state.

Equation of state allows to calculate changes in density due to fluxes of heat and water through sea surface and uses thermodynamic processes.

The global model consisted of horizontal and vertical grid points at a resolution of 2°, with 12 level in vertical. The coarse spatial resolution incorporated viscosity and western boundary currents and mesoscale eddies. With the improvement of computer technology application, the resolution has been increased to 0.03° in North Atlantic model.

Simulation models developed at the Geophysical Fluid Dynamics Laboratory in Princeton. The models provide three dimensional ocean flow. The models used approximations of Boussinesq, equations of continuity, momentum and hydrostatic. These models are called primitive equation models.

The first simulation model was regional.

27.2 Global Ocean Models

Grid spacing 0.1° horizontal and vertical resolution about 30 levels. Models include – (i) realistic coasts, bottom features, (ii) heat & water fluxes through the (sea) surface, (iii) eddy dynamics and (iv) the meridional overturning circulations. Many of these models assimilate data received from satellites and floats. These models must run to calculate variability for about two decades to simulate the ocean. This is termed spin-up. Spin-up is required because initial conditions for density, fluxes of momentum and heat through the sea-surface. The Bryan-cox models are widely used, provide excellent views of the global ocean circulation. Developed as extension of Bryan-cox model, the MOM-Moduler ocean model of Geophysical Fluid Dynamics laboratory most widely used. MOM used for climate studies and ocean circulation studies over wide range of space and time scales.

Main features of recent ocean models:

(i) Models are used to simulate oceanic flows which are nearly correct and useful.

(ii) Models are incorporating or including heat fluxes, wind forcing, mesoscale eddies, orographic features of coasts and seafloor. In vertical resolution contain more than 20 levels and horizontal resolution 0.1°.

(iii) Models still (however) cannot reproduce all turbulence because grid points spaced 10-100 km apart.

(iv) Assimilating ships based data, satellite & drifts data models are predicting correct oceanic conditions which include ENSO in the Pacific and state of Gulf stream in Atlantic.

(v) Ocean-atmospheric coupled models with coarse spatial resolution simulate natural variability of climate system with its increased fluxes of GHGs in the atmosphere.

27.3 NWP in India

A major improvement to NWP in India took place in 1988, with the establishment of NCMRWF with the aid of CRAY-XMP 14 super computer.

Initial variables (boundary conditions) are very critical for improving forecasts. In this regard remote sensing technologies, namely satellites, Doppler radars and Automatic weather stations provided data which had high impact in forecasting weather events by way of assimilation data from these sources. Ministry of Earth sciences (MOES) fully supported to observational and communication networks in India. As a result MOES encouragement Climate Research Centre was established at IITM Pune. At present a number of operational/Research centres and university/institutions are playing important roles in Research and NWP Modelling. The following are important in this research. Andhra University, IIT Delhi, SAC Ahmedabad, IISc Bangalore, ISRO Bangalore, CUSAT Cochin, N10 Goa, N10T Chennai, INC01S Hyderabad, IAF C-DAC Pune, SASE Chandigarh etc.

Some research areas in progress are:

(i) Quasi Geostrophic limited Area Barotropic Model

(ii) Quasi Geostrophic Multivariate Omega equation

(iii) Two level and Four level Quasi Geostrophic Models

(iv) Non-linear Balance Omega equation, using wind as input

(v) Objective Analysis using Cressman Technique

(vi) Decoding meteorological data received through GTS under WWW (WMO), digital gird point data of MRWF produced by Global Meteorological centres and NCEP Washington are widely used. SANS products are also used under WAFS. Under WWW, IMD designated RSMC to provide Cyclone advisories and Warning to the North Indian Ocean.

Recent developments: Input data "Huge data extracted from GTS besides National and Regional data of synop, ship, TEMP/Upper air. Assimilation of this data together with Pseudo observations with main input forecasts produced at the NCMRWF, New Delhi.

***Forecast Model*:** IMD uses limited Area Forecast Model (LAM), which is based on Semi-input Semi- Lagrange, Multilayer Primitive equations (sigma coordinate) Model and Arakawa C-grid in the horizontal which was developed by Florida State University Limited Area Model (FSULAM). The model has a horizontal resolution of 0.75° × 0.75° lat/long (domain 30° S to 50°N and 25°E to 130°E) and 16 vertical sigma levels (1.0 to 0.05). From the operational NCMRWF global spectral model initial boundary conditions are taken. Initial moisture field (from INSAT IR data) over Bay of Bengal and Arabian seas are used. Using this LAM model heavy rainfall events, tropical cyclones and Western Disturbances (Including snow-storms), Monson activity simulated. The model is flexible to run NCEP Global Forecast System (GFS) outputs as initial boundary conditions.

***MM5 Model*:** Mesoscale Models5 developed at NCAR for Atmospheric Research, USA & Pennsylvania State University USA. This model successfully employed for prediction of snow-storm events of Western Disturbances that affect the Western Himalayas.

The MM5 model at IMD has horizontal resolution of 45 km with 23 Sigma levels in the vertical and the integration is carried up to 72 hours over single domain lat 30°S to 45°N and long 25°E to 125°E. Initial and boundary conditions extracted from the NCEP Global Forecast system (GFS). The boundary conditions updated every six hours interval.

***Quasi Lagrangian Model (QLM)*:** IMD operates QLM for track and intensity prediction of tropical cyclones up to 72 hours. The QLM horizontal resolution 40 km and 16 sigma levels in the vertical. The initial and boundary conditions extracted from GFS of NCEP.

27.4 Numerical Models in Oceanography

Type (1) Mechanistic models, (2) simulation models Ocean and Atmospheric Models use different spacing of grid points. Dominant Ocean eddies are $\frac{1}{30}$ the size of dominant atmospheric eddies. Also Ocean features evolve at a rate which is $\frac{1}{30}$ the rate in the atmosphere. As a result, for the same number of grid points, ocean models run 30 times slower than atmospheric models.

27.5 HYCOM (Hybrid Coordinate Ocean Model)

Coordinate (x,y,ρ) where ρ is density. This model is called an isopycnal model. ρ(z) is replaced by z(ρ). Isopycnal surfaces are surfaces of constant density. Primitive equation model driven by wind stress and heat fluxes.

ROMS (Regional Oceanic Modeling System): It is based on the Hybrid coordinate Ocean Model. It is widely used for studying coastal current systems closely connected (knitted) to flow further off shore. For example the Califormia current. In RoMs, Hydrostatic primitive equation (in HYCOM driven by surface fluxes of momentum, heat and water)

Coastal Models: There are many different numerical models for describing (1) Coastal currents, (2) tides and (3) storm surges. The models incorporate (i) beach, (ii) continental slope (iii) a free surface, (iv) realistic coasts and bottom features, river run off and (v) atmospheric forcing. Of many models Princeton Ocean Model is important. It is derived from Bryan-cox model.

This model uses thermodynamic processes, turbulent mixing, and Boussinesq & hydrostatic approximation. Blumberg and Mellor used a vertical coordinate σ scaled by the depth of the water

$$\sigma = (z - \eta) / (H + \eta)$$

Where $z = \eta\,(x, y, t)$ is the sea surface

and $z = H(x, y)$ is the bottom of sea.

The model is driven by wind stress and heat, water fluxes (from meteorological models). The model uses known geostrophic tidal and Ekman currents at the outer boundary. The model usefully applied to calculate three dimensional distributions of velocity, salinity, sea level, temperature and turbulence up to period of 30 days over a region 100-1000 km with grid spacing 1-50 km. The model shortcomings-different results which are adjusted by using vertical & horizontal mixing, spacial & temporal resolution.

27.6 Storm Surge Models

During storms (Cyclones, Hurricanes, Typhoons) wind blowing over shallow, continental shelves (of coasts) water piles up against coast. The rise in sea level is called storm surge. The following are the important processes.

1. Ekman transport by winds parallel to the coast. This transports water toward the coast and causes rise in sea level
2. Wind blowing toward the coast pushes water toward the coast
3. Edge waves are generated by wind travel along the coast
4. The low pressure at the centre of the storm creates inverted barometer effect, which raises sea level at the rate of 1cm rise/1hpa pressure fall
5. Storm surge adds to the routine astronomical tides. High tides trigger large sea level rise even for a weak surge.

The sea level rise and wind stress relation is given by

$$\frac{\partial \zeta}{\partial x} = \frac{T_0}{\rho g H},$$

where ζ = sea level

x = horizontal distance

H = water depth

T_0 = wind stress at the sea surface

Storm surges cause severe damage to coast and structures over it. Storm surges are very high in Bay of Bengal as compared other places over the globe.

In numerical prediction models (1) wind speed calculated at a constant pressure surface. For storm surge, wind strength calculated at 10 m height. (2) The shoreward extent of the model changes with time. (3) The drag coefficient of wind on sea surface is not well known in storms. The drag coefficient of water on the sea floor is also ambiguous. (4) Models include waves and tides that influence sea level in shallow-waters with several draw-backs, the storm-surge heights calculated are very useful which may vary (change) by about 1 meter over a distance of tens of kilometers, overall accuracy is ± 20%.

27.7 Assimilation Models

Model outputs depend on current velocity or surface topography, depend on ocean observations which are calculated. Models that use these type observations are called Assimilation models.

Data Assimilation: (1) A finite number of observations are used. By finite difference method, values of various parameters (observation values) interpolated at infinite number of grid points.

The parameters observed values like wind speed, SST etc interpolated at various grid points are used in ocean dynamics formulae and obtain solutions at some future time scale. Most data now secured from satellites, and other observations

1. Data required for initial conditions for the model.
2. Data introduced into numerical model, which interpolates and smoothes to get best estimate (of density, velocity, SST) etc.
3. The model integrate forward for one week (future time)
4. The new data introduced in the model as in the first step above and the processes is repeated.

The model grid points will have horizontal resolution ($\frac{1}{2}^0$ lat $\frac{1}{2}^0$ long) and 20 levels or fied in the vertical.

Summary

Numerical models are used to simulate oceanic flows. Use: (1) heat fluxes through the sea surface, wind forcing, mesoscale eddies, realistic coastal boundaries and sea-floor features. More than 20 levels in the vertical. Present grid points in horizontal with resolution 0.1° (for ocean circulation). Models provide good estimates but not perfect. Primitive equations are linearised by scale analysis.

Models cannot reproduce all turbulence of ocean

Models can be forced by real time oceangraphic data from ships & satellites.

The large-scale ocean circulation is mainly driven by atmospheric wind stress and heat exchange with atmosphere including insolation. Further the oceanic processes are affected by salinity, evaporation, precipitation or and river run off. Numerical models are reliably used to simulate large scale ocean circulation.

Numerical Weather Prediction (NWP)

The basic or primitive equations are:

(a) Equations, of motion:

$$\frac{du}{dt} = -\alpha\frac{\partial p}{\partial x} + fv + Fx$$

$$\frac{dv}{dt} = -\alpha\frac{\partial p}{\partial y} - fu + Fy$$

$$\frac{dw}{dt} = -\alpha\frac{\partial p}{\partial z} - g$$

(b) Equation of continuity

$$\frac{\partial p}{\partial t} + \nabla \cdot (\rho \vec{V}) = 0$$

(c) Equation of state

$$p\,\alpha = RT$$

(d) First law of thermodynamic

$$C_p dT - \alpha dp = dQ$$

These six equations contain six variables u,v,w,p,α,T which are unknown. By solving these we can predict the weather. The above equations are non-linear differential equations and in general don't have analytical solutions. In order to solve, we have to resort to numerical methods of solution.

Charney's scale analysis shows that atmosphere is in hydrostatic state and nearly in geostrophic balance. Vertical pressure gradient is roughly balances the force of gravity and horizontal pressure gradient force roughly balances the coriolis force.

Filtering: The solutions of primitive equations include Rossby waves (which are significant in meteorology) and unwanted gravity & sound waves. It becomes necessary to eliminate gravity & sound wave effects. Sound waves can be eliminated by the use of hydrostatic relation. Gravity waves can be eliminated by the judicious use of geostrophic approximation.

SOME BASIC CONCEPTS IN MODELING

1. Taylor Series

$$f(x+h) = f(x) + h.f'(x) + \frac{h^2}{L^2} f''(x) + \frac{h^3}{L^3} f'''(x) - \ldots\ldots +$$

where $h > 0$

$$f'(x) = \frac{f(x+h) - f(x)}{h} + R \text{ , where } R = -\frac{h}{2} f''(x)$$

Forward difference or $f'(x) = \dfrac{f(x+h) - f(x)}{h}, R \simeq 0$

$$f(x+h) = f(x) + h\,f'(x) + \frac{h^2}{\underline{|2}} f''(x) + \frac{h^3}{\underline{|3}} f'''(x) + \ldots \qquad \ldots..(1)$$

$$f(x-h) = f(x) - h\,f'(x) + \frac{h^2}{\underline{|2}} f''(x) - \frac{h^3}{\underline{|3}} f'''(x) + \ldots \qquad \ldots..(2)$$

Eq. 1 – Eq. 2 gives

$$f(x+h) - f(x-h) = 2hf'(x) + 2\frac{h^3}{6} f''(x)$$

Centered difference

$$f'(x) = \frac{f(x+h) - f(x-h)}{2h} + O(h^2)$$

Eq. 1 + Eq. (2) gives

$$f(x+h) + f(x-h) = 2f(x) + h^2 f''(x) + \ldots\ldots\ldots\ldots$$

$$f''(x) = \frac{f(x+h)+f(x-h)-2f(x)}{h^2} + O(h^2)$$

$$= \frac{f(x+h)-2f(x)+f(x-h)}{h^2} + O(h^2)$$

Finite differences

2. Newton-Gregory Formula for Forward Interpolation

Let $y = f(x)$ and let consecutive values of x be

$a, a+h, a+2h, a+3h, \ldots\ldots\ldots a+\overline{n-1}h + a + nh$

Then corresponding to these values of x of $y = f(x)$ are

$f(a), f(a+h), f(a+2h), f(a+3h) \ldots\ldots\ldots f(a+\overline{n-1}\,h), f(a+nh)$

The first difference of the function $y = f(x)$ is defined as

$$\Delta f(a) = f(a+h) - f(a)$$

$$\Delta f(a+h) = f(a+2h) - f(a+h)$$

$$\Delta f(a+2h) = f(a+3h) - f(a+2h) \qquad \ldots\ldots(A)$$

$$\ldots\ldots\ldots\ldots\ldots\ldots\ldots\ldots\ldots\ldots\ldots\ldots$$

$$\Delta f(a+\overline{n-1}h) = f(a+nh) - f(a+\overline{n-1}h)$$

where Δ is called the forward or decreasing difference operator.

The second difference (Δ^2)

The differences of the first forward differences of (A) given above are called second forward differences, denoted by Δ^2 f(a), Δ^2 f(a + h), Δ^2 f(a + 2h) etc....

where

$$\Delta^2 f(a) = \Delta f(a+h) - \Delta f(a)$$

$$= [f(a+2h) - f(a+h)] - [f(a+h) - f(x)]$$

$$= f(a+2h) - 2f(a+h) + f(a)$$

$$\Delta^n f(a+h) = \Delta f(a+2h) - \Delta f(a+h)$$

$$= f(a+3h) - 2f(a+2h) + f(a+h)$$

In general the nth forward difference is given by

$$\Delta^n f(x) = \Delta^{n-1} f(x+h) - \Delta^{n-1} f(x)$$

Note: Δ is an operator, Δ^2 is not square of the operator Δ, but it is the repetition of the operator Δ.

Backward or ascending differences (∇)

The differences: $\nabla f(a+h) = f(a+h) - f(a)$

$$\nabla f(a+2h) = f(a+2h) - f(a+h)$$

$$\nabla f(a+3h) = f(a+3h) - f(a+2h)$$

$$\ldots\ldots\ldots\ldots\ldots\ldots\ldots\ldots\ldots\ldots$$

$$\nabla f(a+nh) = f(a+nh) - f(a+\overline{n-1}h)$$

Second backword difference are denoted by ∇^2 given by

$$\begin{aligned}\nabla^2 f(a+2h) &= \nabla f(a+2h) - \nabla f(a+h)\\ &= [f(a+2h) - f(a+h)] - [f(a+h) - f(a)]\\ &= f(a+2h) - 2f(a+h) + f(a)\end{aligned}$$

Similarly third backward difference dented by ∇^3, given by

$$\nabla^3 f(a+3h) = \nabla^2 f(a+3h) - \nabla^2 f(a+2h)$$

and so on

3. Divided differences

Let y = f(x) be a function and let y_0, y_1, y_2, …… y_n (or) $f(x_0)$, $f(x_1)$, $f(x_2)$ …… f(x) be the values of corresponding to any values x_0, x_1, x_2, ….. x_n of the argument. Then the divided differences of y is ascending order are defined as follows.

Fist order divided differences

$$\delta(x_1, x_0) = \frac{y_1 - y_0}{x_1 - x_0}$$

$$\delta(x_2, x_1) = \frac{y_2 - y_1}{x_2 - x_1}$$

$$\ldots\ldots\ldots\ldots\ldots\ldots\ldots\ldots$$

$$\delta(x_n, x_{n-1}) = \frac{y_n - y_{n-1}}{x_n - x_{n-1}}$$

[where Notation: $f(x_1, x_0) = \delta(x_1, x_0)$ etc]

Second order divided difference

$$\delta(x_2, x_1, x_0) = \frac{\delta(x_2, x_1) - \delta(x_1, x_0)}{x_2 - x_0}$$

$$\delta(x_3, x_2, x_1) = \frac{\delta(x_3, x_2) - \delta(x_2, x_1)}{x_3 - x_1}$$

etc

Third order differences

$$\delta(x_3, x_2, x_1, x_0) = \frac{\delta(x_3, x_2, x_1) - \delta(x_2, x_1, x_0)}{x_3 - x_0}$$

$$\delta(x_4, x_3, x_2, x_1) = \frac{\delta(x_4, x_3, x_2) - \delta(x_3, x_2, x_1)}{x_4 - x_1}$$

etc

Symmetry of divided difference

$$\delta(x_1, x_0) = \frac{y_1 - y_0}{x_1 - x_0} = \frac{y_0 - y_1}{x_0 - x_1} = \delta(x_0, x_1)$$

4 Newton's General Interpolations formula for unequal intervals of argument.

1. $\delta(x, x_0) = \dfrac{y - y_0}{x - x_0}$
2. $\delta(x, x_0, x_1) = \dfrac{\delta(x, -x_0) - \delta(x_0, x_1)}{x - x_1}$
3. $\delta(x, x_0, x_1, x_2) = \dfrac{\delta(x, x_0, x_1) - \delta(x_0, x_1, x_2)}{x - x_2}$
4. $\delta(x, x_0, x_1, x_2, x_3) = \dfrac{\delta(x, x_0, x_1 x_2) - \delta(x_0, x_1, x_2, x_3)}{x - x_3}$

   From (1) we have

5. $y = y_0 + (x - x_0)\delta(x, x_0)$

   From (2), $\delta(x, x_0) = \delta(x_0, x_1) + (x - x_1)\delta(x, x_0, x_1)$

   using above expression in (5) we get

   $y = y_0 + (x - x_0)\left[\delta(x_0, x_1) + (x - x_1)\delta(x, x_0, x_1)\right]$

6. $y = y_0 + (x - x_0)\delta(x_0, x_1) + (x - x_0)(x - x_1)\delta(x, x_0, x_1)$

   From (3), $\delta(x, x_0, x_1) = \delta(x_0, x_1, x_2) + (x - x_2)\delta(x, x_0, x_1, x_2)$

   Substituting this in (6) we get

7. $y = y_0 + (x - x_0)\delta(x_0, x_1) + (x - x_0)(x - x_1)\delta(x_0, x_1, x_2) +$

   $(x - x_0)(x - x_1)(x - x_2)\delta(x, x_0 x_1, x_2)$

   Continuing in this manner, the general **Newton formula with divided difference is**

$$y = y_0 + (x - x_0)\delta(x_0, x_1) + (x - x_0)(x - x_1)\delta(x_0, x_1, x_2)$$
$$+(x - x_0)(x - x_1)(x - x_2)(x - x_2)\delta(x_0, x_1, x_2, x_3) + \ldots\ldots\ldots$$
$$+(x - x_0)(x - x_1)(x - x_2)\ldots\ldots\ldots(x - x_{n-1})\delta(x_0, x_1, x_2\ldots\ldots\ldots x_n)$$
$$+R_{n+1}$$

where

$$R_{n+1} = (x - x_0)(x - x_1)(x - x_2)\ldots\ldots\ldots(x - x_n)\delta(x, x_0, x_1\ldots\ldots\ldots x_n)$$

5 Lagrange's Interpolation Formula

Let $f(x) = y$, denote a polynomial of nth degree which takes values $y_0, y_1, y_2, y_3, \ldots\ldots y_n$ for the argument $x = x_0, x_1, x_2, \ldots\ldots\ldots x_n$ respectively, then

$$y = \frac{(x - x_1)(x - x_2)(x - x_3)\ldots\ldots\ldots\ldots(x - x_n)}{(x_0 - x_1)(x_0 - x_2)(x - x_3)\ldots\ldots\ldots(x_0 - x_n)} y_0$$
$$+ \frac{(x - x_0)(x - x_2)(x - x_3)\ldots\ldots\ldots\ldots(x - x_n)}{(x_1 - x_0)(x_1 - x_2)(x_1 - x_3)\ldots\ldots\ldots\ldots(x_1 - x_n)} y_1$$
$$+ \frac{(x - x_0)(x - x_2)(x - x_3)\ldots\ldots\ldots\ldots(x - x_n)}{(x_2 - x_0)(x_2 - x_1)(x_2 - x_3)\ldots\ldots\ldots(x_2 - x_n)} y_2$$
$$+ \ldots\ldots\ldots\ldots\ldots\ldots\ldots\ldots\ldots\ldots\ldots\ldots\ldots\ldots\ldots\ldots$$
$$+ \frac{(x - x_0)(x - x_1)(x - x_2)\ldots\ldots\ldots\ldots(x - x_{n-1})}{(x_n - x_0)(x_n - x_1)(x_n - x_2)\ldots\ldots(x_n - x_{n-1})} y_n$$

APPENDIX

Some definitions and notations used in Geophysical Fluid Dynamics (Meteorology and Oceanography)

1. **Gradient** $(\nabla\phi)$**:** The horizontal gradient (or slope) of a scalar element ϕ is given by the formula

$$\nabla\phi = \left(i\frac{\partial}{\partial x} + j\frac{\partial}{\partial y} \right)\phi = i\frac{\partial\phi}{\partial x} + j\frac{\partial\phi}{\partial y}$$

in three dimensions

$$\nabla\phi = \left(i\frac{\partial}{\partial x} + j\frac{\partial}{\partial y} + k\frac{\partial}{\partial z} \right) = i\frac{\partial\phi}{\partial x} + j\frac{\partial\phi}{\partial y} + k\frac{\partial\phi}{\partial z}$$

where i, j, k are unit vectors along the axis x, y & z respectively.

$\frac{\partial}{\partial x}, \frac{\partial}{\partial y}, \frac{\partial}{\partial z}$ are partial derivatives of the element ϕ and components of the gradient ϕ along the directions x, y and z respectively.

The horizontal gradient is a vector in the direction of the maximum rise (or increase) in magnitude of the element ϕ and is numerically equal to $\frac{\partial\phi}{\partial n}$.

The vertical gradient of the element ϕ is $\frac{\partial\phi}{\partial z}$, where z is the attitude.

2. For the calculations of these derivatives the following approximate formulas used with respect to the figs. as shown below with reference to for Fig. A.1.

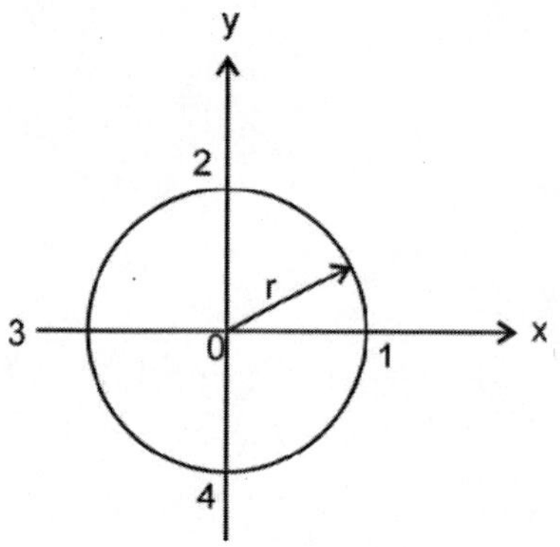

Fig. A.1

The respective values of ϕ at the points 0, 1, 2, 3, 4 denoted b $\phi_0, \phi_1, \phi_2, \phi_3, \phi_4$ respectively

$$\left(\frac{\partial\phi}{\partial x}\right)_0 = (\phi_x)_0 \simeq \frac{\phi_1 - \phi_3}{2r}$$

$$\left(\frac{\partial\phi}{\partial y}\right)_0 = (\phi_y)_0 \simeq \frac{\phi_2 - \phi_4}{2r}$$

$$\left(\frac{\partial^2\phi}{\partial x^2}\right)_0 = (\phi_{xx})_0 \simeq \frac{\phi_1 + \phi_3 - 2\phi_0}{r^2}$$

$$\left(\frac{\partial^2\phi}{\partial y^2}\right)_0 = (\phi_{yy})_0 \simeq \frac{\phi_2 + \phi_4 - 2\phi_0}{r^2}$$

The above method of representation is called central differences.

$$\left.\begin{aligned} \left(\frac{\partial\phi}{\partial x}\right)_0 &\simeq \frac{\phi_1 - \phi_0}{r} \\ \left(\frac{\partial\phi}{\partial y}\right)_0 &\simeq \frac{\phi_2 - \phi_0}{r} \end{aligned}\right\} \quad \text{Method of one sided differences.}$$

3. With reference to Fig. A.2 [circular grid]

$$\left(\frac{\partial\phi}{\partial x}\right)_0 = (\phi_x)_0 \simeq \frac{1}{4r}\left[(\phi_1 - \phi_3) + 0.7(\phi_5 + \phi_8 - \phi_6 - \phi_7)\right]$$

$$\left(\frac{\partial\phi}{\partial y}\right)_0 = (\phi_y)_0 \simeq \frac{1}{4r}\left[(\phi_2 - \phi_4) + 0.7(\phi_5 + \phi_6 - \phi_7 - \phi_8)\right]$$

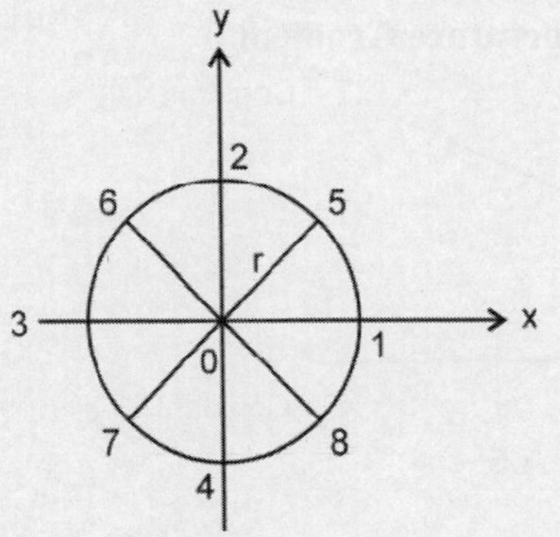

Fig. A.2

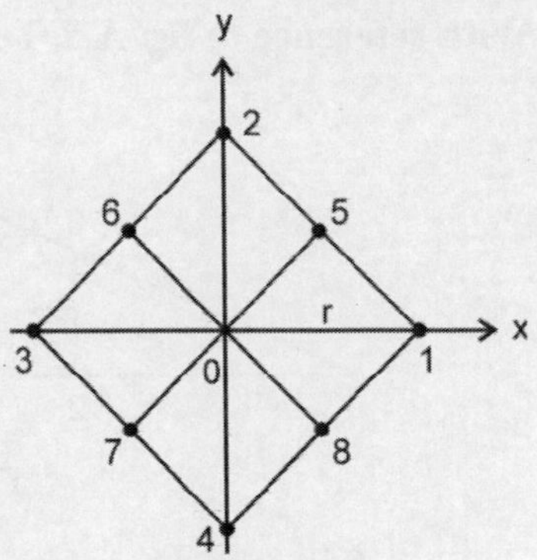

Fig. A.3

With reference to Fig A.3, square grid points

$$\left(\frac{\partial\phi}{\partial x}\right)_0 = (\phi_x)_0 = \frac{1}{4r}[\phi_1 + \phi_5 + \phi_8 - \phi_3 - \phi_6 - \phi_7]$$

$$\left(\frac{\partial\phi}{\partial y}\right)_0 = (\phi_y)_0 = \frac{1}{4r}[\phi_2 + \phi_5 + \phi_6 - \phi_4 - \phi_7 - \phi_8]$$

4. With reference to Fig. A.4, Hexagonal

$$\frac{\partial\phi}{\partial x} = \frac{\phi_1 + \phi_2 + \phi_6 - (\phi_3 + \phi_4 + \phi_5)}{4r}$$

$$\frac{\partial\phi}{\partial y} = \frac{(\phi_2 + \phi_3) - (\phi_5 + \phi_6)}{2r\sqrt{3}}$$

$$\nabla^2\phi = \frac{\partial^2\phi}{\partial x^2} + \frac{\partial^2\phi}{\partial y^2} = \frac{2}{3r^2} = \left[\phi_1 + \phi_2 + \phi_3 + \phi_4 + \phi_5 + \phi_6 - 6\phi_0^1\right]$$

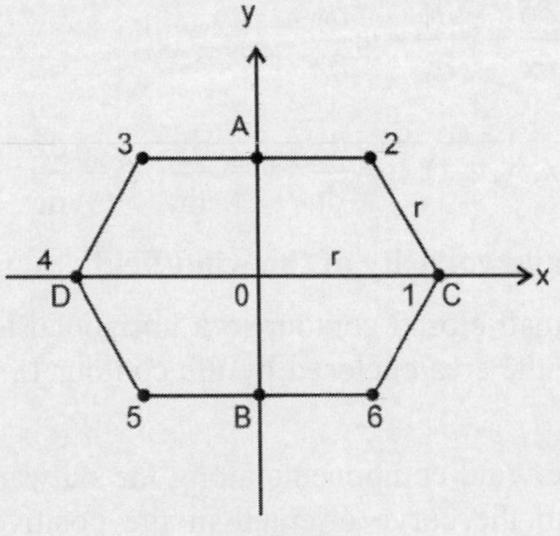

Fig. A.4

5. With reference to fig. A.5, Temperature Gradient

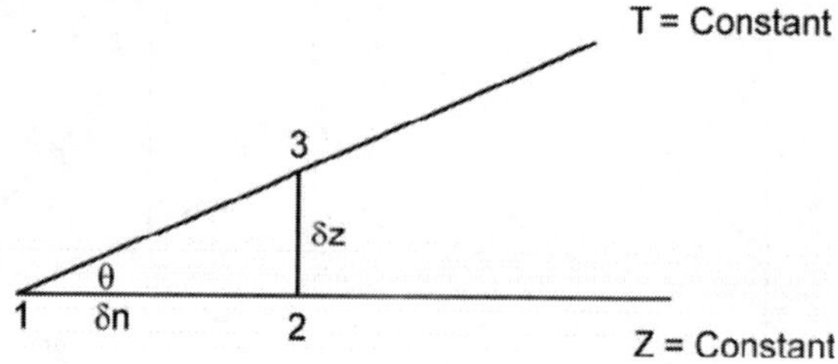

Fig. A.5

$$\text{Tan}\theta = \frac{\delta z}{\delta n}$$

$$\frac{\partial T}{\partial z} = \frac{T_3 - T_2}{\delta z} \quad \text{or} \quad \delta z = \frac{T_3 - T_2}{\left(\frac{\partial T}{\partial z}\right)}$$

$$\frac{\partial T}{\partial n} = \frac{T_2 - T_1}{\delta n} \quad \text{or} \quad \delta n = \frac{T_2 - T_1}{\left(\frac{\partial T}{\partial n}\right)}$$

$$\text{Tan}\theta = \left(\frac{T_3 - T_2}{T_2 - T_1}\right)\left[\frac{\frac{\partial T}{\partial n}}{\frac{\partial T}{\partial z}}\right]$$

6. **Relation between individual and local time derivatives**

The relation between individual and local time derivatives of an element ϕ is given by the formula

$$\frac{d\phi}{dt} = \frac{\partial \phi}{\partial t} + u\frac{\partial \phi}{\partial x} + v\frac{\partial \phi}{\partial y} + \omega\frac{\partial \phi}{\partial z}$$

where $\phi = \phi(x, y, z, t)$, $u = \frac{dx}{dt}$, $v = \frac{dy}{dt}$, $\omega = \frac{dz}{dt}$

7. **Divergence and vorticity of the wind field**

Let L be a small closed contour in a horizontal isobaric or isentropic surface, S be the area enclosed by the contour L, V be the velocity of the field.

V_n, V_L be the wind components along the outward normal and along the tangent to the curve oriented in the positive direction (counter clockwise direction)

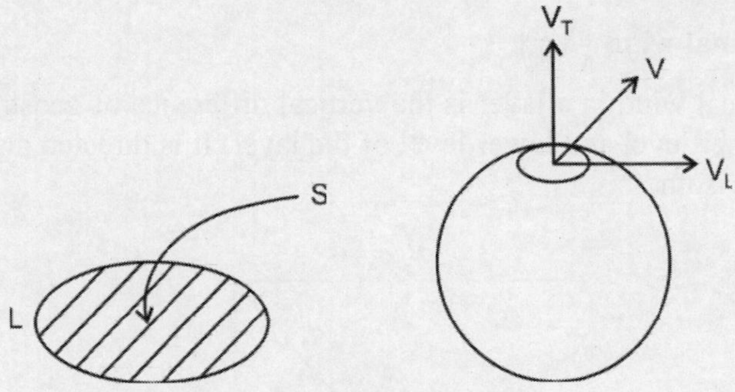

then $D = \text{Divergence} = \underset{s \to 0}{Lt} \frac{1}{S} \oint_L V_n dl$

$$\zeta = \text{Vorticity} = \underset{s \to 0}{Lt} \frac{1}{S} \oint_L V_i dl$$

where S = the area enclosed by the closed contour L

In the horizontal field with u, ν velocity components in x, y directions (in Cartesian coordinates)

$$D = \frac{\partial u}{\partial x} + \frac{\partial \nu}{\partial y}, \zeta = \frac{\partial \nu}{\partial x} - \frac{\partial \mu}{\partial y}$$

In natural coordinates

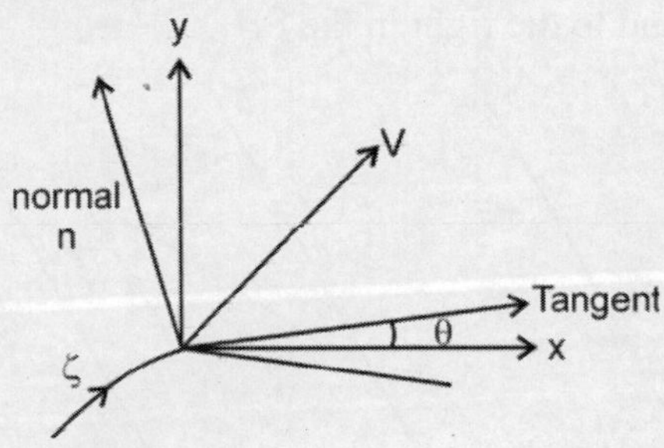

$$D = \frac{\partial v}{\partial s} + V\frac{\partial \theta}{\partial n}$$

$$\zeta = -\frac{\partial v}{\partial n} + V\frac{\partial \theta}{\partial E}$$

where

S = is a natural coordinate along positive direction of the streamline

n = is along the normal to the streamline directed to the left of the S direction

V = wind speed

θ = is the angle between the tangent to the streamline and an arbitrary fixed direction (East or x axis)

8. Thermal wind $\left(\vec{V}_T\right)$

(i) Thermal wind in a layer is the vertical difference of geostrophic wind at upper level and lower level of the layer. It is directed along tangent to the isotherms.

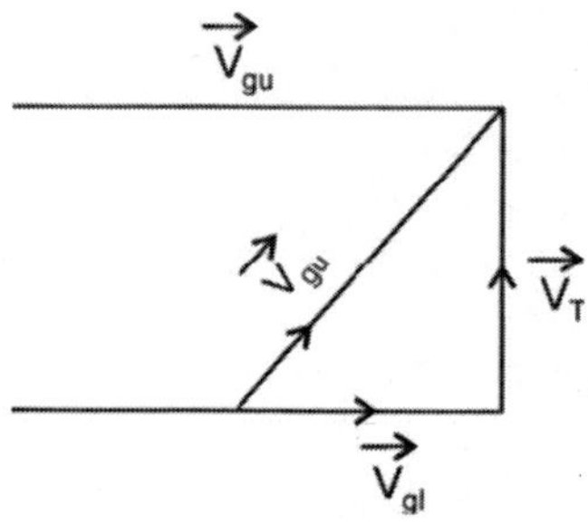

$$\vec{V}_T = \vec{V}_{gu} - \vec{V}_{gl}$$

where

$\vec{V}_T$ = Thermal wind vector

$\vec{V}_{gu}$ = Geostrophic wind vector in the upper level of the layer

$\vec{V}_{gl}$ = Geostrophic wind vector in the lower level of the layer

(ii) $\vec{V}_T$ will be parallel to the isotherms with low temperature to the left of it in the NH and to the right in the SH

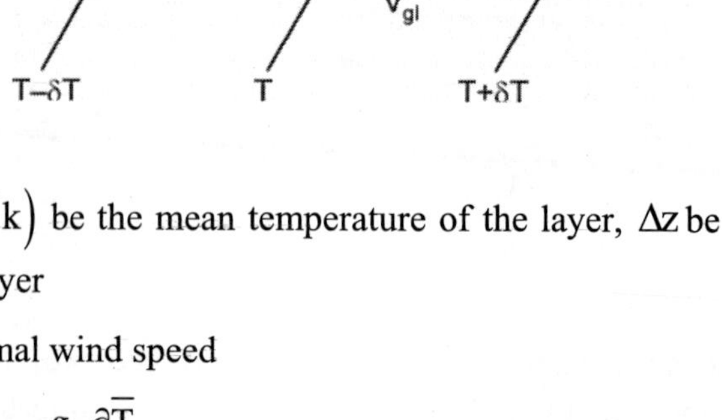

(iii) Let $\overline{T}\left({}^0k\right)$ be the mean temperature of the layer, Δz be the thickness of the layer

$\vec{V}_T$ thermal wind speed

then $\vec{V}_T = \dfrac{g}{f\overline{T}} \dfrac{\partial \overline{T}}{\partial n} \Delta z$

where $\frac{\partial \bar{T}}{\partial n}$ = temperature gradient in the layer

(iv) Thermal wind in an isobaric layer

Let $\frac{\partial H}{\partial n}$ = the horizontal gradient of the relative geopotential

where H is in gpm (geopotential meters)

then $\bar{V}_T = \frac{9.8}{f}\frac{\partial H}{\partial n}$

where $\bar{V}_T$ is thermal wind speed.

9. The geostrophic advection of temperature

The geostrophic advection of temperature is defined as the local change of temperature per unit time caused by the advection of temperature with the geostrophic wind. It is given by the formula.

$$\left(\frac{\partial T}{\partial t}\right)_a = -V_g \frac{\partial T}{\partial n}\cos\alpha = -\frac{1}{pf}\frac{\partial p}{\partial n}\frac{\partial T}{\partial n}\sin\delta$$

$(\ )_a$ = Subscript for advection

where $\left(\frac{\partial T}{\partial t}\right)_a$ = geostrophic advection of temperature

α = angle between the geostrophic wind and the temperature gradient

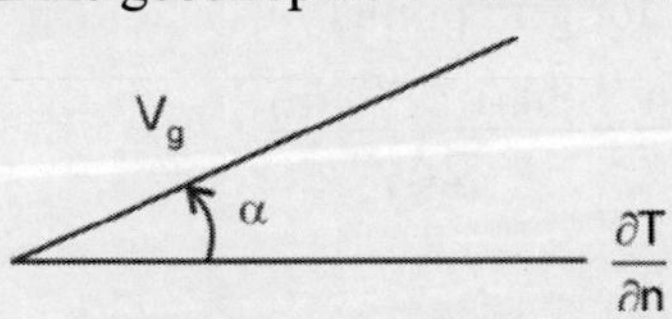

δ = angle from the pressure gradient to temperature gradient counted positive with anticlockwise direction

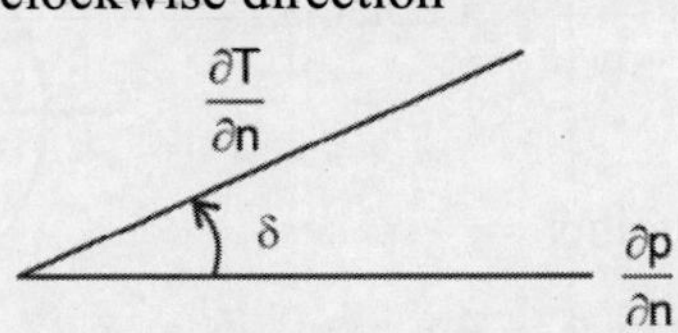

Approximate geostrophic advection is given by $\left(\frac{\partial T}{\partial t}\right)_a \simeq \frac{f}{g}TV_g^2\frac{\partial \delta}{\partial z}$

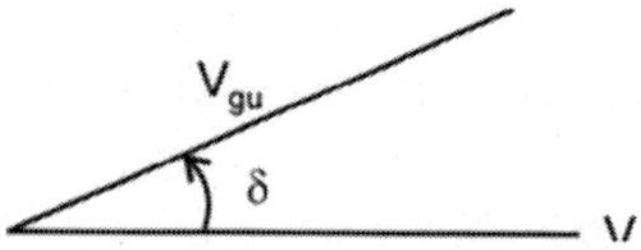

where

δ = angle between the wind at lower level and upper level, counted positive in the anticlockwise direction.

Note: In NH, the veering of geostrophic wind with height indicates warm air advection and backing of geostrophic wind with height indicates cold in advection.

Definition: ***Entropy:*** The physical quantity which describes the ability of the system to do work is called entropy of the system.

If $\oint \frac{dQ}{T} = o$ or $\oint d\phi = o$, where $\phi = \frac{dQ}{T}$, then ϕ is called entropy.

Definition: *Isentropic process:* A thermodynamic process in which entropy of the system remains constant (does not change) is called isentropic process.

10. The vorticity of the geostrophic wind (ζ_g)

$$\zeta_g = \frac{\partial v_g}{\partial x} - \frac{\partial u_g}{\partial y}, \text{ where } \vec{V}_g = iu_g + jv_g$$

Relative vorticity for grid points

$$\zeta = \frac{v_{(i+1,j)} - v_{(i-1,j)}}{2\Delta x} - \frac{u_{(ij+1)} - u_{(i,j-1)}}{2\Delta y}$$

Also $\zeta_g = \frac{1}{pf} \nabla^2 P$

where $\nabla^2 = \frac{\partial^2}{\partial x^2} + \frac{\partial^2}{\partial y^2}$ Laplacian operator

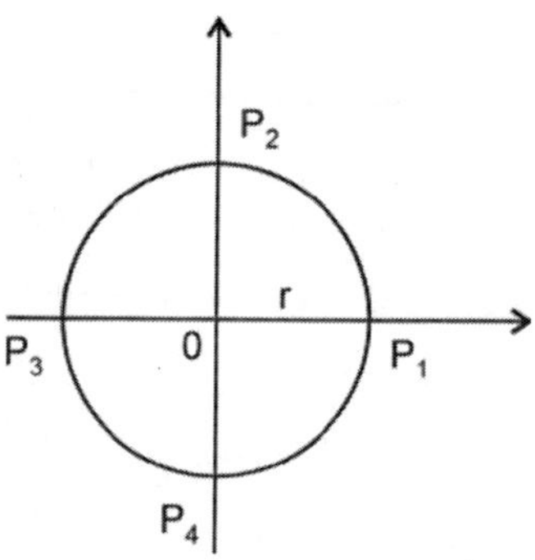

For circular grid points

$$\nabla^2 p = \frac{p_1 + p_2 + p_3 + p_4 - 4p_0}{r^2}$$

Expression in gpm

$$\zeta_g = \frac{9.8}{f} \nabla^2 H$$

11. Divergence of the geostrophic wind

$$\text{Div. } \vec{V}_g = \nabla \vec{V}_g = \left(i\frac{\partial}{\partial x} + j\frac{\partial}{\partial y} \right)(iU_g + j\nu_g)$$

$$= \frac{\partial u_g}{\partial x} + \frac{\partial \nu_g}{\partial y}$$

Also $\text{Div. } \vec{V}_g = \frac{V_g}{R \tan \phi}$

where R = Radius of the earth

ϕ = Latitude

Note: Geostrophic divergence is found between ridge and trough (V_g is negative) and Geostrophic convergence is found between trough and ridge (V_g is positive)

For grid point estimation

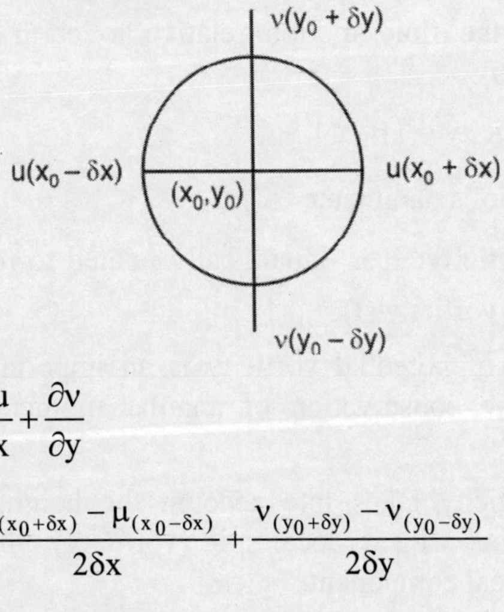

$$\text{Div. } \vec{V} = \frac{\partial u}{\partial x} + \frac{\partial \nu}{\partial y}$$

$$\text{Div. } \vec{V} = \frac{\mu_{(x_0+\delta x)} - \mu_{(x_0-\delta x)}}{2\delta x} + \frac{\nu_{(y_0+\delta y)} - \nu_{(y_0-\delta y)}}{2\delta y}$$

If $\delta x = \delta y = d$, then

$$\text{Div. } \vec{V} = \frac{u_{(x_0+d)} - u_{(x_0-d)}}{2d} + \frac{\nu_{(y_0+d)} - \nu_{(y_0-d)}}{2d}$$

12. Vorticity

The rotation of fluid is vorticity

Vorticity $= 2 \times$ angular velocity at a point in a field

$= 2\vec{\Omega}$

Definition: $\vec{W} = \nabla \times \vec{V}$

where $\overrightarrow{W}$ = vorticity, $\nabla = i\frac{\partial}{\partial x} + j\frac{\partial}{\partial y} + k\frac{\partial}{\partial z}$

Velocity of the fluid = $\overrightarrow{V} = iu + iv + k\omega$

$$\overrightarrow{W} = i\left(\frac{\partial v}{\partial z} - \frac{\partial \omega}{\partial y}\right) + j\left(\frac{\partial \omega}{\partial x} - \frac{\partial u}{\partial z}\right) + k\left(\frac{\partial v}{\partial x} - \frac{\partial u}{\partial y}\right)$$

$$= i\,\xi + jn + k\zeta$$

$$\left.\begin{array}{l}\text{Relative vorticity or local vorticity}\\ \text{or vertical component of vorticity}\end{array}\right\} = \zeta = \frac{\partial v}{\partial x} - \frac{\partial u}{\partial y}$$

Fluids have vorticity because of earth's rotation.

This is called planetary vorticity = 2Ω

$$\text{where } \Omega = \frac{2\pi}{\text{day}} = \frac{2\pi}{86160\,\text{sec}} = 7.293 \times 10^{-5}\ /\ \text{sec}$$

Planetary vorticity $2\Omega = 1.4586 \times 10^{-4}\ /\ \text{sec}$

Vorticity of the fluid motion relative to earth's surface is called relative vorticiy

Absolute vorticity = $(f + \zeta)$

where f = coriolis parameter

Potential vorticity: It is dynamically related to relative vorticity (ξ) and planetary vorticity (f).

Conservation of potential vorticity is an important concept in fluid dynamics (like conservation of angular momentum in solid body mechanics.)

Potential vorticity takes into account the height 'H' of the water column together with its local spin (vorticity) when considered only the local vertical component.

$$\text{Potential vorticity} = \frac{f + \zeta}{H}$$

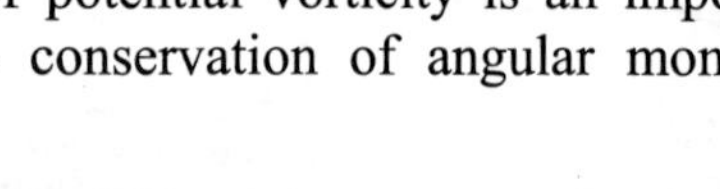

13. Vorticity of the Geostrophic wind

The vorticity of the geostrophic wind is approximately given by the formula

$$\zeta_9 = \frac{1}{pf}\nabla^2 p$$

where $\nabla^2 = \nabla.\nabla = \frac{\partial^2}{\partial x^2} + \frac{\partial^2}{\partial y^2}$ two dimensional laplacian operater

In the field of geopotential, the geopotential vorticity is given by

$$\zeta_g = \frac{9.8}{f}\nabla^2 H$$

where H is the thickness of fluid column measured in gpm.

Application of the vorticity equation

The absolute vorticity is conserved at the level of non-divergence (about 500 hpa). At this level (LND) one may write the vorticity equation in the form

$$\frac{\partial \zeta}{\partial t} + u\frac{\partial \zeta}{\partial x} + v\frac{\partial \zeta}{\partial y} + \beta v = 0$$

where $\beta = \frac{df}{dy}$, meridional gradient of coriolis parameter

= Rossby parameter.

The term $-\left(u\frac{\partial \zeta}{\partial x} + v\frac{\partial \zeta}{\partial y}\right)$ is called vorticity advection (with negative sign).

14. Rossby waves: The waves which are greatly influenced by the meridional gradient of coriolis parameter is called Rossby waves.

The differential equation of the Rossby waves may be expressed as:

$$\frac{\partial \nabla^2 H}{\partial t} + \frac{u\partial \nabla^2 H}{\partial x} + \beta\frac{\partial \nabla H}{\partial x} = 0$$

where H is the geopotential of an isobaric surfaces is = (equal) zonal flow in subtropics.

$\nabla^2 = \frac{\partial^2}{\partial x^2} + \frac{\partial^2}{\partial y^2}$, where x-axis denotes east and y-axis denotes north

Rossby parameter = $\beta = \frac{\partial f}{\partial y} = \frac{2\Omega}{R}\cos\phi$

ϕ= latitude

R = Radius of the earth.

The phase velocity of simple Rossby waves is given by

$$C = u - \frac{\beta L^2}{4\pi^2},$$

where

L = wave length

Regelation: Ice melts under pressure (that is, when pressure increases) and resolidifies when pressure is decreased. This phenomena is called Regelation.

State Variable: A quantity is called state variable if the integral of that quantity around any closed path is zero.

If $\oint d\phi = 0$, then the quantity ϕ is a state variable.

A necessary and sufficient condition that a differential be exact is that its integral around a closed path is zero.

15. **Convection level:** The height δz, which is the ascending air parcel, warmer than the surrounding will attain the temperature of the surrounding is called convection level.

$$\delta z = \frac{T_p - T_3}{\frac{\partial \overline{T}}{\partial z} - \frac{\partial T}{\partial z}}$$

where

T_p = Temperature the parcel (particles)

T_s = Temperature of the surrounding

$\frac{\partial \overline{T}}{\partial z}$ = Temperature gradient in the atmosphere

$\frac{\partial T}{\partial z}$ = Temperature gradient of the parcel particles

If the motion of parcel is assumed to be adiabatic, then

$$\delta z = \frac{T_p - T_s}{\gamma_a + \frac{\partial \overline{T}}{\partial z}}$$ where γ_a = adiabatic lapse rate

16. **Momentum equation**

$$\frac{d\vec{v}}{dt} = -\alpha \nabla_p - 2\vec{\Omega} \times \vec{v} + \vec{g} + \overline{F}_n$$

where $\alpha = \frac{1}{\rho}, \vec{v}$ = velocity vector

∇_p = pressure gradient

$2\vec{\Omega} \times \vec{V}$ = coriolis force

$\bar{g}$ = gravity (acceleration)

$\bar{F}$ = Friction

$\bar{\Omega} = 7.292 \times 10^{-5}$ radians/second

In Cartesian coordinates $\vec{V} = iu + j\nu + k\omega$

$$\frac{\partial u}{\partial t} + u\frac{\partial u}{\partial x} + \nu\frac{\partial u}{\partial y} + \omega\frac{\partial u}{\partial z} = -\alpha\frac{\partial p}{\partial x} + 2\Omega\nu\sin\phi + F_x$$

$$\frac{\partial \nu}{\partial t} + u\frac{\partial \nu}{\partial x} + \nu\frac{\partial \nu}{\partial y} + \omega\frac{\partial \nu}{\partial z} = -\alpha\frac{\partial p}{\partial y} + 2\Omega u\sin\phi + F_y$$

$$\frac{\partial \omega}{\partial t} + u\frac{\partial \omega}{\partial x} + \nu\frac{\partial \omega}{\partial y} + \omega\frac{\partial \omega}{\partial z} = -\alpha\frac{\partial p}{\partial z} + 2\Omega u\cos\phi - g + F_z$$

Note

$$\frac{df}{dt} = \frac{\partial f}{\partial t} + u\frac{\partial f}{\partial x} + \nu\frac{\partial f}{\partial y} + \omega\frac{\partial f}{\partial z} \qquad f = f(x, y, z, t)$$

Hydrostatic Equation $\delta p = g\rho\delta z$

17. Equation of Continuity

$$\frac{1}{\rho}\frac{d\rho}{dt} + \frac{\partial u}{\partial x} + \frac{\partial \nu}{\partial y} + \frac{\partial \omega}{\partial z} = 0$$

or

$$\frac{\partial \rho}{dt} + \frac{\partial \rho u}{\partial x} + \frac{\partial \rho \nu}{\partial y} + \frac{\partial \rho \omega}{\partial z} = 0$$

For incompressible fluid flows

$$\frac{\partial u}{\partial x} + \frac{\partial \nu}{\partial y} + \frac{\partial \omega}{\partial z} = 0$$

18.

(i) Thermal expansion Coefficient $(\propto T)$

$$\propto T = -\frac{1}{\rho_0}\frac{\partial \rho}{\partial T} \text{ (where salinity (=s), and pressure (p) kept constant)}$$

P_0 = reference density

$$\propto T = 1\times 10^{-4} / c^{o}$$

(ii) Dependence of density, salinity (β_s) is given by

$$\beta_s = \frac{1}{\rho_0}\frac{\partial \rho}{\partial s} \qquad \text{(where T, p kept constant)}$$

β varies very little $\simeq 7.6 \times 10^{-4}$ ps u^{-1}

(iii) The approximate dependence of σ on T, s, p is given by

$$\sigma = \sigma_0 + \rho_0 \left[\propto_T (T - T_0) + \beta_s (S - S_0)\right]$$

where $\sigma_0(T_0, S_0)$ is density anomaly at T_0,S_0

$$\propto_T = \propto_T (p)$$

19. Equation of state [Temperature (T), salinity (S) and density (ρ) relation]

Temperature changes by heating, cooling and diffusion.

Temperature change + temperature advection/convection

= heating/cooling term + diffusion

(i) $$\frac{dT}{dt} = \frac{\partial T}{\partial t} + u\frac{\partial T}{\partial x} + v\frac{\partial T}{\partial y} + \omega\frac{\partial T}{\partial z} \qquad \text{(where t is time)}$$

$$= \frac{Q_H}{\rho C_p} + \frac{\partial}{\partial x}\left(K_H \frac{\partial T}{\partial x}\right) + \frac{\partial}{\partial y}\left(K_H \frac{\partial T}{\partial y}\right) + \frac{\partial}{\partial z}\left(K_V \frac{\partial T}{\partial z}\right)$$

Salinity changes by addition or removal of fresh water (which changes the dilution of salts)

(ii) $$\frac{ds}{dt} = \frac{\partial s}{\partial t} + u\frac{\partial s}{\partial x} + v\frac{\partial s}{\partial y} + \omega\frac{\partial s}{\partial z}$$

$$\frac{ds}{dt} = Q_s + \frac{\partial}{\partial x}\left(K_H \frac{\partial s}{\partial x}\right) + \frac{\partial}{\partial y}\left(K_H \frac{\partial s}{\partial y}\right) + \frac{\partial}{\partial z}\left(K_v \frac{\partial s}{\partial z}\right)$$

where Q_H = heat source (positive for heating, negative for cooling) applied only near sea surface

C_p = Specific heat of sea water (at constant pressure)

Q_s = Salinity source (positive for evaporation, brine rejection, negative for precipitation & run off) applied near sea surface.

K_H = horizontal eddy diffusivity

K_v = Vertical eddy diffusivity

(iii) Density $\rho = \rho$ (S, T, p) is a function of S, T & p

Equation of state – dependence of density on salinity,

Temperature and pressure.

Density change + density advection/convection

= Density sources + diffusion

(iv) $$\frac{d\rho}{dT} = \frac{\partial \rho}{\partial t} + u\frac{\partial \rho}{\partial x} + \nu\frac{\partial \rho}{\partial y} + \omega\frac{\partial \rho}{\partial z}$$

$$= \frac{\partial \rho}{\partial s}\frac{ds}{dt} + \frac{\partial \rho}{\partial T}\frac{\partial T}{\partial t} + \frac{\partial \rho}{\partial p}\frac{\partial p}{\partial t}$$

20. Bulk Formulas – Indirect calculations of fluxes.

For fluxes of sensible and latent heat and momentum, the formulas are:

(i) $T = \rho_a C_D U^2_{10}$

(ii) $Q_s = \rho_a C_p C_s U_{10} (t_c - t_a)$

(iii) $Q_L = \rho_a L_e C_L U_{10} (q_s - q_a)$

where

t_a = temperature of air 10 m above sea

t_s = surface temperature through satellite (instrument)

Q_s = sensible heat flux

Q_L = Latent heat flux

T = Wind temperature

ρ_a = density of air

C_D = Drag coefficient

U_{10} = Wind speed at 10 m above sea

C_p = Specific heat capacity of air

C_s = Sensible heat transfer Coefficient

L_e = Latent heat of evaporation

C_L = Latent heat transfer Coefficient

q_s = Specific humidity of air at the sea surface

q_a = Specific humidity of air at 10m above sea.

q = Specific heat at saturation, which depends on SST

(iv) $Q_s = \rho_a C_p C_s U_{10} [SST - T_{air}]$

(v) $Q_L = \rho_a L_e C_L U_{10} [q_x (SST) - q_{air}]$

21.

(i) Meridional ocean heat transport $\left(\vec{H}\right)$ is given by

$$H_{ocean}^{-\lambda} = \rho_r C_w a \cos\phi \int_{\lambda\,west\,coast}^{\lambda\,east\,cost} \int_{ocean\,bottom}^{ocean\,top} vTd_z d_\lambda$$

where a = radius of the earth

ρ_r = reference density (also denoted as ρ_0)

ϕ = Latitude, λ = longitude

a cos ϕ dλ = distance along a latitude circle over an arc dλ

T = potential temperature

(ii) Meridional ocean heat transport at latitude ϕ is given by

$$H_{ocean}^{-\lambda}(\phi) = -a^2 \cos\phi \int_{\phi_1}^{\phi_2} \int_{\lambda\,west\,cost}^{\lambda\,east\,cost} Q_{net} d_\lambda d\phi$$

where Q_{net} = air – sea flux magnitude

$= Q_{sw} + Q_{LW} + Q_s + Q_L$

22.

(i) Dynamic Energy Equation

$$\frac{dw}{dt} = \frac{d}{dt}(T + \phi + E) + \iiint (u\,F_x + v\,F_y) d_x\,d_y\,d_z$$

where

W = Amount of heat added to the system

T = Total KE of the system

ϕ = Potential energy of the system

E = Internal energy of the system

(ii) Dynamic meter (D) is given by

$$D = \frac{1}{10}\int_0^z g d_z$$

where

z = Geometrical depth

$$Z = 10\int_0^D \frac{1}{g} dD$$

(iii) Hamilton operator $\nabla = i\frac{\partial}{\partial x} + j\frac{\partial}{\partial y} + k\frac{\partial}{\partial z}$

∇ is a vector differential

(iv) If U is a scalar quantity, then

$$\text{grad } U = \nabla U = i\frac{\partial U}{\partial x} + j\frac{\partial U}{\partial y} + k\frac{\partial U}{\partial z}$$

where grad U = gradient of U

(v) $\text{Div } \vec{U} = \nabla \cdot \vec{U} = i\frac{\partial u}{\partial x} + j\frac{\partial \nu}{\partial y} + k\frac{\partial \omega}{\partial z}$

$\text{Div } \vec{U}$ = Divergent of $\vec{U}$

$\vec{U} = iu + j\nu + k\omega$

(vi) $\text{Curl } \vec{U} = \text{rot } \vec{U}$

$$= \nabla \times \vec{U}$$

$$= i\left(\frac{\partial \omega}{\partial y} - \frac{\partial \nu}{\partial z}\right) + j\left(\frac{\partial u}{\partial z} - \frac{\partial \omega}{\partial x}\right) + k\left(\frac{\partial \nu}{\partial x} - \frac{\partial u}{\partial y}\right)$$

where $\vec{U} = iu + j\nu + k\omega$

(vii) rot grad $U = \nabla \times \nabla U = 0$

(viii) div rot $\vec{U} = \nabla \cdot \left(\nabla \times \vec{U}\right) = 0$

(ix) $\nabla \cdot \nabla \phi = \nabla^2 \phi$ = Laplacian ϕ

Div of grad ϕ = Laplacian ϕ

where ϕ is a scalar.

23.

(i) **Ekman layer:** The wind driven ocean friction layer is called Ekman layer, which accounts only eddy viscosity and coriolis acceleration.

(ii) A constant vertical eddy viscosity acts in the fluid in the form:

$$\left.\begin{aligned} T_{xz} &= \rho A_z \frac{\partial u}{\partial z} \\ T_{yz} &= \rho A_z \frac{\partial v}{\partial z} \end{aligned}\right\} \qquad \text{.....(A)}$$

where A_z = eddy viscosity or eddy diffusing

T_{xz}, T_{yz} are components of wind stress in x, y directions (that is zonal and meridional directions)

(iii) The momentum equations for homogeneous steady state boundary layer.

$$\rho f v + \frac{\partial T_{xz}}{\partial_y} = 0 \qquad \text{or} \qquad fv + A_z \frac{\partial^2 u}{\partial z^2} = 0$$

$$\rho f u + \frac{\partial T_{yz}}{\partial_z} = 0 \qquad \text{or} \qquad fu - A_z \frac{\partial^2 v}{\partial z^2} = 0$$

where $f = 2\Omega \sin n\phi$, coriolis parameter

$$T_{xz} = \rho A_z \frac{\partial u}{\partial z}$$

$$T_{yz} = \rho A_z \frac{\partial v}{\partial z}$$

Ekman vertical eddy viscosity components, assumed to be constant

The solution of differential equations (A) are of the form

$$u = V_0\, e^{az} \cos\left(\frac{\pi}{4} + az\right)$$

$$v = V_0\, e^{az} \sin\left(\frac{\pi}{4} + az\right)$$

where $V_0 = \dfrac{T}{\sqrt{\rho\omega^2 + A_z}}$ and $a = \sqrt{\dfrac{f}{2A_z}}$

V_0 = velocity of the current at the ocean surface, z = 0

gives $u_0 = V_0 \cos\dfrac{\pi}{4}$

$$v_0 = V_0 \sin\frac{\pi}{4}$$

and $\sqrt{u_0^2 + v_0^2} = V_0 e^{az}$

24. Ekman Mass Transport (M_{EK}) defined as below

x - component M_{EKx}, y - component M_{EKy}

$$M_{EKx} = \int_{-d}^{0} \rho U_{EK} d_z,$$

$$M_{EKy} = \int_{-d}^{0} \rho V_{EK} d_z$$

[M_{EKx} = mass of water passing through a vertical plume yz = 1 meter wide and perpendicular to the transport from z = 0 to z = –d]

For homogeneous steady turbulent incompressible flow in boundary layer, the momentum equations are:

$$f \rho v + \frac{\partial T_{xz}}{\partial z} = 0$$

$$f \rho u - \frac{\partial T_{yz}}{\partial z} = 0$$

this gives

$$M_{EK} = \int_{-d}^{0} \rho \vec{U} dz$$

$$= \frac{T \times \hat{K}}{f}$$

where $\hat{K}$ = unit vector in z - direction

$$\vec{U} = i\, U_{EKx} + j\, V_{EKy}$$

25. Ekman Pumping

Since mass is conserved, the spatial variability of the transport lead to vertical velocities at the top of the Ekman layer.

Integrating the continuity equation in the vertical between

$$z = 0 \text{ to } z = -d$$

$$\rho \int_{-d}^{0} \left(\frac{\partial u}{\partial x} + \frac{\partial v}{\partial y} \right) dz = -\int_{-d}^{0} \frac{\partial \omega}{\partial z} dz$$

or $$\frac{\partial}{\partial x} \int_{-d}^{0} \rho u\, dz + \frac{\partial}{\partial y} \int_{-d}^{0} \rho v\, dz = -\rho \int_{-d}^{0} \frac{\partial \omega}{\partial z} dz$$

using Ekman transport definition

$$\frac{\partial}{\partial x}(M_{EKx}) + \frac{\partial}{\partial y}(M_{EKy}) = -[W(o) - W(-d)]$$

$$= -\rho\, W_{EK}(0)$$

or $$\nabla_H\, M_{EK} = -\rho\, W_{EK}(0)$$

where $M_{EK} = i\, M_{EKx} + j\, M_{EKy}$

This equation states that horizontal divergence of the Ekman transport lead to vertical velocity in the upper boundary layer of the ocean. This process is called Ekman pumping.

26. Sverdrup's theory of the ocean circulation or depth integrated circulation.

Sverdrups assumptions.

(i) The flow is stationary

(ii) Lateral friction and molecular viscosity are small

(iii) Non-linear terms (like $u\frac{\partial u}{\partial x}$ etc) are small

(iv) Turbulence near the sea surface can be described by the use of vertical eddy viscosity

(v) Wind driven circulation becomes zero at some depth of no motion.

Under the above assumptions, the momentum equations are

$$\frac{\partial p}{\partial n} - f \rho v = \frac{\partial T_{xz}}{\partial z}$$

$$\frac{\partial p}{\partial y} + f \rho u = \frac{\partial T_{yz}}{\partial z}$$

Using the Ekman transport definition & notation

$$M_{EKx} = M_x = \int_{-d}^{0} \rho(U_z) dz$$

$$M_{EKy} = M_y = \int_{-d}^{0} \rho(V_z)dz$$

and denoting $\frac{\partial \mathbf{P}}{\partial x} = \int_{-d}^{0} \frac{\partial p}{\partial x} dz$

$$\frac{\partial \mathbf{P}}{\partial y} = \int_{-d}^{0} \frac{\partial p}{\partial y} d_z$$

we get

$$\frac{\partial \mathbf{P}}{\partial x} - f M_x = T_x \qquad \text{.....(a)}$$

$$\frac{\partial \mathbf{P}}{\partial y} + f M_y = T_y \qquad \text{.....(b)}$$

Integrating equation of continuity between z = –d to z = 0

$$\int_{-d}^{0} \frac{\partial \rho u}{\partial x} dz + \int_{-d}^{0} \frac{\partial \rho \nu}{\partial y} dz = \int_{-d}^{0} \frac{\partial \rho \omega}{\partial z} dz$$

which reduces to

$$\frac{\partial Mx}{\partial x} + \frac{\partial My}{\partial y} = 0 \quad \text{since} \quad \int_{-d}^{0} \frac{\partial \rho \omega}{\partial z} dz = 0$$

Differentiating (a) with respect y and (b) with respect x and subtracting we get

$$0 = \beta \; My + f\left[\frac{\partial My}{\partial x} + \frac{\partial Mx}{\partial y}\right] + \frac{\partial Tx}{\partial y} - \frac{\partial Ty}{\partial x}$$

where $\frac{\partial f}{\partial y} = \beta$, $\frac{\partial f}{\partial x} = 0$

we get $\beta \, My = -\left(\frac{\partial Tx}{\partial y} - \frac{\partial Ty}{\partial x}\right)$

$$\beta \, My = \hat{k}.\nabla \times T \qquad \text{where} \quad T = i\,\tau_{n+j}\,\tau_y$$

This equation states that northward mass transport of wind driven current is equal to curl of the wind stress. This is Sverdrups relation.

This relation violates observed western boundary currents.

27. Stommel's theory of Western Boundary currents

Sverdrups relations do not give the vertical distribution of current and fails to answer observed western boundary current. Stommel added in boundary conditions a bottom stress proportional to the velocity and retained other equations:

Sverdrups momentum equation

$$\frac{\partial p}{\partial x} = f\rho v + \frac{\partial T_{xz}}{\partial x}$$

$$\frac{\partial p}{\partial y} = -f\rho u + \frac{\partial T_{yz}}{\partial z}$$

The Mass transport

$$M_x = \int_{-d}^{0} \rho u dz$$

$$M_y = \int_{-d}^{0} \rho v dz$$

Integrated pressure gradients

$$\frac{\partial \mathbf{P}}{\partial x} = \int_{-d}^{0} \frac{\partial p}{\partial x} d_z$$

$$\frac{\partial \mathbf{P}}{\partial y} = \int_{-d}^{0} \frac{\partial p}{\partial y} d_z$$

and horizontal boundary conditions at sea surface wind stress

$T_{xz}(0) = T_x, T_{xz}(-d) = 0$

and $T_{yz}(0) = T_y, T_{yz}(-d) = 0$

Stommel added bottom stress

$$\left(A_z \frac{\partial u}{\partial z}\right)_0 = -T_x = -F\cos\left(\frac{\pi y}{b}\right); \left(A_z \frac{\partial u}{\partial z}\right)_{-d} = -Ru$$

$$\left(A_z \frac{\partial v}{\partial z}\right)_0 = -T_y = 0; \left(A_z \frac{\partial v}{\partial z}\right)_{+d} = -Rv$$

when F and R are constants,

with rectangular basis $0 \leq' y \leq b$, $0 \leq x \leq \lambda$, with depth 'd' constant density (ρ), constant earth's rotation (Ω), variation of coriolis force with latitude.

This explained the cause of crowding of stream lines in the western boundary.

Note: Wind driven flow must reach the ocean bottom, which appears to be unrealistic.

28. Munk's Solution: Western Boundary Currents

Munk added lateral viscosity to Sverdrups solution within an ocean basin.

Munk used lateral eddy friction with constant

$A_H = A_x = A_y$

Under these assumptions the Sverdrups momentum equations become.

$$\frac{1}{\rho}\frac{\partial p}{\partial x} = f\nu + \frac{\partial}{\partial z}\left(A_z \frac{\partial u}{\partial z}\right) + A_H \frac{\partial^2 u}{\partial x^2} + A_H \frac{\partial^2 u}{\partial y^2}$$

$$\frac{1}{\rho}\frac{\partial p}{\partial y} = -fu + \frac{\partial}{\partial z}\left(A_z \frac{\partial \nu}{\partial z}\right) + A_H \frac{\partial^2 \nu}{\partial x^2} + A_H \frac{\partial^2 \nu}{\partial y^2}$$

Munk integrated from z = –d to (surface) Z = Z_0 and currents vanish at depth '–d'. A_H = constant

The above equationss can be written as

$$\frac{1}{\rho}\frac{\partial p}{\partial x} = f\nu + \frac{\partial}{\partial z}(T_{xz}) + A_H\left(\frac{\partial^2 u}{\partial x^2} + \frac{\partial^2 u}{\partial y^2}\right)$$

$$\frac{1}{\rho}\frac{\partial p}{\partial y} = -fu + \frac{\partial}{\partial z}(T_{yz}) + A_H\left(\frac{\partial^2 \nu}{\partial x^2} + \frac{\partial^2 \nu}{\partial y^2}\right)$$

Using stream function ψ, $u \equiv \frac{\partial \psi}{\partial y}, \nu \equiv -\frac{\partial \psi}{\partial x}$

The above equations can be written as:

$$\frac{1}{\rho}\frac{\partial p}{\partial x} = -f\frac{\partial \psi}{\partial x} + \frac{\partial}{\partial z}(T_{xz}) + A_H\left[\frac{\partial^3 \psi}{\partial x^2 \partial y} + \frac{\partial^3 \psi}{\partial y^3}\right] \quad \text{.....(a)}$$

$$\frac{1}{\rho}\frac{\partial p}{\partial y} = -f\frac{\partial \psi}{\partial y} + \frac{\partial}{\partial z}(T_{yz}) + A_H\left[-\frac{\partial^3 \psi}{\partial x^3} - \frac{\partial^3 \psi}{\partial y^2 \partial x}\right] \quad \text{.....(b)}$$

Differentiating (a) with y, and (b) with x and subtracting we get

$$0=-\frac{\partial\psi}{\partial x}\beta+\frac{\partial}{\partial z}\left(\frac{\partial T_{xz}}{\partial y}-\frac{\partial T_{yz}}{\partial x}\right)$$

$$+A_H\left[2\frac{\partial^4\psi}{\partial x^2\partial y^2}+\frac{\partial^4\psi}{\partial x^4}+\frac{\partial^4\psi}{\partial y^4}\right]$$

where $\beta=\frac{\partial f}{\partial y},\frac{\partial f}{\partial x}=0$

$$0=-\beta\frac{\partial\psi}{\partial x}+\frac{\partial}{\partial z}(T)+A_H\nabla^4\psi,$$ where

$$\nabla^4=\left(\frac{\partial^4}{\partial x^4}+2\frac{\partial^4}{\partial x^2\partial y^2}+\frac{\partial^4}{\partial y^4}\right)$$ bihormonic operator.

$$A_H\nabla^4\psi-\beta\frac{\partial\psi}{\partial n}=-\text{curl}_z T,\ A_H\nabla^4\psi=\text{function}$$

$$\beta\frac{\partial\psi}{\partial x}=-\text{curl}_z T$$ Sverdrup balance.

Munk assumed

$$\psi_b=0,\ \left(\frac{\partial\psi}{\partial n}\right)_b=0$$

where b = boundary

n = normal to the boundary

Basic x = 0 to x = r and y = –s to y = +s

wind stress zonal in the form T = acosny + b sinny + e

where $n=\frac{j\pi}{s},\ i=1,2$

29. **Ageostrophic Ocean flow:** Ocean currents generally concentrate in the top one kilometer depth due to wind force. These currents are not basotropic and not independent of depth and the flow has to account friction from frictional drag.

Let $\vec{F}$ = frictional drag = $-\frac{K\vec{U}}{\delta}$

where $\overline{K}$ = drag coefficient

$\vec{U}$ = Ageostrophic wind

$Z=\delta=$ depth of friction layer (Ekman layer)

Ageostrophic momentum equations are:

$$\frac{1}{\rho}\frac{\partial p}{\partial x}=f\nu-\frac{\bar{k}u}{\delta} \quad \text{.....(a)}$$

$$\frac{1}{\rho}\frac{\partial p}{\partial y}=-fu-\frac{\bar{k}\nu}{\delta} \quad \text{.....(b)}$$

Differentiating (b) with x, and (a) with y and subtracting we get

$$0=f\left(\frac{\partial \nu}{\partial y}+\frac{\partial u}{\partial x}\right)+\nu\frac{\partial f}{\partial y}+\varepsilon\left(\frac{\partial u}{\partial y}-\frac{\partial \nu}{\partial x}\right)$$

$$0=-f\frac{\partial \omega}{\partial z}+\nu\beta+\varepsilon\zeta$$

where $\frac{\partial f}{\partial x}=0,\ \frac{\partial f}{\partial y}=\beta$ and $\frac{\partial u}{\partial x}+\frac{\partial \nu}{\partial y}=-\frac{\partial \omega}{\partial z}$

$$\zeta=\frac{\partial \nu}{\partial x}-\frac{\partial u}{\partial y}$$

Or $f\frac{\partial \omega}{\partial z}=\beta\nu+\varepsilon\zeta$

using streamline $\psi,\ u=\frac{\partial \psi}{\partial y},\ \nu=-\frac{\partial \psi}{\partial x}$

we get $f\frac{\partial \omega}{\partial z}+\beta\frac{\partial \psi}{\partial x}=\varepsilon\zeta$

This equation is similar to the Munk's flow equation.

30. Taylor - Proudman Theorem

For geostrophic flow of a fluid, the influence of vorticity due to earth's rotation is most striking if $\rho=\rho_0$ is constant &

$f=f_0=$ constant.

Geostrophic equations

$$f_0\ \nu_g=\frac{1}{\rho_0}\frac{\partial p}{\partial x} \quad \text{.....(a)}$$

$$f_0\ u_g=-\frac{1}{\rho_0}\frac{\partial p}{\partial y} \quad \text{.....(b)}$$

$$g=-\frac{1}{\rho_0}\frac{\partial p}{\partial z} \quad \text{.....(c)}$$

Differentiating (a) with respect to z

$$f_0 \frac{\partial v_g}{\partial z} = \frac{1}{\rho_0}\frac{\partial}{\partial z}\left(\frac{\partial p}{\partial x}\right) = \frac{\partial}{\partial x}\left(\frac{1}{\rho_0}\frac{\partial p}{\partial z}\right)$$

$$= -\frac{\partial g}{\partial x} = 0 \qquad \text{using (c)}$$

$$\Rightarrow \frac{\partial v_g}{\partial z} = 0$$

Similarly differentiating (b) w, r, t, z we get

$$\frac{\partial u_g}{\partial z} = 0$$

It follows that the vertical derivation of the horizontal velocity field is zero viz $\frac{\partial u_g}{\partial z} = \frac{\partial v_g}{\partial z} = 0$

This is the Taylor – Proudman Theorem

31. Fluid Dynamics on the Beta Plane

Taylor - Proudman theorem says that geostrophic motion does not change in the direction $f\hat{k}$.

[Equation of continuity for incompressible], gestrophic flow is

$$\frac{\partial u_g}{\partial x} + \frac{\partial v_g}{\partial y} + \frac{\partial \omega_g}{\partial z} = 0 \qquad \text{.....(a)}$$

From T – P theorem $\frac{\partial u_g}{\partial x} = \frac{\partial v_g}{\partial y} = 0$, implies $\frac{\partial \omega_g}{\partial z} = 0$

That is flow cannot expand or contract in the vertical direction.

Consider the flow on Beta plane

Let $f = f_0 + \beta y$

Consider $\frac{\partial u}{\partial x} + \frac{\partial v}{\partial y} = \frac{-\partial}{\partial x}\left(\frac{1}{f\rho_0}\frac{\partial p}{\partial y}\right) + \frac{\partial}{\partial y}\left(\frac{1}{f\rho_0}\frac{\partial p}{\partial x}\right)$, using geostrophic equation on $f = f_0 + \beta_y$

$$= \frac{1}{f\rho_0}\left(\frac{-\partial^2 p}{\partial x \partial y} + \frac{\partial^2 p}{\partial y \partial x}\right) - \frac{\beta}{f}\frac{1}{f\rho_0}\frac{\partial p}{\partial x}$$

$$= 0 - \frac{\beta}{f}\nu$$

i.e., $f\left(\frac{\partial u}{\partial x} + \frac{\partial v}{\partial y}\right) = -\beta\nu$

or $f_0 \frac{\partial \omega_g}{\partial z} = \beta\nu$ [using (a)]

32. Heat Capacity (C) is defined by the relation

$C = \frac{\delta Q}{\delta t}$ where δQ = a very small amount of heat (energy) transferred to the body

δt = small change in temperature of the body T to $T + \delta t$

33. Specific heat c is the heat capacity of a unit mass of homogeneous substance

$$c = \frac{C}{M} = \frac{\text{heat capacity}}{\text{Mass of the substance}}$$

$$c = \frac{1}{M}\frac{\delta Q}{\delta t}$$

34. First law of thermodynamics

The quantity of heat absorbed by a substance

= External work done by the system + increase in its internal energy

$\delta Q = \delta w + di$

$\delta Q = C_v\, dT + p d\alpha$

or $\frac{dQ}{dt} = C_v \frac{dT}{dt} + p\frac{d\alpha}{dt}$ (Gaslaw $p\alpha$=RT)

where δw = work done

di = Increase in internal energy

and $H = \frac{\delta Q}{\delta t}$

35. Potential temperature (θ): It is the temperature attained by a sample of dry air when the sample is compressed or expanded adiabatically from a given state to pressure of p_0 = 1000 h P_a

$$\theta = T\left(\frac{p}{P_0}\right)^k, \quad \text{where} \quad K = \frac{R}{C_p}$$

36. Another form of thermodynamic equation

$$\frac{d}{dt}(l_n \theta) = \frac{1}{C_p}\frac{H}{T}$$

37. **Entropy:** The physical quantity which describes the ability of a system to do work is termed entropy of the system.

If $\oint \frac{dQ}{T} = 0$ or $\oint d\phi = 0$

where $d\phi = \frac{dQ}{T}$

Then ϕ is called entropy.

38. Change in entropy of any reversible process from an equilibrium state 'a' to 'b' is

$$\Delta\phi = \int_a^b \frac{dQ}{T} = \int_a^b d\phi$$

or $\Delta\phi = \phi_b - \phi_a = \frac{Q}{T}$

where $\Delta\phi$ reduced heat delivered to the system.

39. **Relation between Entropy (ϕ) and potential temperature (θ)**

$$\phi = C_p \, l_n \, \theta + \text{Constant}$$

40. **The operator ∇ (del)**

$$\nabla = i\frac{\partial}{\partial x} + i\frac{\partial}{\partial y} + k\frac{d}{\partial z}$$

(i) If ϕ is any arbitrary scalar quantity

$$\nabla\phi = i\frac{\partial\phi}{\partial x} + j\frac{\partial\phi}{\partial y} + k\frac{\partial\phi}{\partial z}$$

where i, j, k are unit vectors in x, y, z directions (Cartesian orthogonal system –

$\nabla\phi$ is called gradient of ϕ, also written as "grad ϕ'.

(ii) If $\vec{V} = iu + jv + k\omega$ is a velocity vector, then the divergence of $\vec{V}$ is denoted by div $\vec{V}$ is given by

$$\text{div }\vec{V} = \nabla\cdot\vec{V} = \left(i\frac{\partial}{\partial x} + j\frac{\partial}{\partial y} + k\frac{\partial}{\partial z}\right)\bullet(iu + jv + k\omega)$$

$$= \frac{\partial u}{\partial x} + \frac{\partial v}{\partial y} + \frac{\partial \omega}{\partial z}$$

(iii) The rotation or curl of a vector $\vec{V}$ is denoted by rot $\vec{V}$ or Curl $\vec{V}$ and is given by

$$\text{rot }\vec{V}\text{ or Curl }\vec{V} = \nabla\times\vec{V} = \begin{vmatrix} i & j & k \\ \frac{\partial}{\partial x} & \frac{\partial}{\partial y} & \frac{\partial}{\partial z} \\ u & v & \omega \end{vmatrix}$$

$$= i\left(\frac{\partial \omega}{\partial y} - \frac{\partial v}{\partial z}\right) + j\left(\frac{\partial u}{\partial z} - \frac{\partial \omega}{\partial x}\right) + k\left(\frac{\partial v}{\partial x} - \frac{\partial u}{\partial y}\right)$$

$$= i\xi + j\eta + k\zeta$$

where

$$\xi = \frac{\partial \omega}{\partial y} - \frac{\partial v}{\partial z}$$

$$\eta = \frac{\partial u}{\partial z} - \frac{\partial \omega}{\partial x},$$

$$\zeta = \frac{\partial v}{\partial x} - \frac{\partial u}{\partial y}$$

(iv) $\nabla\cdot\nabla = \nabla^2 = \frac{\partial^2}{\partial x^2} + \frac{\partial^2}{\partial y^2} + \frac{\partial^2}{\partial z^2}$ is called Lapalacian.

41.

(i) Equation of Continuity

$$\frac{\partial \rho}{\partial t} + \nabla\cdot\left(\rho\vec{V}\right) = 0$$

or

$$\frac{\partial \rho}{\partial t} + \vec{V} \cdot \nabla \rho + \rho \nabla \cdot \vec{V} = 0 \quad \text{or} \quad \frac{d\rho}{dt} + \rho \nabla \cdot \vec{V} = 0$$

$$\text{Since} \left[\frac{d\rho}{dt} = \frac{\partial \rho}{\partial t} + \vec{V} \cdot \nabla \rho \right]$$

$$\frac{d}{dt} = \frac{\partial}{\partial t} + \vec{V} \cdot \nabla$$

$\nabla \cdot \left(\rho \vec{V}\right)$ = mass divergence in three dimensional space

$\nabla \cdot \vec{V}$ = dialation or expansion

$\frac{d}{dt} = \frac{\partial}{\partial t} + \vec{V} \cdot \nabla$ denotes that, the individual time derivative is equal to local time derivative +(plus) the convective acceleration.

$\frac{d}{dt}$ = rate of change or individual time derivative

$\frac{\partial}{\partial t}$ = local change or local time derivative

$\vec{V} \cdot \nabla$ = Velocity advection or convection acceleration

(ii) Equation of continuity for incompressible fluid (ρ = constant)

$$\nabla \rho = 0, \quad \frac{\partial \rho}{\partial t} = 0$$

is given by $\nabla \cdot \vec{V} = 0$ or $\frac{\partial u}{\partial x} + \frac{\partial v}{\partial y} + \frac{\partial \omega}{\partial z} = 0$

42. Hydrostatic equation

$$\delta p = -g\rho \delta z$$

Hydrostatic equation for incompressible fluids

$$p = p_o - g\rho z \quad \text{or} \quad p_2 - p_1 = -g\rho(z_2 - z_1)$$

43. Equations of motion in Cartesian Coordinates

$$\frac{du}{dt} = -\alpha \frac{\partial p}{\partial x} + fv - e\omega + \alpha F_x$$

$$\frac{dv}{dt} = -\alpha \frac{\partial p}{\partial y} - fu + \alpha F_y$$

$$\frac{d\omega}{dt} = -\alpha \frac{\partial \rho}{\partial z} + eu - g + \alpha F_z$$

where

$\vec{V} = iu + j\nu + k\omega$ = velocity

$f = 2\Omega \sin \phi$ = coriolis parameter

$\rho = 2\Omega \cos \phi$

$\alpha = \frac{1}{\rho}$ = specific volume

$\vec{F} = iF_x + jF_y + kF_z$ = Friction

g = gravity force

$\vec{\Omega}$ = Angular velocity of the earth

ϕ = latitude

u, v, w velocity components in x, y, z directions respectively

p = pressure

44. Bernoulli's equation

$$p + \frac{1}{2}\rho V^2 + \rho gh = \text{constant}$$

where

p = static pressure & p + ρgh = is called static pressure

ρ = density

V = Velocity

$\frac{1}{2}\rho V^2$ = dynamic pressure

The above equation states that for steady, non-viscous incompressible fluid flow:

Static pressure + dynamic pressure = constant.

45. Geostrophic wind

The horizontal wind occurring in a frictionless flow, when the pressure gradient force is in balance with coriolis force is called geostrophic wind.

In geostrophic balance $\alpha\nabla p = -2\vec{\Omega} \times \vec{V}_g$

where $\vec{V}_g = iu_g + jv_g$ = geostrophic wind vector

$$\alpha\frac{\partial p}{\partial x} = fv_g$$

$$\alpha\frac{\partial p}{\partial y} = -fu_g$$

46. **The Gradient wind**

The wind occurring in a frictionless flow when there is a balance between pressure force $(\alpha\nabla\rho)$, the coriolis force $(2\vec{\Omega}\times\vec{V})$ and the centrifugal force $(\Omega^2\vec{R})$

i.e., $\alpha\nabla p = -2\vec{\Omega} \times \vec{V} - \Omega^2 \vec{R}$

47. (i) Divergence of Geostrophic wind

$$\text{Div } \vec{V} = \frac{-V_g}{R \tan\phi}$$

(ii) Vorticity of the Geostrophic wind

$\zeta_g = \kappa.\nabla \times \vec{V}_g$, where $\vec{V}_g = iu_g + jv_g$

$$= \frac{\partial v_g}{\partial x} - \frac{\partial u_g}{\partial y}$$

(iii) Thermal wind or vertical shear of the geostrophic wind

$$\vec{V}_T = \vec{V}_{gu} - \vec{V}_{gL}$$

Thermal wind vector in a layer

= Geostrophic wind vector in upper level – geostrophic wind vector in the lower level

48.

(i) Divergence of the wind field (D) is given by

$$D = \underset{S\to 0}{\text{Lt}} \frac{1}{S}\oint_L V_n dl$$

where L is a closed curve in a horizontal isobaric surface

s = Area occupied by the contour (L)

V_n = Wind component along the outward normal to the curve

V_L = Wind component along the tangent to the contour (in positive direction)

$$D = \frac{\partial u}{\partial x} + \frac{\partial v}{\partial y}$$

(ii) Vorticity of the wind field (ζ) is given by

$$\zeta = \underset{s \to 0}{Lt} \frac{1}{S} \oint_L V_L \, dL$$

$$= \frac{\partial v}{\partial x} - \frac{\partial u}{\partial y}$$

49. The convection level

The δz (be the height) at which the ascending air warmer than the surrounding will get a temperature equal to the temperature of the surrounding is called the convective level (CL). The approximate formula is

$$\delta z = \frac{T - T_e}{\frac{\partial T_e}{\partial z} = \frac{\partial T}{\partial z}}$$

where T = temperature of the particle/parcel

T_e = temperature of the surrounding

$\frac{\partial T_e}{\partial z}$ = Temperature gradient of the atmosphere

$\frac{\partial T}{\partial z}$ = temperature gradient of the particle/parcel

50. The Geostrophic advection of temperature

The geostrophic advection of temperature is defined as the local change of temperature per unit time caused by the advection of temperature with the geostrophic wind

$$\left(\frac{\partial T}{\partial t}\right)_a = -V_g \frac{\partial T}{\partial n} \cos \alpha$$

$$= -\frac{1}{\rho f} \frac{\partial p}{\partial n} \frac{\partial T}{\partial n} \sin \delta$$

where $\left(\frac{\partial T}{\partial t}\right)_a$ = geostrophic advection

α = angle between the geostrophic wind and the temperature gradient

δ = angle from the temperature gradient to the temperature gradient (+ for cyclonic)

51. The vorticity of the geostrophic wind

$$\zeta_g = \frac{1}{\rho f}\nabla^2 p$$

or
$$\zeta_g = \frac{9.8}{f}\nabla^2 H$$

where
$$\nabla^2 = \frac{\partial^2}{\partial x^2} + \frac{\partial^2}{\partial y^2} \quad \text{laplacian operator}$$

52. Rossby waves: Rossby waves may be defined as waves which are greately influenced by the meridional gradient of the coriolis parameter.

Rossby waves differential equation

$$\frac{\partial \nabla^2 H}{\partial t} + u\frac{\partial \nabla^2 H}{\partial x} + \beta\frac{\partial H}{\partial x} = 0$$

where H = geopotential of an isobaric surface

and
$$\nabla^2 = \frac{\partial^2}{\partial x^2} + \frac{\partial^2}{\partial y^2}$$

$$\beta = \frac{\partial f}{\partial y} = \frac{2\Omega}{R}\cos\phi,$$

R = Radius of the earth

Phase velocity of simple Rossby wave is

$$\sigma = u - \frac{\beta\lambda^2}{4\pi^2}$$

where λ = wavelength

REFERENCES

1. **The ocean,** H.V. Sverdrup, Martin W. Johnson and Richard H. Fleming. Asia Publishing House, Indian Edition – 1961.
2. **Introduction to Physical Oceanography** Robert H. Stewart. Department of Oceanography, Texas A&M: University Sept 2006 Edition.
3. **Atmosphere, Ocean and Climate Dynamics. An introductory Text** John Marshall. R. Alan Plumb, Elsevier Academic Press – 2008
4. Supplementary Material to the Text Book **An Introduction – Descriptive Physical Oceanography** by Lynne D. Tolley, George L. Pikcard William J. Emery and James H. Swift. Elservier Academic Press – Sixth edition 2011, 2012 Reprint
5. **Fundamentals of Geophysical Hydrodynamics** FV. Dolzhansky Encyclopedia of Mathematical Sciences 103 Springer –Verlag Berlin Heidelberg – 2013
6. **Physical Oceanography** Prof. ASN Murthy, Dr. V.S.N Murthy APH Publishing Corporation
7. **An Elementary Hydrodynamics** By. Bansilal – 1961 Atmaram & Sons, Delhi
8. **WMO. No. 226. TP 150, Compendium of Lecture Notes for training class IV Meterological Personnel. Vol I Earth Science & Vol II Meteorology**
9. **WMO No. 364, Compendium of Meteorology for use by class I and class II Meteorological personnel. Part I Dynamic Meteorology**

10. **WMO No. 261, TP. 146 Problems in Dynamic Meteorology**
11. **Handbook of Physics.** B. Yavorsky and A. Detlaf. Mir Publishers – Moscow – 1975
12. **Numerical Prediction and Dynamic Meteorology.** George J. Haltiner, Roger Terry Williams. Second Edition 1980 John Wiley & Sons, NewYork
13. **Atmospheric Sciences,** I.N. James Camebridge Books Online
14. **Five Decades of Numerical Weather Prediction (NWP) – in India. IMD – 2008**
15. **A Course in Dynamic Meteorology** Navale Pandharinath B.S. Publication - 2007, Hyderabad -500095
16. **Aviation Meteorology** Navale Pandharinath B.S. Publication – 2012, Hyderabad – 500095
17. **The Atmosphere of the Living Planet Earth.** WMO. No 735, 1990

INDEX

E

F

G

P

R

S

T